Information Systems: Failure Analysis

NATO ASI Series

Advanced Science Institutes Series

A series presenting the results of activities sponsored by the NATO Science Committee, which aims at the dissemination of advanced scientific and technological knowledge, with a view to strengthening links between scientific communities.

The Series is published by an international board of publishers in conjunction with the NATO Scientific Affairs Division

A Life Sciences **B Physics**	Plenum Publishing Corporation London and New York
C Mathematical and Physical Sciences	D. Reidel Publishing Company Dordrecht, Boston, Lancaster and Tokyo
D Behavioural and Social Sciences **E Applied Sciences**	Martinus Nijhoff Publishers Boston, The Hague, Dordrecht and Lancaster
F Computer and Systems Sciences **G Ecological Sciences** **H Cell Biology**	Springer-Verlag Berlin Heidelberg New York London Paris Tokyo

Series F: Computer and Systems Sciences Vol. 32

Information Systems: Failure Analysis

Edited by

John A. Wise

Westinghouse Research and Development Center
1310 Beulah Road, Pittsburgh, PA 15235, USA

Anthony Debons

University of Pittsburgh, Pittsburgh, PA 15260, USA

Springer-Verlag
Berlin Heidelberg New York London Paris Tokyo
Published in cooperation with NATO Scientific Affairs Divison

Proceedings of the NATO Advanced Research Workshop on Failure Analysis of Information Systems held in Bad Windsheim, Federal Republic of Germany, August 18–22, 1986

ISBN-13: 978-3-642-83093-8 e-ISBN-13: 978-3-642-83091-4
DOI: 10.1007/978-3-642-83091-4

Library of Congress Cataloging in Publication Data. NATO Advanced Research Workshop on Failure Analysis of Information Systems (1986 : Bad Windsheim, Germany) Information systems. (NATO ASI series. Series F, Computer and systems sciences ; vol. 32) "Published in cooperation with NATO Scientific Affairs Division" "Proceedings of the NATO Advanced Research Workshop on Failure Analysis of Information Systems held in Bad Windsheim, Federal Republic of Germany, August 18–22, 1986"—T.p. verso. Includes index. 1. Electronic data processing—Congresses. 2. System failures (Engineering)—Congresses. 3. System analysis—Congresses. I. Wise, John A., 1944- . II. Debons, A. III. North Atlantic Treaty Organization. Scientific Affairs Division. IV. Title. V. Series: NATO ASI series. Series F, Computer and systems sciences ; vol. 32. QA75.5.N38 1986 003 87-12818

Softcover reprint of the hardcover 1st edition 1987

2145/3140-543210

This proceedings is dedicated to the late

Dr. Mario di Lullo

Program Director
NATO Scientific Affairs Division

A champion of the new idea
from whatever quarter,
and an unflinching supporter
of its definition and development.

PREFACE

Although system analysis is a well established methodology, the specific application of such analysis to information systems is a relatively new endeavor. Indeed, it may be said to be still in the trial-and-error stage. In recent years, such analysis has been given impetus by the numerous accounts of information system failures, some of which have led to serious consequences - e.g., the accident at Three Mile Island, the chemical spills at Bophal, India, and at Institute, West Virginia, and the loss of the space shuttle Challenger. Analysis of the failure of the W. T. Grant Company, the third largest retail organization in the United States, indicated that improper use of the available information was a significant factor in that failure.

In spite of these incidents and their widespread impact, only meager attempts have been made to develop an effective methodology for analyzing the information systems involved in such incidents. There have been no well developed guidelines for determining the causes of such events and for recommending solutions so that similar failures could be avoided.

To address the need for such a methodology, the North Atlantic Treaty Organization (NATO) sponsored an Advanced Research Workshop attended by a group of 32 scientists, scholars, and expert investigators, representing a variety of disciplines and countries. Additional funding was provided by the Aluminum Company of America and the U. S. Army Research Institute for the Behavioral and Social Sciences - European Office. The Workshop meetings were held in Bad Windsheim, West Germany, August 18 to 22, 1986.

This Proceedings document includes the papers prepared by the Workshop participants, along with summaries of the conclusions reached by the group. The material is intended to provide basic guidelines for both the design of information systems and the prevention of future information system failures. The Proceedings will also serve as a basis for generating Research and Development programs that are directed specifically toward establishing new methodologies for the investigation of information system failures.

Meeting in small groups and plenary sessions, the Workshop participants addressed the following specific topics:

- Investigation techniques. Identification of techniques used in the evaluation of system failures.

- System issues. Effects of the functional relationship among components on system failure.

- Hardware/software. Factors related to hardware and software contributions to system breakdown.

- Decision-making and problem-solving issues. Factors related to human decision-making and problem-solving during crisis events. The human ability to deal with options before and after a crisis event.

- Organizational issues. Personnel management factors in system breakdown. The role of communications during pre- and post-events. Causes of communications breakdown.

- Human/computer interface. Basic human factors considerations in data acquisition from the system. Loss of an effective communication between user and computer.

In examining these topics, the group identified and evaluated various investigative techniques and methods of converting results into design and test of procedures and standards verification. They also identified case histories that can be used as models, resources required for investigation of information system failure, and new methods that are information system specific.

The opinions expressed in this volume are those of the participants and do not necessarily represent those of any of the sponsors.

ACKNOWLEDGMENTS

The editors would like to acknowledge the work of those individuals whose untiring effort and dedication to the Advanced Research Workshop made possible the publication of this Proceedings. In a certain sense, each of them should have his or her name on the cover, because without any one of them this volume would not exist.

For the success of the Workshop, we must thank our co-director, Dr. Gianni Casale. He did all of the leg work and a significant amount of the coordination that was required to set up and run a meeting of this magnitude. A very special thanks must be extended to Mrs. M. Barbara Gibson, who handled the finances and acted as the general manager during the event. Her business sense was a key reason for the successful operation of the Workshop. We also owe a significant debt of gratitude to Dr. Mary C. Culver, who served as our technical writer and editor to ensure the high quality of the Proceedings.

We want to thank Ms. Thelma Crum, Ms. Michele Ginley, Ms. Lisa Girty, and Mr. Mark Wise for their technical support in getting the papers into the word processing system, and especially Ms. Gerry Markle for her untiring work in processing and preparing the papers for publication.

While our names are on the cover, we respectfully share whatever credit arises from this volume with all of the above.

John A. Wise

Anthony Debons

EXECUTIVE SUMMARY

During the week of August 18-22, 1986, thirty-four scientists and interested participants met in an Advanced Research Workshop to discuss the nature of information system failure and to explore whether investigation methods are available that could provide understanding of the nature of these failures and how they could be prevented. These methods, it was hoped, could then be formalized as tools for investigating teams that would be asked to assess the causes of a failure and how it could be prevented.

Several working groups were formed to address specific aspects of the topic: human factors methodological issues, systemic issues, hardware-software, decision making, and organizational aspects. The total workshop met in plenary session for open discussion of these issues, and attendees then met in working groups to identify and detail technical aspects of the issues. Presentation of papers was discouraged, but the workshop did entertain statements by some of the participants, who presented their experiences, views, and suggestions in plenary session.

WORKSHOP BACKGROUND

In proposing and conducting the Advanced Research Workshop, the directors were motivated by three existing situations:

1. Contemporary events of considerable significance to national and public welfare suggest that information was a significant force on the character and outcome of the each event (e.g., the Challenger shuttle accident, the explosion at Bophal). The need to explicate the role of information in these events was deemed to be relevant to national and private interests.

2. Although the failure to respond to such events because of factors attributable to the breakdown of information was explicit, the role of information as a multi-faceted system phenomenon was not. The tendency of the accounts of these events was to attribute the information breakdown to things (i.e., computers, humans, warning devices) and not to a process of the many components that make up an information system. The Workshop Directors were interested in learning whether (and if so, how) a mind set could influence the understanding of an information system failure and thus negate the possibility of establishing investigative procedures.

3. A disarming aspect of system failures that can be directly or hypothetically linked to information failure is the lack of any accepted concept of an information system. The ubiquity of the term "information" compounds the difficulty inherent in the term "system."

A structure for an information system based on a physiological metaphor was proposed to serve as a guidepost for the workshop in its attempt to assess the state of affairs in regard to information system failure. The directors were interested in learning the usefulness of the model in developing insight to current problems and in provoking research and study.

WORKSHOP POINTS OF EMPHASIS

The reports of the special working groups (included in this volume) serve to bring together many of the technical statements contained in the papers prepared by the scientists who were invited to present their views prior to the workshop. These statements can be summarized briefly as follows:

1. There are several methods that can be applied directly to the investigation of information system failure -- some quantitative and some descriptive. In the absence of any clear definition of an information system (and possibly the inclination to resist the acceptance of such a definition), the tendency to accept a component, such as a computer or a human, as the information system itself rather than as part of the system becomes the focus of the failure analysis. Some aspects of a component may be difficult to test because of their nature (e.g., software) or changing structure, as pointed out by A. Bilinski in this volume. However, among methods available in the rest of the overall system, data flow analysis, which is the practice of system analysis, is applicable. Simulation is suggested as an alternative to operational testing and verification (see Ehrenberger and Yarman in this volume). Diagnostic methods as part of expert systems development are relevant within the contemporary interests of those engaged in research on artificial intelligence (see Hollnagel).

2. The utilization component of the information system, i.e., the human, is a key element in information system failure. As to conditions that precede failure (pre-mortem), ergonomic considerations are explicit to the extent that statistical data are available that indicate the predisposing factors of human failure. Data on interpretive, analytical, and synthesizing capabilities (cognitive element) are less available. Methods have been proposed, however, that indicate ways of identifying these predisposing factors and their containment (see Janis, Reason).

3. An often neglected but important element in information system functioning is the organization in which the information system is encased. The organization, which is itself an information system, embraces a multitude of other information systems that aid in the achievement of objectives. In this hierarchy of information systems, the information system at whatever level is a source of power even to those individuals who themselves represent information systems with desires and objectives.

These various information entities with related powers are all potential sources of failure. For example, individuals in an organization can induce failure of other information systems in the hierarchy (see Weiner, Conway). Centralization and decentralization of completed data flow in an organization (top-down, bottom-up) as a system organizational philosophy can be the stimulus for failure (see Westrum). Thus, any investigation of information system failure needs to account for the rationality (and the related organizational culture or belief systems) of those who direct the activities of an information system (see Westrum).

Each of these dimensions of the workshop is discussed in greater detail in the individual concept papers prepared prior to the convening of the workshop and to a lesser degree in the working group reports.

IMPLICATIONS FOR FUTURE RESEARCH

Were the aspirations of the directors met? Most certainly. But more important, what are the implications for future research on information system failure? Several areas that need to be studied are summarized below.

1. There is a tacit recognition that the concept of an information system favors a reductionist component-oriented "mind set". Despite the power of analogic, metaphorical applications (see Reason, Debons), the utility of such thinking in the analysis of failure remains to be questioned. It is doubtful, however, that information failure analysis can proceed without some standard by which an information system can be conceived. In this connection, it would appear

that Janis' reference to Scriven's modus operandi deserves careful attention in regard to "post mortem" analysis. Janis' pre-mortem analysis reference to barometers could provide insight to the state of data flow throughout the system and the likelihood of possible failure. These barometers can suggest the state of the interaction between the components of the system and provide concepts for system experimentation and simulation.

2. It appears that the life cycle for a system provides the basis for concepts concerning the level of efficiency and effectiveness and the degradation of the system. These concepts, at present, could be considered hypothetical and need to be studied in detail.

3. Concerning the utilization of a human component in an information system, the question should be raised of whether a taxonomy of human error (see Empson) can be established for the cognitive domain. Within the Janis (Scriven) framework of symptoms, perhaps both Reason's identification of ergonomic factors and Yovits' drive to obtain a quantitative measure of information can be integrated into a checklist for the containment of information system failure.

   Psychologists and ergonomists are eminently qualified to realize that a human information system is not simply a matter of brain functioning but, on the contrary, the interaction of emotion and effector subsystems in their total functioning. Likewise, a human information system that is technologically augmented brings sensor, transmission, processing, utilization, and transfer functions within a concept that integrates the functions of these individual components. Thus, an investigative checklist must explicitly recognize the breakdown of cognitive factors influenced by the breakdown of the respective subsystems.

4. The role of management in past incidents of system failure is clear (see Mason). Understanding the role that centralization and decentralization of data can play in these failures needs to be studied. Of critical importance is the use of information systems by management to induce the destination of their organization for their own purposes or for the purposes of others (see Conway).

5. The documentation and centralization of data on information system failures are important factors in the understanding and curtailment of such failures. Study should be conducted on the resources presently available for this purpose and on the advisability of establishing a centralized data base.

6. Investigations require individuals trained in careful observation and rational thinking. Considering the important role these individuals play, research should be conducted to identify the related background and skills required to undertake the task of information system failure analysis (see Petrie).

7. Lastly, it remains to be established whether the investigative methods and concepts be formally integrated in standards which allow for identifying and preventing information system failure.

A final comment is in order to further highlight the importance of recognizing information system failure. It can be surmised from the Tower Commission Report that many information system issues, including President Reagan's management style and the decentralization of data caused by the numerous individuals involved, contributed to the Iran weapons sale crisis in the United States. It is significant to note that these are among the issues addressed by the authors in this Proceedings.

CONTENTS

SYSTEMIC ASPECTS OF INFORMATION SYSTEM FAILURE

DISASTER WARNING SYSTEMS: LEARNING FROM FAILURE

Harold D. Foster

University of Victoria
Victoria, British Columbia, Canada

INTRODUCTION

Disaster warning networks are a particularly significant subgroup of information systems because if they fail to operate as anticipated, serious life loss, injury, and property damage almost inevitably occur. For example, at 11 p.m. on 12 November 1970, a major cyclone struck the coast of Bangladesh. The associated storm surge, which drowned some 225,000 people, reached a maximum of almost seven metres above normal high tide. However, landfall had been predicted for several days, since this storm had been tracked by satellite as it swept inexorably up the Bay of Bengal (Burton, Kates and White, 1978). Similarly, on November 13, 1985, lahars from Nevado del Ruiz, in west central Columbia, inundated the towns of Armero, Tolima and Chinchina, killing at least 22,000 inhabitants. Despite widespread fatalities, it could not be claimed that these volcanic mudslides were unexpected. Indeed, in October a hazard map of Ruiz volcano had been issued that accurately predicted the scale of the November 13 eruption. Prior to the destruction of Armero, both its mayor and the Red Cross had pleaded for evacuation (Herd, 1986).

These two catastrophes have been chosen to illustrate a significant point: most disaster-related life loss now occurs despite, rather than without, warning. However, such warnings are often ineffectual, failing to result in the desired response. This problem stems, in large part, from the complexity of the warning process. Although warning systems vary in size, scope, and effectiveness, reflecting both the hazard being monitored and its social setting, their operation normally involves sixteen sequential steps (Foster, 1980). Experience has shown that failure to provide a warning or to respond adequately to it can occur at any one, or more, of these steps. Table 1 shows how the EATPUT model is related to this conceptualization of the typical warning system.

RECOGNITION OF DANGER

The most obvious reason any organization or community fails to receive advance notice of an impending disaster is that no warning system has been set up. Certain hazards are, as yet, unpredictable. There are, for example, no known methods to forecast submarine earthquakes, although once tsunamis have been generated, their movements can be predicted with considerable accuracy. However, many warning systems are not put in place because of an unwillingness to accept the existence of a particular hazard. This reticence can occur for a variety of reasons, including the realization that such an acknowledgment carries with it a variety of economic and social costs. Denial, therefore, may be seen as the best short-term strategy.

The implications of such an unwillingness to face up to reality were illustrated on 11 May 1953, when Waco, Texas, was struck by a tornado that caused 114 fatalities and over 1000 injuries, some 145 of them serious (Moore, 1958). The great majority of these casualties could have been avoided if as little as five minutes advance notice had been given of the approaching threat. However, despite the fact that there had been storm and tornado warnings throughout the day for the general area, little attention had been paid to them in Waco, which had no effective local response system (Moore and Crawford, 1955). Vigilance was not considered

NATO ASI Series, Vol. F32
Information Systems: Failure Analysis
Edited by J. A. Wise and A. Debons

Table 1
Sixteen Steps Required for the Successful Operation of a Disaster Warning System, Considered in Relation to the EATPUT* Model.

STEPS	E	A	T	P	U	T
1. Recognition by decision makers that there is the possibility of danger from a particular source.	•	•		•		
2. Design of a system to monitor changes in the hazard and to issue warnings if danger crosses threshold.		•	•			
3. Installation and operation of the system.		•	•			
4. Education of user group so that responses will be appropriate.		•			•	
5. Testing of the system, when there is little danger, to ensure that it is technically sound and that those who issue and receive warnings act as required.		•	•	•	•	•
6. Modifying the system if test results indicate that changes are necessary.		•	•	•	•	•
7. Detection and measurement of changes in the hazard that could result in increases in death, injury, or property damage.	•	•				
8. Collation and evaluation of incoming data.				•		
9. Decisions as to who should be warned, about what damage, and how.					•	
10. Transmission of warning message(s) to those whom it has decided to warn.						•
11. Interpretation of the warning messages and action by recipients.					•	•
12. Feedback of data about the actions of message recipients to the issuers of the warning.		•				
13. Transmission of further warning messages, corrected in terms of the user's responses to the first and subsequent messages, and noting any secondary threats						•
14. Transmission of All Clear when danger has passed.						•
15. Hindsight review of the operation of the warning system and implementation of necessary improvements.		•	•	•	•	•
16. Testing and operation of revised system.		•	•	•	•	•

*EATPUT components: Event world, Acquistion, Transmission, Processing, Utilization, Transfer.

necessary since most residents believed an Indian legend, widely publicized by the Chamber of Commerce two years previously, that the city was immune to tornadoes (Dacy and Kunreuther, 1969).

DESIGNING THE WARNING SYSTEM

At least nine disaster agent characteristics have serious implications for warning network design (McLuckie, 1970): frequency, physical consequences, speed of onset, length of possible forewarning, duration, scope of impact, destructive potential, gross predictability, and gross controllability. Space does not permit a detailed description of how each of these can affect the success or failure of the warning network, although some comments are obviously pertinent.

The frequency with which a disaster agent is likely to strike greatly affects the efficiency of any warning system. Rare events are often the most devastating, since affected communities lack the comprehensive knowledge or organizational structure to deal with them. It is no coincidence that some of the worst volcanic disasters have occurred during the eruption of volcanoes, such as Mount Pelee and Krakatoa, that had long been considered extinct (Macdonald, 1972). In such locales, it may also be difficult to generate public interest even when warnings have been issued. Darwin, Australia, has often received tropical cyclone alerts or warnings. Normal storm paths, however, have saved the city from disaster on many occasions. Since little cyclone damage had been suffered since 1927, when cyclone warnings were issued on Christmas Eve 1974, most residents believed that, as usual, the city would be spared and so continued their normal festivities (Haas, Cochrane, and Eddy, 1976). This proved to be an expensive error since the city, on this occasion, suffered extensive hurricane damage, which led to its eventual evacuation.

Andow (1986) has shown that the reverse situation may also undermine the effectiveness of a warning system. In many process plants, alarms occur with great regularity indicating only very minor problems, most of which may be corrected automatically. As a consequence, real alarms are easily lost in the noise created by other alarm signals.

The design of any warning network must take great cognizance of the hazards to which the network may be subjected. Obviously, the warning system itself must not be seriously damaged by the event it seeks to predict and monitor. Unfortunately, this has frequently not been the case. To illustrate, the 28 March 1964 Alaskan earthquake destroyed the control tower at Anchorage International Airport, breaking vital communication links between the nerve center of the Seismic Sea Wave Warning System in Honolulu, the College and Sitka seismic observatories, and the Kodiak, Sitka, and Unalaska tidal stations. As a result, the receipt of data for these locations was delayed and the functioning of the system was impeded (Spaeth and Berkman, 1967).

Such breakdowns, resulting from the initial consequences of disaster, need not be technical. During the Bhopal gas disaster of December 1984, for example, virtually all workers in the Union Carbide Plant panicked, fleeing the scene. There is considerable doubt as to whether the community warning siren was ever sounded. Clearly, this warning system was very ineffective since a large cloud of methyl isocyanate and phosgene gases was allowed to asphyxiate over 2,500 sleeping inhabitants in the early hours of the morning (Lyer, 1984).

The current degree of gross controllability also influences the warning process. If an agent is potentially controllable, there may be a reluctance to warn and alarm people. This occurred in the case of the Vaiont Dam, where on 9 October 1963, a huge landslide fell into its impounded reservoir. The displaced water overflowed the dam and engulfed the town of Longarone and more than a dozen other hamlets in northeastern Italy. Some 2600 lives were lost (Kiersch, 1965). That the slide was imminent was well known, but it had been erroneously believed that the reservoir could be emptied before a slide occurred. For this reason, no general alarm had been raised.

INSTALLATION, OPERATION, AND BACKUP

Clearly, the type and distribution of sensors used to collect data are critical. There is inevitably a minimum number below which the system is ineffective. Many sensors may be damaged by the impact of the hazard, by unrelated technical problems, or by sabotage. During the severe flooding in the midwestern United States in 1969, some of the worst losses were suffered by Minot, North Dakota. Many of the data necessary to predict the flood levels of the Mouse (Souris), the river responsible for damaging two schools, 15 churches, 212 businesses, and the homes of 11,864 people, were unavailable because Canadian gauging networks in the headwater of the river had been drowned out (White, 1969).

Ideally, resilience should be provided by an independent backup system that is capable of functioning in isolation. The author, for example, suggested that the tsunami warning system in British Columbia, which is based on a radio and telephone fan out, should have as its backup system aircraft flying above the speed of sound, followed by slower planes dropping flares into the ocean (Sewell and Foster, 1975). Such a system could be used to alert the population over a wide area very quickly, even if the primary means of warning malfunctioned.

EDUCATION OF THE USER GROUP

When disaster strikes, there is no substitute for knowledge. However, ensuring that the public response to disaster warnings is optimal is no simple matter. When flash flooding devastated the Big Thompson Canyon, Colorado, on 31 July 1976, some of those threatened tried to escape by driving out of the canyon. The high velocity flood waters overtook many of these, drowning them almost instantaneously. Had they abandoned their cars and climbed 25 feet up the canyon wall, they would most certainly still be alive today, little the worse for their experience (Gruntfest, 1977).

Public education is complicated by great differences in the experience, age, intelligence, race, health, and interests represented by the resident population. It must be remembered that the deaf cannot hear warnings, nor the blind see signals. In addition, in tourist areas, many people may be transients, completely unaware of local hazards. Any effective education program must accept these difficulties and try to minimize their negative impacts.

TESTING AND MODIFYING THE SYSTEM

Testing through the use of gaming, field exercises, or "dummy warnings" may dramatize several types of weakness, the most common of which stem from an inadequate data collection and processing base, staff shortages, and interpersonal or interagency rivalry. The destruction caused by the November 1970 Bangladesh cyclone, which has already been described, resulted from several factors, the most significant of which was the failure of initial warnings to be passed on by the local radio station, which was closed for the night. Moreover, a newly adopted streamlined system of warning was resented by officials, high and low, who deliberately blocked it (Burton, Kates and White, 1978).

DETECTION OF THE THREAT

Two opposing requirements, the need for speed and the need for accuracy, complicate threat detection. This implicit conflict makes it essential to establish, well in advance of any danger period, just what constitutes sufficient evidence of increased threat to issue warnings. This may prevent disruptive disagreement during those crucial periods when the potential for disaster is uncertain.

Another characteristic of threat detection is that the groups and agencies involved in monitoring the environment are often primarily interested in hazard identification (McLuckie, 1970). This can lead to warning signs being ignored since they are detected by individuals involved with pure research, product design, or some other problem.

COLLATION AND EVALUATION

However elaborate the technology of the warning network, it will not function effectively if the significance of the data is inadequately assessed. The chief problems to be faced at this stage of the warning processes are a) evaluation of the reliability of the data; b) estimating its precise implications; c) deciding whether danger is indicated and whether a warning must be issued, and if so, to whom, how and when. Any contradictions between data must be resolved. Such lack of consistency creates a common problem in predisaster situations that has often led to a risk being underestimated.

There is a well-established psychological principle that, when an individual is faced with conflicting statements, he or she is likely to accept as more valid those which are least threatening (McDavid and Harari, 1968). This principle also appears to apply to organizations. For this reason, evaluating groups faced with data inconsistencies tend to underestimate threats.

An example of this reaction to uncertainty occurred during the early phases of the 1972 Wyoming Valley Flood, caused by Hurricane Agnes (Massari, 1974). When the flood forecasters of the U.S. National Weather Service predicted a record-breaking 40-foot crest a few hours away from inundating Luzerne County, local officials refused to accept the prediction. Such a flood could have been expected to overflow every dike in the area. They therefore recomputed their available data and predicted a 38 foot crest. Civil defense officers used this more acceptable level to prepare for flooding in Wilkes Barre, assuming that the federal information was in error. Enormous destruction followed this decision (Cullen, 1978).

THE DECISION TO WARN

It has been estimated that the Mississauga derailment of 1979, which required the rapid evacuation of 226,000 people, cost $68 million (University of Toronto, 1983). Clearly, such decisions cannot be taken lightly. Indeed, once a major warning has been issued, the organization involved is unlikely to be unaffected regardless of the ultimate validity of the message.

Faced with potential danger, four general scenarios can be envisaged. In the first, although some apparent threat clues are collated, no decision to warn is taken. If disaster does not follow, the information is unlikely to be made public and the agency or organization will continue to operate as before, perhaps in a more conservative manner.

Conversely, disaster may in fact strike, as was the case on 9 October 1963, when the Vaiont Dam was overtopped (Kiersch, 1965). In such circumstances, deaths, injuries, and property losses will generally be attributed to the negligence of the officials involved. Loss of employment, status, and potential support is likely to follow, and legal action may be taken by victims or their relatives (Lewis and Clark Law School, 1977).

A third scenario takes place when the decision to warn is made but the envisaged disaster does not follow. This situation recently occurred in British Columbia when several coastal settlements were unnecessarily evacuated as the result of a tsunami warning. Such an error typically results in resented anxiety, inconvenience, and loss of time and money by those warned. Since hindsight will reveal that this was unnecessary, those responsible may be subjected to bitter criticism and even legal action.

Such mistakes undermine the future effectiveness of the warning system in two ways: those in charge may become too cautious and fail to respond to future threat clues; and conversely, people who have wasted time and money as the result of an earlier false alarm are probably less likely to respond to future warning messages. This is particularly true if the area has never suffered devastation from the predicted disaster agent.

A fourth scenario can be constructed in which warning is given and the anticipated disaster subsequently strikes. Under these circumstances, the adverse side effects are likely to be ignored and the value of the warning network confirmed.

TRANSMISSION OF WARNING MESSAGES

No warning system will function effectively if its messages, however logically arrived at, are ignored, disbelieved, or lead to inappropriate action. To avoid such pitfalls, warning messages must be designed and transmitted with the greatest of care.

All warnings must be specific, clearly identifying who is expected to respond. They should also include details of the expected time of impact of the disaster agent, in addition to the area threatened. Messages must also be specific in terms of the consequences.

The setting of warning message is also very critical. Clifford (1955), for example, records the unfortunate case of Piedras Negras, a Mexican town threatened by the flooding Rio Grande. Since disaster was imminent, two loudspeaker cars were coopted from the local cinema to supplement four official units. It is said that one of these vehicles cruised through the city repeating the following message:

> An all-time record flood is going to inundate the city. You must evacuate immediately. (Pause) The . . . theater is presenting two exciting features tonight. Be sure to see these pictures at the . . . theater tonight.

Such a warning message is unlikely to result in anything but total bewilderment.

INTERPRETATION OF WARNING MESSAGES BY RECIPIENTS

From the sociological and psychological research, three important generalizations can be drawn concerning the way individuals and families respond to disaster warnings (Killian, 1952; Fritz, 1957; Drabek, 1969). First, even though many people listen to the same message, each hears and believes different things. Secondly, people respond to warnings on the basis of how what they hear encourages them to behave. Thirdly, individuals are stimulated differently depending on who they are, whom they are with, and whom and what they see (Mileti, 1975). An organization's bureaucratic structure, especially if it is accustomed to centralized decision-making, may be slow to react to warnings. This seems particularly true if the required action would severely disrupt normal organizational activities. For example, competition among steamship lines in their attempts to deliver "express train service" encouraged the captain of the Titanic to ignore iceberg warnings and continue steaming at full speed, an error that ultimately resulted in loss of his ship and more than 1500 passengers and crew (Jones, 1976.

It is clear from this complexity that the behavior of those at risk cannot be easily predicted. Their behavior will reflect present activity, personal prejudices, beliefs, education, experience, and health. Response can be improved, however, by the type of educational programs already discussed. The relative unpredictability of public reaction also necessitates the next step in the design of warning systems, the planning for feedback.

FEEDBACK TO ISSUERS OF WARNINGS

On 23 May 1960, a tsunami inundated the lower sections of Hilo, Hawaii, killing 61 people and injuring several hundred others (Lachman, Tatsuoka and Bank, 1961). This loss occurred despite the fact that local officials had known the arrival time of this seismic sea wave for ten hours. Soon after this disaster, 327 survivors were interviewed in an attempt to discover why they had not evacuated the threatened area in spite of prolonged warning. It was found that, although the seismic sea wave siren, which had sounded for a 20-minute period more than 4 hours prior to the impact of the first wave, had been heard by 95% of the survivors, only 40% actually evacuated their homes. Further questioning revealed that the threatened population had been very confused about the the significance of those sirens. Numerous erroneous meanings had been attached to this signal to evacuate; these included a belief that the sirens were an alert, a preliminary signal preceding official evacuation, a notice to await further instructions or to make preparations to leave the area.

It is clear from this example that information must flow in both directions through warning systems, both to and from those at risk. Whenever the speed of onset allows, if those for whom the warning is intended do not respond as anticipated, new information must be provided. Further actions can also be taken to reinforce the message.

TRANSMISSION OF FURTHER WARNING MESSAGES

Warnings need to be repeated frequently, since many individuals will not react until the fourth or fifth time that they are exposed to such a message. Response is more likely to occur if the decreasing time available is stressed in later messages. The need for such confirmation was demonstrated during the Darwin cyclone disaster (Haas, Cochrane, and Eddy, 1976). It is in this context that feedback is most valuable since it permits changes in the nature and urgency of the information relayed. Later warnings can be structured so as to reduce existing misconceptions and prevent any further undesirable responses.

ISSUING OF ALL CLEAR MESSAGES

There are many widely believed myths about natural and man-made hazards. For example, during the 1964 Good Friday disaster in Alaska, three Aleuts were killed in Kaguyak because of their belief that the second wave of a tsunami was always the highest (Engle, 1966). After this had passed without causing damage to their homes, they returned, only to be killed by a bigger third wave.

Because of such widespread superstitions about many types of disaster agents, it is essential that subsequent warning messages stress that it is too early to return to the threatened area. It is also necessary, when the danger has passed, to issue an all clear. Widespread occurrences of premature returns necessitate the guarding of perimeters until such an all clear has been issued.

HINDSIGHT REVIEW AND TESTING OF THE REVISED SYSTEM

No matter how well a disaster warning system has apparently worked, it can always be improved. The necessity for many alterations becomes most obvious during actual disaster situations. It is therefore essential that the operation of the system be carefully monitored.

Such a process of hindsight review is likely to be most successful if detailed records are kept of the wording and time of arrival of all data, watch or warning messages, the system's response to them, and public reaction to this information. Such material is also very beneficial in the subsequent production of public education programs and in the training of new personnel. It is also necessary to interview those served by the system to determine when and how the warnings were heard, and when and what actions were taken as a result.

The major purpose of such a hindsight review is to use experience to achieve improvement. While any resulting changes might be expected to be beneficial, it cannot be assumed that this will be the case. For this reason, it is necessary to test the network after any significant alterations have been made.

CONCLUSION

As the preceding discussion illustrates, warning systems can fail to operate satisfactorily during any one or more of the sixteen steps outlined in Table 1. Most of the actions taken to make such networks reliable tend to be designed to enable them to withstand calculable, predictable types of technical or social failure. While this is essential, it is inadequate. Ours is a world of wars, strikes, sabotage, embargoes, corruption, mental illness, drug abuse, interagency competition, shifting social values, and dramatic technological change. Unfortunately, many of today's disaster networks, and indeed information systems as a whole, are not designed to accommodate stresses of this nature. Traditional analyses of the reliability of disaster warning systems tend to assess the probability and consequences of failure and to take appropriate steps against it. However, the most serious and unacceptable types of failure are often difficult to predict or to quantify, since many of these, such as sabotage, depend on the virtually unpredictable actions of a small group of malcontents. Even petty human weaknesses, such as alcoholic abuse, have frequently undermined apparently well-designed warning systems.

In reality, the total vulnerability of complex warning systems cannot be predicted completely. Since this fact has been repeatedly demonstrated, it becomes apparent that most such systems fail because they lack resilience, that is, they do not have the capacity to absorb change without catastrophic failure. To paraphrase Kahn (1978), what are needed are systems with the capacity to absorb shocks gracefully.

In an earlier report, the author (Sewell and Foster, 1983) sought to define resilience and identified 32 elements that assist in providing such flexibility to any system. While space does not permit detailed description of the characteristics of resilience (Table 2), it is suggested here that many of these should be used in any checklist designed to determine why a particular warning system failed to operate as anticipated. For example, did it have functional redundancy? Was it efficient? Was it too site specific? Did it lack in fail-safe design? In this way, principles might be developed that could be applied to reduce the vulnerability of information systems as a whole.

TABLE 2

THE ELEMENTS OF RESILIENCE

SOCIAL
1. Compatibility with diverse value system
2. Capacity to satisfy several goals
3. Equitable distribution of benefits and loss
4. Generous loss compensation
5. Accessibility

SYSTEM CHARACTERISTICS
1. Internal variables paramount
2. Capable of withstanding large external variable fluctuations
3. Generates rather than diminishes options
4. Diversity of components
5. Functional redundancy

ECONOMIC
1. Incremental funding possible
2. Wide range of potential financial support
3. High benefit/cost ratio
4. Early return on investment

ENVIRONMENTAL
1. Minimum adverse effects
2. Reversibility of impacts
3. Replenishable or extensive resource base

TIME AND TIMING
1. Short lead time
2. Responds rapidly
3. Operational life span open-ended

OPERATIONAL
1. Efficient
2. Hierarchical embedding present
3. Incremental operation possible

PHYSICAL
1. Not site-specific
2. Fine grained and modular
3. Prefabricated and standardized
4. Mobile
5. Esoteric components unnecessary
6. Unique skills not necessary
7. Stable
8. Fail-safe design
9. Early fault detection

REFERENCES

Andow, P. (1986). Failure of process plant monitoring systems. This volume.

Burton, I., R.W. Kates, and G.F. Wite. (1978). *The Environment as Hazard.* New York: Oxford University Press.

Clifford, R. A. (1955). *Informal Group Actions in the Rio Grande Disaster.* National Academy of Sciences-National Research Council, Washington, D.C.

Cullen, J. M. (1978). *A Comparison of Lifeline System Vulnerability in Two Large Regional Disasters: The Wyoming Valley Flood and the Projected Puget Sound Earthquake.* Report prepared for the Federal Disaster Assistance Administration, Region Ten, Seattle, Washington.

Dacy, D.C., and H. Kunreuther. (1969). *The Economics of Natural Disasters: Implications for Federal Policy.* New York: Free Press.

Drabek, T.E. (1969). Social process in disaster: family evacuation. *Social Problems.* 16, pp. 336-349.

Engle, E. (1966). *Earthquake: The Story of Alaska's Good Friday Disaster.* New York: John Day.

Foster, H. D. (1980). *Disaster Planning: The Preservation of Life and Property.* New York: Springer-Verlag.

Fritz, C.E. (1957). Disasters compared in six American communities. *Human Organization.* 16 (Summer), pp. 6-9.

Gruntfest, E. C. (1977). *What People Did During the Big Thompson Flood.* Working Paper No. 323, Institute of Behavioral Science, University of Colorado, Boulder, Colorado.

Haas, J.E., H.C. Cocrane, and D.G. Eddy. (1976). *The Consequences of Large-Scale Evacuation Following Disaster: The Darwin, Australia Cyclone Disaster of December 25, 1974.* Working Paper No. 27, Institute of Behavioral Science, University of Colorado, Boulder, Colorado.

Herd, D.G. (1986). The 1985 Rujiz volcano disaster. *EOS.* 7 (19), pp. 457-460,

Iyer, P. (1984). India's night of death: more than 2500 people killed in the worst industrial disaster ever. *Time* , 124 (25), pp. 14-21.

Jones, M.V. (1976). *Deadline Disaster.* London: David and Charles.

Kahn, E. (1978). *Reliability Planning in Distributed Electrical Energy Systems.* LBL-7877 Lawrence Berkeley Laboratory. Berkeley, California.

Kiersch, G.A. (1965). The Vaiont Reservoir disaster. *Mineral Information Service.* 18(7), pp. 129-138.

Killian, L.M. (1952). The significance of multi-group membership in disaster. *American Journal of Sociology.* 57, pp. 309-314.

Lachman, R., M. Tatsuoka, and J.W. Bank. (1961). Human behavior during the tsunami of May 1960. *Science.* 133, pp.1405-1409.

Lewis and Clark Law School. (1977). *Legal Constraints on the Planning and Development of Disaster Home Warning Systems.* Report to the National Science Foundation Research Applied to National needs Directorate.

Macdonald, G.A. (1972). *Volcanoes.* Englewood Cliffs, New Jersey: Prentice-Hall.

McDavid, J., and H. Harari. (1968). *Social Psychology.* New York: Harper and Row.

McLuckie, B.F. (1970). *The Warning System in Disaster Situations: A Selective Analysis.* Ohio State University, Disaster Research Center Report Series No. 9, prepared for Office of Civil Defense, Office of Secretary of the Army, Washington D.C.

Mileti, D.S. (1975). *Natural Hazard Warning Systems in the United States: A Research Assessment.* Program on Technology, Environment and Man, Institute of Behavioral Science, University of Colorado, Boulder, Colorado.

More, H.E. (1958). *Tornadoes over Texas.* Austin, Texas: University of Texas Press.

Mussari, A.J. (1974). *Appointment with Disaster.* Wilkes-Barre, Pennsylvania: Northeast Publishers.

Sewell, W.R.D., and H.D. Foster. (1975). *Disaster Scenario British Columbia's Tsunami Warning System.* Paper presented at the Environmental Hazards and International Development Session. Pacific Science Congress, Vancouver, British Columbia, August 1975

Sewell, W.R.D. and H.D. Foster. (1983). *Energy Policy Formulation for an Uncertain Future.* Renewable Energy Division, Energy, Mines and Resources Canada.

Spaeth, M.G., and S.C. Berkman. (1967). *The Tsunami of March 28, 1964 as Recorded at Tide Stations.* U.S. Department of Commerce, Enviornmental Science Services Administration, Coast and Geodetic Survey, Technical Bulletin No. 33.

University of Toronto, Institute for Environmental Studies. (1983). *The Mississauga Evacuation: Final Report.* Toronto: University of Totonto.

White, T.A. (1969). Satellites gave warning of midwest floods. *National Geographic.* 136(4), pp. 574-592.

White, W.R.H. (1966). The Alaska earthquake - its effects in Canada. *Canadian Geographical Journal.* LXII, pp. 210-219.

SYSTEM ISSUES IN INFORMATION SYSTEM FAILURE

Anthony Debons[1]

University of Pittsburgh
Pittsburgh, Pennsylvania, USA

INTRODUCTION

Recently several major events (e.g., Three Mile Island, the Challenger Shuttle accident, Chernobyl) have shocked man's consciousness as to the possible consequences that advanced technology could impart. Unfortunately, some interpretations of these events as technological failures (accidents) have masked the important role that information plays in such occurrences. On a more positive note however, these occurrences provide the basis for studies that are directed specifically to the significant role that information plays in such events.

Our task is to determine whether or not it is possible to establish methods that would determine the causal factors leading to the failure of information to meet the objectives that they have in regard to institutions, organizations, projects, and other entities in the social structure.

In this task, a given is that an information system consists of a number of identifiable components, the function of which aid or augment the human component of the system. People use information systems to contend with the diversity of conditions and events that influence and direct their lives.

The question of information system failures has received some notice in the literature. For the most part this treatment centers around those organization structures, habits, and practices, indulged or not indulged, that lead to the mitigation of proper data flow in an organization.

The work of Laplat (1984) and his colleagues is particularly pertinent. Laplat's research has dealt with the application of systems theory to occupational accidents. Laplat refers to an accident as "a consequence of dysfunctioning in the system which does not work as planned." In this connection, the terms "accident" and "failure" may be considered related, if not synonymous:

Accident: Something that happens without a cause that can be seen at once. (*Oxford Advanced Dictionary of Current English*, 1974).

Failure: Something that comes to an end while still needed or expected. (*Oxford Advanced Dictionary of Current English*, 1974).

Laplat cites a number of factors that can lead to the breakdown of a component of a system.

1. There is no clear indication of what the component or subcomponent of a system is to do.
2. Improper data flow between components of a system

These two concepts can be directly applied to the task of determining whether tools can be established to aid in investigating information system failure.

[1]This paper was written to provide guidance to those individuals who prepared working papers concerned with systemic issues.

NATO ASI Series, Vol. F32
Information Systems: Failure Analysis
Edited by J. A. Wise and A. Debons

INFORMATION SYSTEM MODEL

An information system consists of sensors whose function is to capture signals from the external world and to transport (transmit) such signals to a processing component (machine, brain, etc.) and ultimately to an agent (machine, human) that interprets such data for possible action (transfer). The question arises as to whether it is possible to develop an overall parameter of system activity that would indicate the state of these components and the stability of the overall system. Each of the components can be said to be governed by certain principles or laws related to its function:

Event World	Physics
	Language
Acquisition:	Threshold
	Physical loss of energy
Transmission:	Energy propagation
	Bandwidth
	Loading temporal phasing (neural)
Processing:	Automata
Utilization:	Cognition
	Decision theory
Transfer:	Language
	Motor responses
	Communication

In system or cybernetic terms, one can say that the principle of feedback can be applied to the entire system as well as to the actions governing the links between the components. But the task is more simply stated than realized. The identification of components is a complex matter deserving significant attention. For example, when one considers the human information system, there are significant questions about the extent to which our brain (processing element) controls what we see, and how much our sensing systems control what is processed in the brain. In technology-aided information systems, the question centers around the fixed capacities of sensor systems, the programming of outputs, the transmission limits that establish the prioritization of signals , the inclusion of noise, the coding properties that influence processing, and ultimately the match of these factors to data storage, retrieval and access. These are critical items that need to be studied before an understanding of the boundaries of information system components can be established.

SYSTEM BAROMETER

If a barometer of systems failure is to be established, other factors need be considered. For example, what function does time play in inducing, disturbing, or accelerating failure within the system? In the dramatic incidents of failure (Three Mile Island, Challenger, etc.), time was of the essence. Time can be considered from several points of view. Failure can occur in the functioning of a component and not be accounted because of the lag time in the reporting (sensor or transmission) mechanism. The specifications of the reporting sensor component can aid in accounting for possible incipient failure.

The time to respond is another factor. In the Challenger situation, events occurred at such a rapid state that operators could not have responded even if they wished to. If time to acquire and respond is a constraint on the system, the question amounts to accounting for such constraints not only on the component in question but on the overall information system. Time is ordinarily a constraint on the human information system to respond adequately, for example, to a possible head-on collision. The time involved in seeing the on-coming vehicle is related to the time involved (reaction time), but the response possible is also constrained by the characteristics of the entire nervous system in collecting and transmitting the signals to be processed by the brain. In machine-augmented information systems, the sensors operate within a particular time envelope (within which they can and can't function in capturing the event). The integration of both the human and machine capacities for realizing failure would be important in developing a barometer of individual component functioning.

We can postulate that the adequacy of whatever barometer of system failure one can come up with would be based on the reliability of the component and the measurement of their reliability as a function of time. It is easy to refer to sensors, both human and machine, because they serve to illustrate the points in question. The reliability of a machine or technological sensor is indicated presumably by the equipment specifications that accompany the sensors. Generally, sensor functions are exponential with established decay rates. Stress on sensors due to environmental conditions can influence such decay rates. Sensor redundancy can ameliorate the effects from such influences. The reliability of human sensor systems is much more difficult to define and acknowledge, but human factors research is available to provide guidance on such matters. Nevertheless, overall component reliability is an important consideration n overall system performance.

Maintenance and reliability are correlated. Ensuring that a fire extinguisher will actually extinguish the fire when it is used depends on how reliable the fire extinguisher is in extinguishing the fire - given its state of efficiency at the time the fire occurs. To ensure that the fire extinguisher will do the job will depend to a large extent on the periodic inspections applied to the extinguisher. Thus, one can postulate that maintenance data on sensor, transmitting, and processing components could constitute an essential aspect of the barometer of overall systems failure we are attempting to devise. From the human point of view, maintenance can be related conceptually to training - the level of human ability to respond to events. In nuclear power plants, control operations are subjected to numerous training exercises via simulation. Performance indications can be a subset of equipment performance indicators as part of the barometric measures of overall information system performance we are seeking.

In several failures of information systems, one aspect is predominant. Individuals fail to tell others (transfer) what has happened or is happening. Instructions delivered to induce actions are not received. When they are received, they are disregarded or discarded. This leads to what Laplat identifies as a "lack of link-up between the elements of the system... Information is indispensable when the actions of an individual depend on what another individual says or does." Furthermore, Laplat clearly indicates that "the greater this distance [referring to the position of an element in a hierarchy], the more different the communication and the higher the level of uncertainty and insecurity" (brackets are mine). How to account for these dimensions in an overall assessment of failure is a difficult task that will be encountered by an investigating team.

As Laplat clearly states, "incidents are only the symptoms of dysfunctioning." The breakdown of a seal in a rocket booster creates a whole series of incidents that represent failure in the overall system. If the barometer we are seeking for information systems failure is to be effective, the barometer should be sensitive enough to pick out the slightest change that can be considered an aberration contributing to system performance. One slight "knock" heard from an internal combustion engine may be insignificant. On the other hand, the knock is an indication and should be heeded as an incident leading to possible failure. But this is undoubtedly more easily said than done. The human tendency is to regard the "knock" as an incident that merits only fleeting attention.

INFLUENCES ON FAILURE INVESTIGATION

At this point, I would like to address the question of those factors that directly influence the investigation of an information system failure. A main question raised by the Workshop is whether information systems can be investigated as other systems are (transportation, management, financial, etc.). Again, Laplat's research is helpful as a frame of reference.

Wise (1977) in a doctoral dissertation cited the importance of change in understanding information systems. Laplat notes that "change and deviation are important concepts both in systems theory and in accident theories," referring also to the work of Kjellen (1983). And "change is conceived as the destabilizing element in a system which has no adequate response for it." The investigating teams that we are proposing will be faced with the issue of determining the role of change in failure directly.

Change is always in terms of some reference point. The elucidation of objects in space through a sensor system is based on the movement of the object within an established reference of azimuth and range. When a blip on a radar scope disappears, the change is based on a determined, predicted course, which is cause for alarm.

But these are situations that are not as clear cut. Deadly poison gas seeps through the air ducts of a conference room. If the change in the atmosphere is not detected by the occupants of the room, it can lead to their demise. The failure of the system in this last case can be relegated to the failure of the human sensory system in not detecting the change in the chemical content of the atmosphere. But the cause of the failure of the human information systems could be related to the possible failure of machine sensors (if any) that would have detected the change in the atmosphere in the conference room.

There are methods for the study of the changes that are alluded to briefly here. The INRS method proposed by Cuny (1977) and further illustrated by Johnson (1980) attempts to provide a sociotechnical classification based on individual, task, equipment, and environmental considerations. Rasmussen (1983) and Laplat later have attempted to use these methods in practical situations. I believe our present effort should include a careful consideration of such efforts.

Last, but certainly not least, is the role of the organization in information system failure and how this role can be accounted for in the barometer we are seeking. By organization, I mean that formal and informal structure, consisting of individuals, technology, and procedures, that guides systems to an objective or objectives. In this sense, the investigating team we are envisaging is also an organization (and both are political entities).

If we are to apply our own dictum of systems thinking, an accounting of group processes would appear important and relevant. Unfortunately, group processes encompass a wide area of theoretical and operational considerations that cannot be included in this paper (see Westrum, pp. 109-127 and Janis, pp. 129-162 for more thoughts on this matter). But there are certain fundamentals that merit a note of reference in this paper.

Referring again to constraints, organizations impose constraints on cost and time. Moreover, there are regulatory constraints imposed by the organization and by other constituencies like the government and the public. Often the internal and external constraints are not well understood until they have been initiated as the result of detailed assessment of a possible failure. The level of understanding of such regulatory issues could be part of the barometric instrument we are seeking.

Quite often the argument is made that an information system is as good as the statement of objectives it serves. Presumably, well defined goals or objectives and methodologies could seem to reduce failure. On the other hand, fixed ways of doing things foster boredom, leading to inattention and, possibly more important, loss of initiative. In some systems, failures, breakdown could have been avoided if the individual's own initiative had been exercised - an initiative that would have gone beyond the bounds of established policy.

In all organizations, segments (units, divisions, etc.) exist that tend to protect their boundary areas (turf). Perhaps the greatest source of misinformation, non-information, occurs at these boundaries. Considering the power spectrum that exists in hierarchially-oriented organizations, boundary conditions propagate practices that are conducive to system failure. Attempts to divest oneself of responsibility to an act that has gone wrong can influence the entire organization.

The organization is itself an information system made up of sensor, transmitting, processing, utilizing, and transfer elements supported by information systems infrastructure. The possibility of induced, self-directed as well as non-induced failures in information system functioning is enormous. Several years ago W. T. Grant, the third largest retailer in the Untied States, filed for bankruptcy under Chapter 11. Investigation of this incident as an information system failure revealed that certain segments of the management formulated decisions in the absence of well established information (data). Although not confirmed, it was alleged that certain segments of the company directed their activity towards negating the influence of the information systems to serve their purposes (Naharoni, 1980). It is clear in this case that the role of personal as well as collective motivation can provide a basis for understanding the interrelationship of three types of fundamental information systems: organizational, technologically aided, and the human information system.

The enormity of the task that will be faced by an investigating team in determining the cause of the failure of information systems is realized in light of what has been presented. Of course, there are other problems as well. For example, who will constitute the investigating team? What is the relationship of the investigating team to the organization in which the failure occurred? But these issues must be addressed by other working groups.

REFERENCES

Biddle, W. (1986). What destroyed the Challenger. *Discover.* April, pp. 40-47.

Cuny, X. (1977). An accident analysis method. In *Research on Occupational Accident.* Stockholm: Swedish Work Environment Fund, pp. 37-41.

Johnson, W. G. (1980). *MORT Safety Assurance Systems.* New York: Marcel Dekker, p. 525.

Kjellen, U. (1983). *The Deviation Concept in Occupational Accident Control - Theory and Method.* Royal Institute of Technology, Report No. Trita- A06-0019, Stockholm.

Leplat, J. (1984). Occupational accident research and systems approach. *Journal of Occupational Accidents*, 6, 77-89.

McKean, K. (1986). They fly in the face of danger. *Discover.* April, pp. 48-58.

Nahouraii, A. (1980). *An Investigation of W. T. Grant as Information System Failure.* Ph.D.dissertation, University of Pittsburgh, Pittsburgh, PA.

Oxford Advanced Dictionary of Current English. (1974). London: Oxford University Press.

Rasmussen, J. (1985). *Risk and Information Processing.* Riso National Laboratory, Roskilde, Denmark.

Wise, J. A. (1977). *Change as a Concept in Information System Theory.* Ph.D. Dissertation, University of Pittsburgh, Pittsburgh, PA.

SYSTEM FAILURE MODELS AS A RESULT OF DESIGN INADEQUACY

Andrew E. Bilinski

United States Air Force
Washington, D. C., USA

A PERSPECTIVE

Western society has been increasingly putting its way of life, recreational pursuits, economic systems, and national defense into the hands of complex information systems. These systems, composed of micro, mini, and mainframe computers, make use of thousands or millions of lines of computer code. Their use requires complex operating instructions or even advanced technical training. Basically, these systems include hardware, software, and human factors elements.

During a typical day, we use information systems to wake us in the morning and to cook our food. A micro computer in our automobile engages our ignition system and regulates fuel into our engine as we drive to work. Still another system serves as a security system that allows us access to our offices. For many people, work is stored by a computer, and even their rate of work is managed by an information system. We communicate with our peers and management using electronic mail. Finally, our paychecks and performance evaluations are calculated by still another complex information system.

While we are at work, our children are in school and may use micro computers to learn French or German. Still another system is available for searching out their reference books in the library. Information systems are used to prepare their bus assignments and report cards.

On a larger scale, complex information systems regulate control over nuclear reactors in power plants and submarines. Complex information systems control domestic and international air traffic, resulting in millions of miles of safe air travel annually. Systems composed of millions of lines of application code provide warning against nuclear attack deep within Cheyenne Mountain, and other systems await activation commands to deliver a sure and swift retaliation to our enemies.

In an advanced society, it is not unreasonable to expect that as long as a complex information system performs without a failure, we will become increasingly dependent on it and begin to take it for granted. However, when our microwave oven reduces an expensive roast to cinders, or our electronic ignition leaves us stranded on the highway on the way to a weekend at the beach, our confidence in these systems is shattered. A more thorough investigation might reveal that we misread the microwave instructions or failed to have our automobile serviced for the last 50,000 miles - and subsequently our confidence in these systems is quickly restored and hence our dependence on them.

However, other failures of information systems composed of hardware, software, and human factor elements have more serious consequences and result in comprehensive investigations regarding the cause of the failure and the necessary corrective action. The Air Florida crash in the Potomac River several years ago, the Delta Airlines crash in Dallas, and the recent Space Shuttle explosion all resulted in comprehensive investigations.

This paper, as part of the NATO Advanced Research Workshop in Failure Analysis of Information Systems, looks at the adequacy of design as a contributing factor in information

NATO ASI Series, Vol. F32
Information Systems: Failure Analysis
Edited by J. A. Wise and A. Debons

systems failure. In many cases, proper design of hardware, software, or human factor elements would serve to reduce system failure, prolong usable system life, or prevent catastrophic failure.

INTRODUCTION

The adequacy of the total system design is the fundamental component of a successful implementation of a complex information system. System failures as a result of hardware component failure or poor human factors design clearly contribute to system rejection by users or to catastrophic failure. Software failures resulting from a poor design phase can shorten the usable system life, and clearly accelerate to the point where the level of effort necessary to perform routine maintenance and functional enhancements, as well as to react to "software bugs," will be greater than the effort to redesign still another "new" information system.

The combined or simultaneous effects of hardware component failure, software failure, and human factors aspects can serve to continually give users the impression that the "bugs" have never been worked out, or in the worst case, can cause an information system failure resulting in tragedy.

Consider a typical implementation of a new information system (i.e., a system never installed before) from the perspective of the three major components. For emphasis, we will illustrate using the Air Florida, Delta Airlines, Space Shuttle, and other tragedies as examples.

HARDWARE COMPONENT

From the information systems hardware aspect, the initial installation and the immediate time-frame around the installation offer the greatest opportunity for unexpected failure. Usually these failures are in direct relationship to the design adequacy, the quality control provided during the hardware manufacturing process, any pre-shipment "burn in" quality control activities, as well as the skill and experience of the installation engineers and operators.

Consider, as an example, the early prototype flights of almost every tactical fighter over the last twenty years. Hardware failure was frequently the cause of early mishaps for the F-111, F-16, and others. Design characteristics were continually modified for these prototypes. Another example of hardware failure as a result of poor design might be the press-speculated "Stealth" fighter aircraft and its unique design characteristics.

Following the initial high-stress period, the system will usually fall into a rather routine, predictable failure posture on a component-by-component basis, with the mechanical components having the least mean-time-between-failures. However, as the system begins to reach the natural end of its life cycle (not easily defined), the frequency of component failures begins to increase, and the predictability of their occurrence becomes more difficult to anticipate.

This is not always the case when a system has latent design flaws, as in the case of the Space Shuttle disaster. Clearly, the design of the O-rings on the solid rocket booster was deficient from its inception. But the design flaw was not of a nature to cause a failure upon the initial shuttle mission. The extremely low temperatures on the launch morning, coupled with other problems, caused the latent design flaw of the O-rings to be manifested in catastrophic proportions. Latent design defects not discovered during or shortly after implementation become almost impossible to predict or discover in and of themselves, as indicated by Figure 1.

This is the case because of the nature of implementing the hardware aspects of our current high technology information systems. The engineers expected hardware failures, including the design defects, to be manifested during initial tests and during the first shuttle missions.

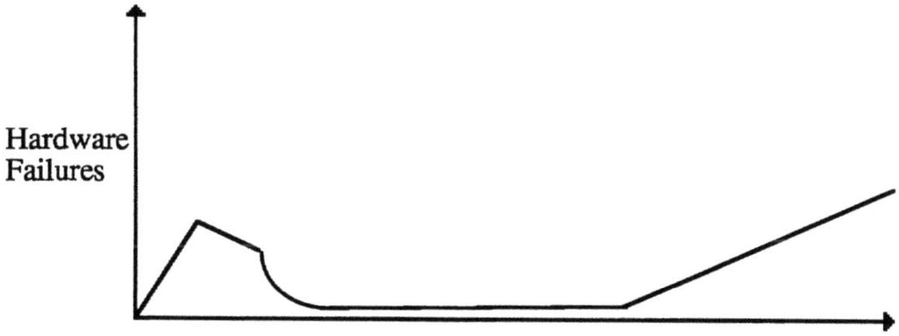

Figure 1. Hardware failure over time.

Although operating temperature warnings were known as far back as 1980, these warnings were ignored by all except a few people as successful shuttle missions were flown.

SOFTWARE COMPONENT

Consider the life cycle elements of the software component of information systems. Historically, the ability to test any complex software system has been recognized as essentially impossible. As a result, the evolving software engineering methodologies and techniques have emphasized processes and procedures that apply software systems at the earliest possible stage of the software development life cycle.

Frequently, inexperienced software designers will not only provide incorrect design of the functional software itself, but will also provide incomplete and insufficient designs for the automatic failure analysis and recovery of selected boundary data conditions that might be encountered. Again, assuming that a 100%, or complete, data condition and data path test program is not practical (the normal situation for large systems), it must be assumed that the initial operational version of the software component will have inherent functional design flaws, and will not be able to automatically recognize and/or recover from data and processing failure conditions. However, as the information system is installed and operated in its intended environment, the identification and correction of these inherent design flaws is initiated. With most new systems, the first two years of operation represent the heaviest period for this level of software failure analysis, identification, and correction.

The software failures as a result of design flaws observed in the initial operation of the Internal Revenue Service's new automation program in 1985 led to mountains of unprocessed tax returns and lost data that had to be manually reconstructed. In 1986, the Army's cancellation of the DIVAD project was largely based on the inability of the software integration to isolate, track, and employ the firing systems in order to hit the targets consistently. These problems were discovered during the prototype testing or during implementation.

The next period of the software life cycle represents a reasonable period of software stability, with most of the failures introduced during the addition of software enhancements as the system is notified and/or extended beyond the original design. Generally, the failures are localized to those elements of the code that were modified during the software enhancement and maintenance activity. However, there is a point in the software life cycle when historically the amount of enhancement/modification begins to exceed the original software design structure, i.e., a specific section of code has been modified several times over, and the result is an increasing number of software failures due to the number of software corrections (sometimes known as "patches") applied to that section of code. This is the twilight period of the software system during which conscious decisions must be made as to the redesign and re-implementation of the software system.

Users of the large IBM mainframe family of operating systems will remember all too clearly the progression of MVT, MVS, and MVS/XA. IBM continually provided upgrades (or "patches," depending on your perspective) to its operating systems. However, every several years, another major new operating system would be released. Clearly, during the period prior to the release of a new system, and after "fixes" to the old operating system were stopped by IBM, system programmers were severely challenged to keep pace with the problems.

Recent thrusts in software engineering methodologies have emphasized new tools and techniques to support improved maintainability of software systems, but the generic problem will continue as long as the system user needs and demands modifications to software systems to extend their utility to the initial and primary functional mission.

Figure 2 shows two curves of the probability of software failures as a function of time. The lower curve depicts the failures likely to result from rigorous and structured design phases. The upper curve shows the failures likely to result over time as a result of a design phase with serious flaws. These flaws could be the result of the infamous "moving baseline," an imprecise design methodology, or a design team working in a vacuum. Nonetheless, we find that the system failures resulting from major design flaws peak at three times the level of the optimum design process. Second, we find that the on-going failures occur at a rate of more than four times our optimum design. Finally, we find that our system never reaches a stable period during the life cycle.

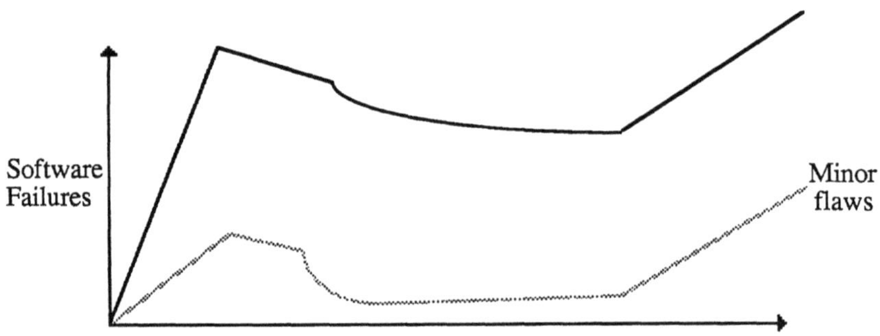

Figure 2. Software failures over time.

HUMAN FACTORS COMPONENT

Shifting now to the human factors element of the information systems, it is not atypical for the operators and end-users of new information systems to receive extensive training on a new system before it becomes operational. However, in many cases, this training is inadequate to provide the level of proficiency necessary for fault-free operation. Many factors come into play: the natural human resistance to change with a new system; the system development cycle, which is always late with the software development, usually resulting in a reduction in the time provided for testing; the end-users' pressure to go into operation prematurely; and frequently the ability of the formal training (if provided at all) to effectively emulate the "real operational world" that the computer system operators and the end-users will face.

The result of these elements is that the initial operational period of any new information system has frequent system or subsystem failures which can be directly attributed to human actions, although they are disguised by the situation to be diagnosed as either hardware or software failures. The worst conditions usually result from the human operator's inexperience with recovery capabilities, resulting in actions that may, in fact, significantly magnify the recovery of the failure condition.

During the second World War, it was not uncommon for fighter training squadrons to experience accident rates of more than 40% during the first several training missions. During the Vietnam conflict, due to improved training methods, the accident rates were cut by more than 50%. Currently, with the use of Computer Based Instruction (CBI) to reinforce textbook and classroom instruction, flight simulators and special training aircraft like the T37 have radically reduced training accident rates.

As the human element gains experience on the system (both by actual individual experience and improvement to formal and on-the-job training programs), the human failure element is reduced to a period of fairly stable and minimal impact on the operational system. However, there is a period subsequent to the implementation life cycle of the system during which the human factor again comes into play. This is the period when the equipment is reaching obsolescence and/or the magnetism and charisma of the system have dwindled for the human operator. During this period, the experienced, well qualified operators seek other employment opportunities, and the resulting work force has neither the experience or, in some cases, the motivation and professionalism to provide a human-failure-free operation.

The Air Florida Flight 90 crash into the icy Potomac River in 1981 is a tragic example of what can happen when on-going training is not provided. Clearly, the airlines spend a great deal of effort in explaining to the pilots the effects of ice and snow accumulation on the aircraft during takeoff. The lift coefficients drop radically with the accumulation of each 1/8 inch of ice or snow. The Air Florida aircraft was properly de-iced at the gate, but due to a long queue of aircraft waiting to depart Washington National Airport, it was estimated that 3/4 of an inch of snow had accumulated at the time of liftoff. The problems were significantly expanded by the fact that the PT2 probe pressure transducer used to measure engine thrust had also iced over. The ice accumulation caused an artificially high reading for engine thrust, which the pilot compensated for by reducing power during takeoff. Finally, the proper procedure would have been for the pilot to request another de-icing, as did several other pilots waiting in the queue.

Figure 3 shows two curves of human factors failures as a function of time. The lower curve depicts the failures likely to result from adequate human factors considerations during the design phases. These considerations include a "user friendly" man-machine interface, adequate training and training material, and a stable operator base. The upper curve shows the system failures as a result of inadequate human factor considerations. Although not as traumatic or alarming as the software example, the curve peaks at twice the optimum level, and the stable period is shorter.

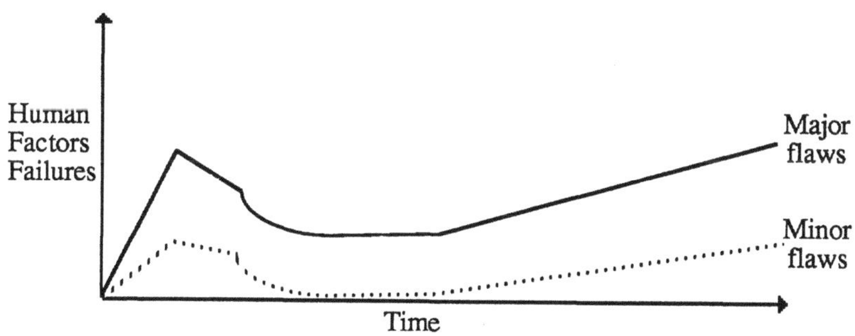

Figure 3. Human Factors failures over time.

COMBINED EFFECTS

The combined effects of the various models show some surprising results (Figure 4). The first is that by summing the failures as a result of hardware, software, and human factors, we find no "flat spot" to any of the models. This is particularly alarming because systems managers and even customers are usually led to believe that things will "settle down" once the initial implementation "bugs" are worked out. Hence, in a non-scientific and crude fashion, we have challenged the age-old myth that after the system implementation is complete, the technical challenge has been overcome. The examples show a distinct challenge for maintaining a competent maintenance staff with increasing numbers over time, ensuring an active training and re-education program, and finally, providing for adequate hardware maintenance support.

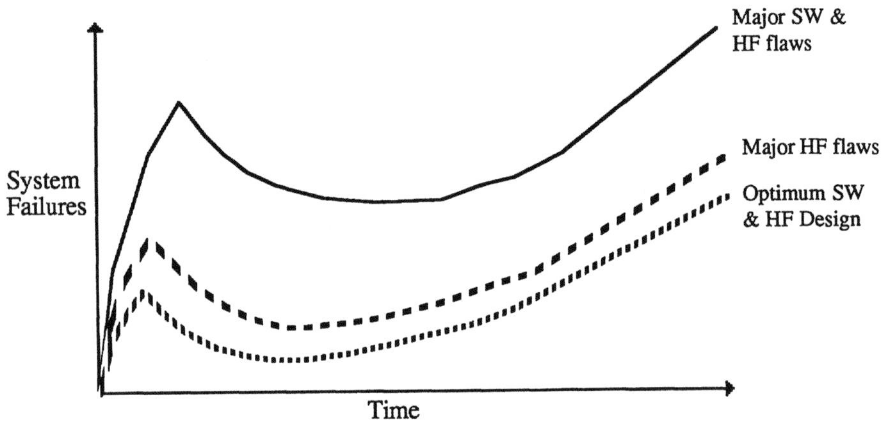

Figure 4. Combined effects over time.

An analysis of the individual curves shows that in the worst case of a flawed software design phase and inadequate human factors considerations, total system failures are more than twice as likely to occur than in the optimum configuration.

The Space Shuttle investigation shows a hardware design failure of the O-rings combined with human factors errors in the decision to launch during extremely low temperatures. The United States Air Force Base-Level Automation upgrade program of the early 1980's was plagued by software conversion problems with three million lines of code as well as training problems. As a result, the Air Force experienced schedule slippages as they overcame these problems.

From examining the models, we find that combined information systems failures continue to increase for a period twice as long in duration as in the optimum situation. This situation can be unnerving for the system manager trying to keep his job, the contractor trying to lower his systems staff in order to maximize his fixed price profits, and the user trying to see the benefits of his expensive new automated system. Clearly, the problems facing senior government leaders, politicians and our military leadership do not have to be expanded further.

It is at best difficult, if not impossible, to predict in absolute terms the number of system failures that will occur as a result of hardware, application software, or human factors considerations. However, it is possible to insist on an adequate, rigorous design phase, to expect system failures after implementation, and finally to be prepared to expend the necessary resources after implementation to respond to these system failures.

BIBLIOGRAPHY

Boston Globe. (1985). Why Weinberger Abandoned the York. 8 September, p. 1.

Government Computer News. (1985). Congress, DOD Disagree Over NORAD Reliability. 11 October, p. 3 .

Information Week. (1986). Federal DP: Red, White and Blues. 4 August, p. 26.

Information Week. (1986). Data Processing Taxes the Internal Revenue Service. 4 August, p. 33.

Los Angeles Times. (1985). HIgh Tech, High-Fail Defense. 16 September, pp. 11-15.

Report on the Presidential Commission on the Space Shuttle Disaster. (1986). June.

Wall Street Journal. (1986). Air Force Is Said To Have Bought 50 Stealth Planes. 25 August, p. 15.

Wall Street Journal. (1986). Recycled Blades Wrecked Jet Engines, Probers Say. 8 August, p. 36.

Washington Post. (1986). 50 Stealth Fighters in Operation." 22 August, p. 1.

FAILURE OF BUSINESS INFORMATION SYSTEMS

John D. Spiers

DCE Computer Consultancy
Working, Surry, U. K.

Traditional concerns over the reliability of computer-based systems have focussed on the real-time area where computers directly interact with a physical system, such as flight management. Increasingly, as commercial business systems acquire a more central role in the strategy and management of business operations, the consequences of failure of such systems become potentially more destructive.

This paper is primarily concerned with computerized information systems used for the operation and management of business and technical processes. Such systems typically provide interfaces with human beings, as well as with automation equipment. Failure issues are discussed with particular reference to the six generic components of the EATPUT model and the techniques of Information Analysis.

COMPONENTS OF A SYSTEM

Systems operate within an environment, sometimes referred to as the Universe of Discourse. To operate within this environment, a system is provided with mechanisms (sensors) for capturing events and information from the Event World or perception space of the system. Such acquisition may well be dependent on people processes (e.g., selection of significant events, actual data entry). Similarly, the total system must be provided with a means to utilize its information to act upon objects in its sphere of influence or action space. Again, these means may well involve people, e.g., in a management decision and the implementation of such a system. It is important to recognize that, except for fully automated closed-loop control systems, people form an essential component of all information systems.

For our purposes, an information system can be viewed as physically made up of the following types of components: sensors, actors, computers, communications links, information storage, and people.

THE UNIVERSE OF DISCOURSE

For a system to perform a useful function, there must be an element of commonality of scope between the perception space and the action space of the system. For example, a system could be constructed to make investment recommendations based on the current state of the weather. More useful, though, would be an investment system based on a monitoring of the financial markets. Similarly, we would expect to find a correlation in size between the action space and the perception space. This avoids far-reaching decisions based on inadequate information, or conversely, the expense of capturing information that cannot be effectively applied. Many, though certainly not all, information systems exhibit an element of feedback from their action space to their perception space.

THE STRUCTURE OF INFORMATION

It is an essential characteristic of all non-trivial information systems that they contain stored information. Typically, information is stored by a Processing unit so that it can be used in subsequent processing and utilization. In simple cases, information may be stored purely as

NATO ASI Series, Vol. F32
Information Systems: Failure Analysis
Edited by J. A. Wise and A. Debons

the state of a processing unit in the way that a switch "remembers" its last input. More generally, however, information is stored on manual or computer-based storage media.

EAR Information Model

Information within the scope of a system or problem has its own structure, which is largely independent of the processes to be performed upon it. Such a structure may be clearly described in terms of entities, attributes, and relationships.

An entity is something of fundamental importance to the area in question. For the organizer of a NATO Workshop, such items as attendee, workshop group, and working paper would be entities. Attributes are descriptive characteristics of an entity; for example ,the entity "attendee" would have such attributes as name and address. Relationships are two-way connections between entities, as shown in the following example:

Attendee	may lead a maximum of one	Working Group
Working Group	must be led by exactly one	Attendee
Attendee	must prepare one or more	Papers
Paper	may have been prepared by (one)	Attendee
Paper	may have been prepared by (one)	Working Group

Entities, attributes and relationships may be summarized diagrammatically in the form of a data model (see Figure 1).

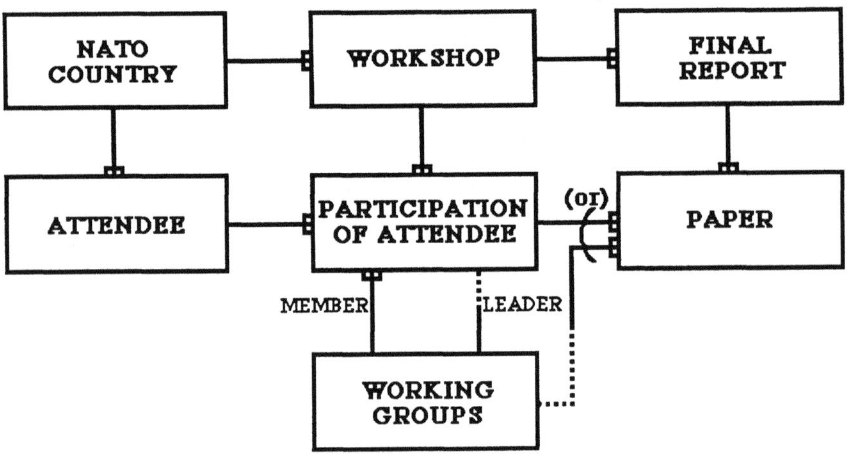

Figure 1. Illustrative data model

The definition of information will include integrity constraints to be imposed. Such constraints can be of 3 main types:

Static:
- An Attribute must take its value from a Domain (of permitted values).
- Interdependence between attributes in one or more entities.
- Referential integrity - relationship rules must be preserved e.g., there can be no Attendee who does not belong to a Working Group.

Dynamic:
- Attribute values can make only certain transitions (e.g., Marital Status can only become Divorced if previous status was Married).

Authorities:

- Operations on the information (e.g., to create and remove entity occurrences, to change attribute values, and to make and break relationships) are restricted to particular people, processes, and circumstances.

It is a function of any information system to protect the integrity of the information that it seeks to manage and use. Unfortunately, data management software is generally not able to enforce all the integrity requirements arising from the nature of the information. Clearly, the design of a system should ensure that integrity enforcement is built into all relevant processing components. In practice, however, many systems are built without even identifying the real integrity requirements.

Static integrity can be verified directly by examination of the state of stored data. Dynamic integrity can only be verified by an examination of the processes by which data has been created and changed, for example from an activity log.

A SYSTEM MODEL

A manual or computerized information processing system can be modeled (See Figure 2) using the techniques of activity analysis and can be represented by the following components:

EVENTS	external to the system which trigger activities
SOURCES	external to the system from which events and/or inward data flows arise
SINKS	external to the system which receive data flows
DATA STORES	in which information is stored internal to the system, corresponding to combinations of entities and attributes
DATAFLOWS	to and from Sources, Sinks and Datastores and between activities
ACTIVITIES	which transform inward data flows to outward dataflows

The overall function of a system is to produce the correct outward dataflows (to people or automated actors) given the historical combination of events and inward dataflows from external sources. Each activity within the system can be viewed as a sub-system, with the function of transforming incoming data to outgoing data. At a detailed level of analysis, each activity would normally serve a single acquisition, transmission, processing, or utilization function.

THE DOMINO EFFECT

Any non-trivial information system involves many activities, generally working in parallel. These activities are interdependent through dataflows and the use of common data stores. A failure may therefore be rapidly propagated through a system, leading to further failures in a cascading process. By the time the system reaches stability (comes to a halt or is otherwise controlled), considerable damage may have taken place.

This model parallels the more physical failure in something like an air crash in which an initial problem or combination of circumstances may lead to the ultimate destruction of the total system. The job of the investigator is to establish how and where the failure chain was initiated.

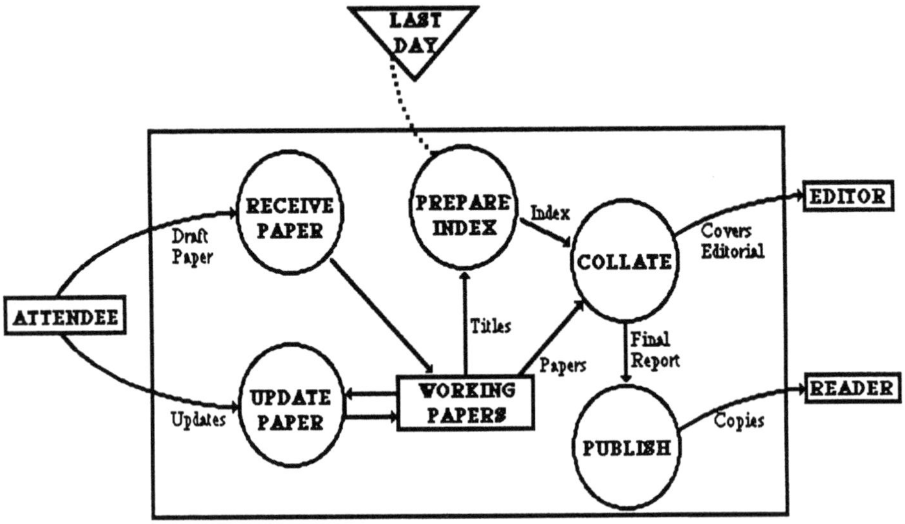

Figure 2. Illustrative activity model

MEASURING THE TYRE MARKS

After a car crash, it is common police practice to examine the wreckage and the environment to assess the circumstances of the accident (e.g., speed, point of impact, direction). Such an accident by its very nature leaves behind a mass of physical evidence that, with careful analysis, provides a record of events. Air crash investigators go to inordinate lengths to retrieve any pieces of wreckage for such examination.

Manual systems, too, provide a form of automatic audit trail, mainly because that they are paper-based and paper is not re-usable. A coroner performing an autopsy has a mass of medical evidence in physical form before him. Computer and people systems do not have such evidence. A failing system does not necessarily leave behind a physical trail of wreckage, nor does it carry the wrinkles and scars of a lifetime's work. Because a computerized system can reuse its storage media, it does not necessarily leave behind an audit trail.

It is a characteristic of many types of failure that through the cascading process they tend to destroy their own evidence. Electronics systems are particularly adept at this.

Software Engineering techniques seek to remove redundancy because of the overheads and the consistency problems that redundancy creates. This, of course, also reduces the opportunities for post-failure investigation. Audit ability is dependent on redundancy and can only be achieved if redundancy is deliberately built into the design of systems.

PROTECTING THE BLACK BOX

Commercial aircraft use flight and cockpit voice recorders to provide an audit trail for use by investigators in the event of accident. These items are clearly redundant and exist only for post-hoc analysis.

It is obvious that where a redundant audit trail is provided it should be as independent as possible of the main systems so that it is shielded from the effects of cascading damage by the failure.

DAMAGE CONTROL

A widely accepted heuristic of systems design is to favor minimum complexity of interfaces between components. This makes it possible to design and build complex systems on the "divide and conquer" principle without the complexity of an increasing number of interfaces overwhelming the development.

Effective damage control requires a similar approach. A failure is isolated by shutting down the damaged component and isolating it from neighboring processes. The system can continue to operate, perhaps in a degraded mode, provided each component has a recognized function that is in some sense redundant. Given a testable interface between sub-systems, it is possible to detect failure of a neighboring component and to close it down or ignore its outputs.

Damage control mechanisms, together with redundancy (backup), can therefore prevent a cascading failure. Even where no recovery is possible, damage control at least has the effect of preserving as much evidence as possible in an undamaged state for investigation.

WATCHING THE LIGHTS

It is well-known that computer operators and programmers become adept at detecting problems by observing the general pattern of activity - for example, through the panel lights. Similarly a motorist who notices a change in engine sound from normal patterns will probably take the car for a check-up.

In these cases, it is important to note that what has been detected is a change from an established pattern, rather than any detailed knowledge of how the system works or of the failure that may be taking place. These examples involve the learning of a normal pattern of activity to allow the first sign of a failure to be recognized. Experiments have been carried out in the practical application of this kind of technique - for example, to detect imminent failure of machine tools by monitoring the vibrations.

In investigating a systems failure, our aim must be to establish the first point at which the system deviated from normal behavior and follow through the cascade to establish in detail the actual failure mechanism. Human experience or artificial intelligence gained by monitoring normal patterns of systems behavior can therefore provide a base for failure analysis. Similarly, systems might effectively monitor themselves through a built-in barometer. It is important to note that such a barometer need only provide a normality rating and is not necessarily dependent on any *a priori* knowledge of the system in question. In principle, therefore, such a general-purpose barometer could be applicable to a wide variety of systems.

THE AUDIT ROLE

It may now be apparent that there are marked similarities between the roles of accident investigator and auditor. Depending on circumstances, an auditor may have responsibilities for examining the following:

- Existence of procedures to prevent and detect failure.
- Presence of conditions that may lead to a failure.
- Inconsistencies that may give early indication of failure.
- Presence and accuracy of the audit trail.
- Causes and lessons of a failure.

An auditor is particularly concerned to remain as far as possible independent of the system being audited. Investigation procedures must similarly ensure that the investigators themselves are not a part of the failed system. By way of example, in the UK investigations of police failures or malpractices are carried out by police from another area. Opinions differ as to whether or not this constitutes independence.

REQUIREMENTS VERSUS TECHNOLOGY

It goes without saying that systems should, in principle, be designed to meet their requirements. However, engineering has traditionally been concerned with building on the basis of what is technically possible. There is a requirement to transport people between locations at high speed; the fact that this requires them to sit in narrow tubes reflects a technical limitation, not an essential characteristic of high-speed transport.

The construction of information systems is subject to much less in the way of physical constraints than, say, the construction of aircraft, but there are normally many practical constraints and pressures. The danger of the pragmatic (technically-driven) approach is that a system can be built to meet a general area of requirement but with its actual operational specification unknown. Clearly, a very important question for an investigator is whether a system was asked to operate outside of its design limits. This requires the boundaries of the perception space and the action space of the system to be identical, together with the assumptions that have been made about them - e.g., the expected event rate.

Unfortunately, many information systems, particularly manual systems that have grown by evolution, do not have identified limits. As an example, companies which outgrow their management systems are clear candidates for failure. To understand how something fails, it is necessary to understand how it is intended to work. Or perhaps we should say that a system with unknown operating parameters cannot by definition fail.

MIS SYSTEMS AND INFORMATION MANAGEMENT

The rise of the management information system reflects the recognition of the value of information as a key business asset, coupled with the availability of cost-effective technology allowing it to be stored and utilized. Like all things of value, information needs to be loved and nurtured if it is to remain in good condition. What value is a wise decision based on bad information? The introduction of information systems needs to be supported by an ongoing Information Resource Management function to protect the value of the information asset by the required completeness, accuracy, and integrity

Increasingly, artificial intelligence techniques will be applied to business models for the purposes of planning, trend analysis, evaluation of alternatives, etc. The term decision support which is already in common use, gives a clue to the future.

With traditional business systems, it has been possible to readily understand the automated processes and to gain reassurance, if necessary, by manual walk-through. As technical automation systems have become more powerful and more automated, however, the key problem has become that of who is the final arbiter, man or system, particularly as man has less and less chance of comprehending the circumstances that have led to an action or recommended action. Business decisions generally need human actors to apply them - people who must be persuaded, cajoled, or pressured into agreement... people to be held accountable when the system fails.

CASE STUDIES

Not surprisingly, most business organizations that suffer loss through the failure of an information systems do not wish to gain publicity for the failing; hence, in general, only spectacular cases come to light. Indeed, failures relating to computer systems seem to receive even less exposure than people failures -- perhaps because it is easier to sack a person than to replace a computer system.

Two examples of information systems failure are outlined below. Both have been chosen from the same organization, a small supermarket chain.

Case A

Over a period of several years, an organized fraud was carried out within the operation, whereby orders were issued for goods that were never received; goods received notes that were fraudulently prepared and were passed for payment. Clearly, such a fraud, involving a conspiracy among many individuals, could take place in many organizations, but in this instance, it lasted for at least five years on a substantial scale.

Ultimately, the problem can be seen as one of information integrity. No controls were in place to establish and audit the connection between the order/receipt/payment process and the existence of the goods themselves. Procedures did not ensure the integrity of the relationship.

Case B

At the same time, the organization used a management accounting procedure based on a supposed sales mix by goods category (e.g., groceries, fresh foods, toiletries). This mix had been calculated several years previously by analysis of sales at each store. Since all internal accounting was performed using the selling price, profitability was derived by dividing turnover into categories and applying current average margins for each goods category. In fact, the turnover mix had changed so substantially over the years that a higher proportion of turnover arose from lower margin goods. The effect was to overstate profits by an increasing amount for several successive years.

The failure here was threefold: the use of what should have been a derived item (trading mix) as though it were a fixed independent item of information; the failure of anyone to remember that such a fixed value would require periodic adjustment; the lack of any effective independent calculation which would have allowed the discrepancy to be detected. The system incorporated assumptions about the real-world which were no longer valid.

CONCLUSIONS

- Information systems need to be audited to ensure that they are operating effectively and are likely to continue to do so.

- Information systems need an audit trail but do not in themselves automatically provide such an trail through their functional operation. The audit trail requires the introduction of controlled redundancy.

- The audit trail needs to be protected against damage during the failing of the information system.

- The components of a system should as far as possible be insulated from other components. This serves both to limit the consequence of a failure and to preserve as far as possible the evidence of the system state in the period leading up to failure.

POSTSCRIPT

The following list summarizes the main failure issues against each of the six elements of the EATPUT model:

EVENT
- Did a relevant event occur for which the system did not provide?

- Was the system subjected to events outside its qualitative boundaries?
- Did the event rate exceed the system's capacity?

ACQUISITION
- Were all events successfully detected and acquired (especially when dependent on manual capture)?

TRANSMISSION (e.g., data entry)
- Was all information triggered by events captured accurately?
- Failure across man-machine interface.

PROCESSING
- Failure of design of activity.
- Failure of a processing unit (hardware / software / human).
- Invalid dataflows to other activity elements.
- Invalid updates to stored data (loss of integrity).

UTILIZATION
- Use of invalid input dataflows (due to loss of integrity).
- Accuracy and effectiveness of outward dataflows.
- Failure of utilization unit (hardware / software / human).
- Failure of utilization algorithm (e.g., inadequate human understanding).

TRANSFER
- Accuracy and timeliness of transfer.
- Failure of transfer unit (hardware / software / human).
- Failure across interface to actor especially where human).

REFERENCES

Jones. (1985). *Air Crash.* Robert Hale.

Rock-Evens. (In Press). *Information Analysis.* New York: Pergamon Infotech

Ward and Mellor. (1985). *Structured Development for Real-Time Systems.* New York: Yourdon Press.

Winter. (In Press). Conceptual modelling for database management systems. In *Fundamentals of Information Technology.* New York: Wiley.

THE RELATIONSHIP BETWEEN INFORMATION AND DECISION MAKING AND THE EFFECT ON THE RELIABILITY AND FAILURE OF INFORMATION SYSTEMS

M.C. Yovits, A. de Korvin, R. Kleyle, L. Medsker, and M. Mascarenhas

Indiana University-Purdue University
Indianapolis, Indiana,USA

INTRODUCTION

Research by the principal investigator and a number of colleagues -- first at Ohio State University and now at the Purdue School of Science at Indianapolis -- has been underway for some time in order to develop a fundamental theory of information flow and analysis and to apply it effectively to practical and useful situations. We attempt to identify, and more important, to quantify the significant variables and parameters in the information flow process, and then to establish relationships among these variables.

All of our work has been directed toward the ultimate goal of applying the theoretical and conceptual base we are developing to practical and useful situations. It is our belief that the models we have developed and the assumptions we have made are descriptive of real situations and are fundamentally useful.

We have now developed a formal and quantitative framework that permits the explicit definition of a number of terms and parameters analytically, measurably, and reliably. Thus, we can use our model and theory to make comparisons among various systems and to identify those components that are most significant in determining the reliability of these systems.

Our approach is probabilistic. It is our belief that because of the extremely large number of extraneous factors that can influence complex information system, any approach must be statistical. Identification of the probabilities and their effects on the system reliability is essential.

Our approach relates information to its effectiveness and its use and value. In other words, information implies application and use. In this regard, we relate use to decision making. This approach has led to the powerful and flexible concept of a Generalized Information System. Theory, measures, and relationships thus derived relate to the way information is utilized by a decision maker and the effects on the decision that is made. Specifically, we treat information conceptually as inversely related to the uncertainty in making a decision. That is, information is a direct function of the certainty involved. Starting with this definition, we are able to define information quantitatively and then to establish some of its properties and interrelationships. This approach also relates to the effectiveness of decision makers in a particular situation; thus, we also define explicitly and quantitatively the term "decision maker effectiveness," or DME.

Information is frequently used rather specifically in the sense of the Shannon and Weaver "information theory" (more accurately called "communication theory"). In this sense, the context of the message is of no significance; the theory is concerned with the probability of the receipt of any particular message for various conditions of the transmission system. While this may be of interest, it is certainly not the major nor even a large part of information flow and analysis. Such a treatment does not consider the vital areas of concern, almost all of which are involved with the context, meaning, and effectiveness of the message. For these reasons, the

NATO ASI Series, Vol. F32
Information Systems: Failure Analysis
Edited by J.A. Wise and A. Debons

Shannon and Weaver approach is generally regarded as too restrictive to be a basis for the reliability analysis of information systems. At the other extreme, the treatment of information as synonymous with knowledge appears to be far too broad to lead to meaningful and useful principles or relationships.

Our approach is quite different from these. It concerns, as stated earlier, the relationship between information and decision making and how important or valuable that information is to the resulting decision. The significance of the information is in its use; information is thus defined explicitly and quantitatively.

Information is related to the uncertainty (more accurately, to the certainty) that exists regarding the choice of a course of action by the decision maker (DM). We have referred to this as "quantity of information." By "certainty," we mean a function of the decision maker's probability distribution for choosing the various courses of action. This probability distribution can be considered to be the internal information state or simply the information state of the decision maker. These terms are used interchangeably. While this formulation may somewhat delimit the total range of interest in an intellectual sense, it does have virtually universal applicability with regard to potential application or use of the information, and we believe this is the important direction in which to proceed.

There is generally available, of course, external documentation, consisting of documents, reports, books, messages, data sets, etc., which may be used by a decision maker. Although these are called information in the lay sense, they are not information in our sense. They give rise to information in that they will change the information state of the decision maker. In fact, the way in which these external documents change the information state and thus relate to a resulting decision has been the main focus of recent work by our group. Although these external documents are not information, they have the potential to become information.

In our approach, we define a so-called average decision maker. This average decision maker is defined as follows. Given a large number of decision makers with the same effectiveness, learning, and confidence, the average of all their outcome values obtained is, in fact, the actual expected value for each course of action chosen. This average decision maker will thus obtain the actual expected value as his outcome value at each execution. By dealing with an average decision maker, we are able to establish unique and useful relationships. (Of course, we always recognize that the behavior of an individual decision maker may differ from the average -- perhaps even significantly.)

We are consequently able to develop a conceptual framework that can be used in a general way to understand the behavior of decision makers and to understand how information affects this behavior. With this approach, we are also able to identify, define, and quantify in a repeatable way the important parameters involved in information flow.

BACKGROUND

We begin our background discussion with the model of a Generalized Information System, shown as Figure 1. This model follows from our formulation relating information to decision making. Information and decision making are intimately connected, and one cannot exist without the other. Decision makers, we suggest, use this model or a similar one (at least in part) in making decisions. On the basis of a current assessment of the total situation, the decision maker must choose a best course of action (COA) from among many available. As the decision maker makes more decisions and obtains feedback from these decisions, he becomes a more effective decision maker. In the process, he learns more about the entire situation, and his confidence increases as well. External documentation, of course, also modifies the decision maker's decisions.

The decision maker, we suggest, chooses all courses of action probabilistically, where the probabilities are some function of his assessment of the expected values of all the courses of action. As he becomes more effective, he tends to choose the course of action with the greatest expected value with greater and greater probability until he becomes completely certain and chooses that course of action virtually with certainty. Even though the finer details of decision making will likely depend on the precise model of optimization used by the decision maker, the broad features, we believe, are still applicable. Although we describe our approach in mathematical terms, most decision makers do not, in general, think in analytical terms but somehow infer a model on which they base their decisions. The decision maker's subjective expected values are generally not quantitative. They may be of the form, "If I do this, such and such will probably result." We believe our approach is essentially descriptive of these decision makers.

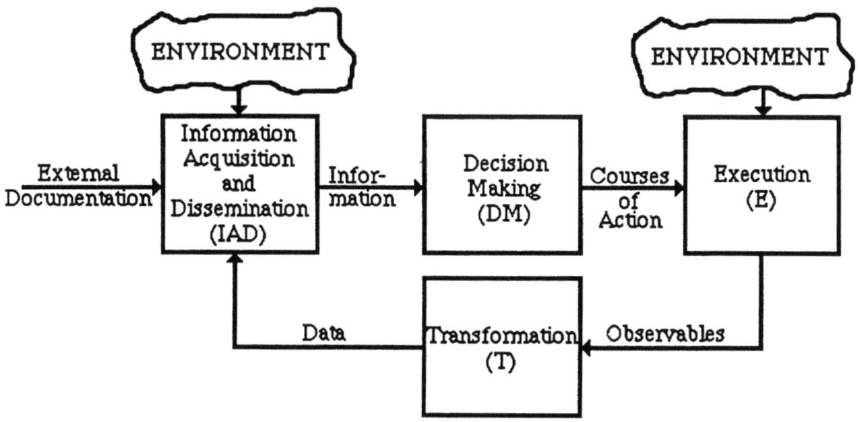

Figure 1. The Generalized Information System.

Our Generalized Information System bears a direct correspondence to the EATPUT Model as shown in Table 1.

We consider an average decision maker, for whom the value obtained from each course of action is the actual expected value for that course of action. By the use of this approach, we are able to develop unique and useful relationships. Although our relationships are developed for the average decision maker, they provide a framework for understanding and quantifying the behavior of the individual decision maker. We believe that these relationships have relatively general application.

Whenever a decision maker executes a course of action, some outcome value results. On the average, this will be the expected value, which we call the actual expected value and which we denote by EV*. Thus, we can define the performance of any decision maker simply as the sum of all the EV*'s weighted by the decision maker's probability of choice for each course of action. We define the decision maker's effectiveness as performance divided by the maximum possible performance, namely by the maximum EV. Thus,

$$DME = \Sigma_i P(a_i)EV_i^*/(EV_k^*)max,$$

where $P(a_i)$ is the subjective probability of choosing the ith course of action, EV* is the actual expected value for the ith course of action, and (EV*) max is the maximum actual expected value, designated here as the kth course of action. We note that decision maker effectiveness is dimensionless and ranges from zero to one.

Table 1. Generalized Information System and EATPUT

EATPUT	Generalized Information System.
EVENT WORLD	Environmental Data Feedback from Previous decisions External Documentation
ACQUISITION	
TRANSMISSION	Transformation
PROCESSING	IAD
UTILIZATION	Decision Making
TRANSFER	Execution (Choosing course of action)

A decision maker effectiveness of particular interest is that of the decision maker who knows the structure but has no knowledge of the relationships involved. Thus, the decision maker would begin with a random guess, and the decision maker effectiveness (DME) for this case, which we call the random point is,

$$DME_{random} = (1/\text{No. of viable COAs})\ \Sigma_i\ EV_i^*/(EV_k^*)\max.$$

We have related information, denoted by I, to the certainty that exists regarding the choice of a course of action by a decision maker. The certainty is some function of the decision maker's probability distribution, which we call his internal information state. For the case of relational uncertainty, we have shown that this function is proportional to the mean square variance of the probabilities of choice. For structural uncertainty, this function is inversely proportional to the number of courses of action remaining. We define the information measure in terms of binary choice units, or BCU's.

We further suggest a unique and monotonically increasing relationship between decision maker effectiveness and I; this we define to be a function, G(DME). Although G(DME) must ultimately be determined by experiment, we have established a number of its important properties.

RECENT ACCOMPLISHMENTS AND RESULTS

In earlier work, beginning with the framework of the Generalized Information System relating information to decision making, we developed a basis for a conceptual model and theory. We have, as a consequence, been able to identify and define quantitatively a number of important parameters and variables and other terms that are involved. Furthermore, we established a number of significant unique relationships and were able to analyze the dependencies among the important variables and parameters. Much of this was due to our consideration of an average decision maker.

In particular, in our earlier work we did the following:

1. Identified and defined the various uncertainties facing a decision maker and related information to the reduction of these uncertainties.

2. Defined a decision state and showed how decision makers modify their decision state by comparison between actual and expected outcomes resulting from various courses of action.

3. Suggested a reasonable selection rule by means of which decision makers can probabilistically select appropriate courses of action depending on their knowledge of the situation and their confidence in their model of this situation.

4. Suggested a reasonable rule showing how decision makers can learn and accordingly update their decision state on the basis of feedback data from decisions already made. (This is the model for a closed system.)

5. Established an average and unique learning rule that applies to decision makers in a given situation as a function of their particular decision state and of two parameters, confidence factor and learning factor, which characterize particular decision makers.

6. Explicitly defined quantity of information, value of information, effectiveness of information, decision maker performance, decision maker effectiveness, and other important quantities.

7. Developed a sophisticated and flexible simulation model that permits us to examine the interrelationship between information and decision making for a wide variety of situations.

We also carried out extensive experimentation that essentially verified our approach and that also established specific values and ranges for the important parameters. In the process, we have collected a great deal of data and have also developed flexible experimental protocols and procedures, which can now be modified and expanded for additional experimentation. We are currently planning further experiments around these procedures and protocols.

More recently we have extended and refined our theory and have developed a number of important results. We provide here a very brief summary of some of our recent results.

1. We have extended the earlier developed theory to a much wider set of situations for which external documentation (such as books, documents, reports, etc., which virtually all decision makers use) plays the dominant information role. This involves the consideration of the way external documentation will change an average decision maker's effectiveness for a particular problem. Since the change in effectiveness is a function of the decision maker effectiveness at the time the document is utilized, we have developed the concept of an information profile. By "information profile" we mean the relationship giving the change in decision maker effectiveness as a function of the decision maker effectiveness for a specific document that contains a given fraction of the solution of a particular problem. Further, we have developed a theory that permits us to define our information profile analytically. We have developed the following analytical relationship for a profile:

$$\Delta DME = \alpha\, DME^{n} (1-DME^{r})/G'\,(DME)$$

where DME is the Decision Maker Effectiveness as previously defined, α is the amount of knowledge available to solve the problem being considered, n is the learning exponent, r is the knowledge exponent, and G is a function relating information to decision maker effectiveness. A typical information profile is shown in Figure 2

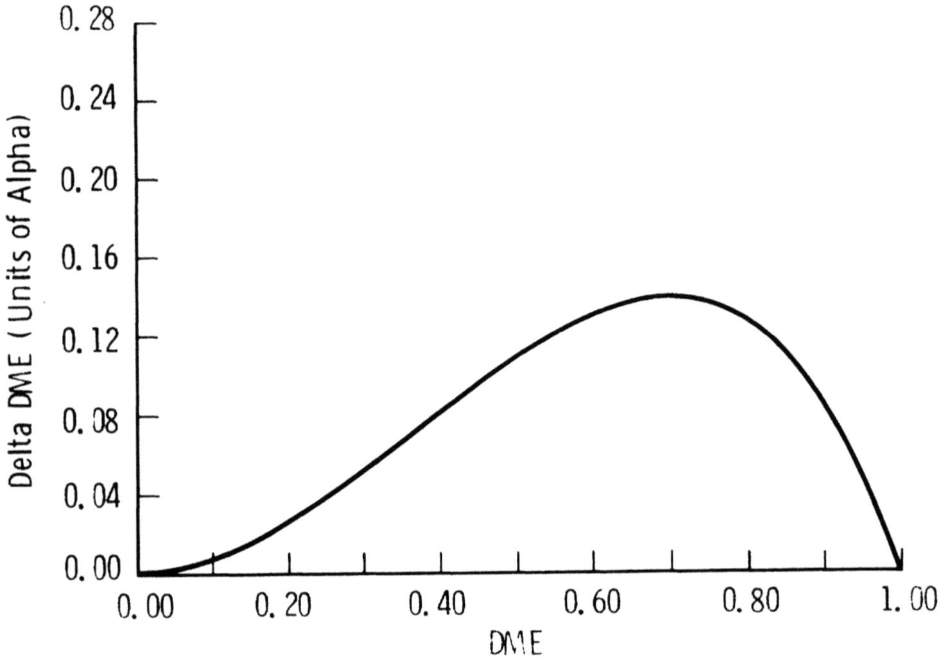

Figure 2. A Typical Information Profile
a=1, mDMER=1.68, n=2, r=3

2. We have related information to decision making and have accordingly developed a measure for the quantity of information that a decision maker has for a specific decision making situation. We have explicitly related information to the internal information state or probability distribution of the decision maker and have defined information to be directly related to the certainty which a decision maker has in making a decision. We have shown that information in our context must be defined as decision maker dependent and, consequently, as a relative rather than an absolute quantity. We have also provided a number of important requirements that an information measure must satisfy.

The resulting measure is a simple one, easy to calculate and to understand. It is significant and of general use in all decision-making situations. We have defined a fundamental unit for this quantity, called a binary choice unit, or BCU, which is the maximum quantity of relational information in a two choice decision problem. We have shown that the quantity of information can be thought of as having two components, structural and relational, so that a certain fraction of the total is due to knowing the structure of the problem and the remainder is due to knowing the probabilistic relationships for choosing the various courses of action. We show in particular that the information due to knowing the relationships among the various courses of action is proportional to the mean square variance of the probabilities divided by the mean of the probabilities. We further show specifically:

$$I_{rational} = 2(\sum_{i=1}^{m} P(a_i) - 1/m)$$

and

$$I_{total} = 2(\sum_{i=1}^{m} P(a_i)^2)$$

in BCUs.

where $P(a_i)$ is the subjective probability for choosing course of action a_i and m is the number of courses of action being considered.

3. In previous work, we considered only the situation for which the decision maker knows all of the viable courses of action but initially does not know the relationships among them. This situation we call "relational uncertainty" or "structural certainty." We have recently extended the model to the situation where the decision maker does not know the various courses of action; this case we call "structural uncertainty."

4. We have defined external documentation in terms of a quantity we call "knowledge." By the term "external documentation", we refer to books, documents, data, references, messages, etc. to which the decision maker has access. Knowledge is the contents of external documentation. It may be converted to information when utilized by a decision maker. It is quantified by the parameter α. This quantity (α) measures the amount of knowledge in an external document. It is the maximum amount of information that could ideally ever be obtained from a document. We have defined (α) to have a value between zero and two.

5. We have defined a number of additional terms explicitly, quantitatively, and measurably. Included are the terms "solve," "know," "learn," "ideal decision maker," "earning exponent," "knowledge exponent," as well as others.

SUMMARY OF IMPORTANT DEFINITIONS AND RELATIONSHIPS

It is appropriate to summarize here the various terms, measures, and parameters that we have used, defined, and developed. These definitions provide insight into our approach and results.

Decision Maker Effectiveness (DME) is the ratio of average performance to maximum performance. This ranges from zero to one and indicates how well a decision maker performs. This is not known by the decision maker and can only be estimated.

DME_{random} is the initial decision maker effectiveness where the decision maker knows the structure of the situation but has no knowledge of the relationships involved. It is simply the ratio of the unweighted mean of all actual expected values of the courses of action to maximum performance.

External Documentation refers to books, documents, data, references, messages, etc. to which the decision maker has access.

Knowledge is the contents of external documentation. It may be converted to information when utilized by a decision maker. It is quantified by the parameter a.

The *Quantity* α measures the amount of knowledge in an external document. It is the maximum amount of information that could ideally ever be obtained from a document.

To Know means that the decision maker has already incorporated into his information state the knowledge contained in external documentation.

To Learn means that the decision maker is incorporating into his information state the knowledge contained in external documentation.

To Solve a problem means that the decision maker always chooses the course of action with the greatest expected value.

The *Learning Exponent n* is the power to which decision maker effectiveness is raised to give the learning capability of the decision maker.

The *Fraction of External Knowledge* that is already known by the decision maker is represented in terms of r. It is the power to which decision maker effectiveness is raised. The symbol r represents the *Knowledge Exponent.*

Average Decision Maker: Given a large number of decision makers with the same effectiveness, learning, and confidence, the average of their expected outcome values is, in fact, the actual expected value for each course of action chosen. The average decision maker will obtain the actual expected value as his outcome value after each execution.

The *Ideal Decision Maker* is one who knows nothing about the problem but who learns very rapidly and converts external knowledge entirely into information.

Information is related to the uncertainty (more accurately, to the certainty) which exists regarding the choice of a course of action by the decision maker. We have referred to this as *Quality of Information.* By *Certainty* we mean a function of the decision maker's probability distribution for choosing the various courses of action.

Decision making is, in fact, the heart of and even the reason for the existence of any information system. By "decision making" we mean the choosing of a course of action from among the number available. The decision maker here may be a human being, a computer, or more likely a combination. Indeed, it may be a group of people and computers. In a somewhat simplified set of circumstances, decision making can be made by a set of rules or algorithms.

Recognizing that complexity necessitates a probabilistic or statistical approach, it is appropriate to indicate some of the areas of potential system failure:

First, there is only a probability that any course of action will obtain the desired outcome. There is always the finite probability that an undesired outcome will result.

Second, the decision maker has only a probability of choosing the best course of action. There is always some finite probability that some other course of action would give better results. In fact, there is a finite probability that for some reason a good course of action has already been eliminated from consideration.

Third, any external documentation utilized by the decision maker will increase his effectiveness only probabilistically, depending on his current effectiveness and his state of knowledge. Since external knowledge must necessarily be used by many different decision makers at many different levels of effectiveness, this change in effectiveness must necessarily be probabilistic.

There are, of course, other errors of a statistical nature which will also cause failures.

Our theory and analysis have provided a quantitative and analytical framework by means of which the effects of the probabilistic errors can be understood and, we hope, dealt with. Knowledge of the various probabilities involved provides the decision maker with a measure of their effects on system performance. The ability to analyze and quantify these effects provides a powerful tool for prediction and comparison among various system designs.

ACKNOWLEDGMENT

Research for this article was supported in part by the National Science Foundation under Grant Number IST 83-12728 and Grant Number IST 83-039900. Any opinions, findings, and conclusions expressed herein are those of the authors and do not necessarily reflect the views of the National Science Foundation.

BIBLIOGRAPHY

Yovits, M. C., C. R. Foulk, and L. L. Rose. (1981). Information flow and analysis: theory, simulations andexperiment. Part 1. Basic theoretical and conceptual development. Part II. Simulation, examples, and results. *Journal of the American Society for Information Science*, 32: 187-20.

Yovits, M. C. and C. R. Foulk. (1985). Experiments and analysis of information use and value in a decision-making context. *Journal of the American Society for Information Science*, 36(2): 63-81.

Yovits, M. C., A. de Korvin, R. Kleyle, and M. Mascarenhas. (1985). Information and is relationship to decision making: The information profile and other quantitative measures; A brief summary. *Proceedings of Third Symposium on Empirical Foundations of Information and Software Science*, Denmark.

Yovits, M.C., A. de Korvin, R. Kleyle, and M. Mascarenhas. (In Press). External documentation and is quantitative relationship to the internal information state of a decision maker: The information profile. *Journal of the American Society for Information Science.*

Yovits, M. C., A. de Korvin, R. Kleyle, and M. Mascarenhas. (Submitted). A quantitative measure of information as related to decision making. *IEEE Transactions on Systems, Man and Cybernetics.*

Yovits, M. C., R. Kleyle, A. de Korvin, and M. Mascarenhas. (summitted). A decision making process for the elimination of low valued courses of action in an unstructured decision making problem: Measuring information and decision maker effectiveness in the early stages of a decision process. *Operations Research.*

FOR THE RECORD--SOME DIFFICULTIES OF GENERATING, DISCERNING, TRANSMITTING AND RECEIVING A SIGNAL

Derek Conway

Consultant
Basingstokes, U. K.

I am starting from the point of view of the aftermath of a disaster, survivors emerging from the scene, figures emerging from a background, perhaps reluctant to communicate, perhaps unable to communicate, being terrorised.

I could have started the argument from the point of view of being old, fatigued, spent and a victim. It would be preferable to choose the scenario of the SPACE AGE, the "Star Trek" context, where we are in a state of suspended animation waiting to be awakened, perhaps as in fairy tales, by a kiss from a prince or princess.

The scenario often depicted in SPACE AGE dramas is that of a signal coming from somewhere in the universe. Upon investigation, it is found that some disaster has overcome a crew, community, or civilization. Here I am reminded also of Nevil Shute's earth-bound novel *On The Beach* . In this particular setting in the novel, a signal is being transmitted from somewhere in the aftermath of a nuclear war or missile exchange. It is received by a patrolling submarine.

So, a signal is transmitted from a noisy background -- in some cases, after a disaster, in other cases before the disaster occurs. At all times the source of the signal could be vulnerable to technological difficulties if it is equipment, or psychological, social, physiological and philosophical hazards if the source is a person.

In order that a signal be transmitted and to initiate action, it is necessary to have received, generated, or transmitted a signal to begin with, and it is this generation and transmission of a signal that could be curtailed or suppressed. An example of what I have in mind is a "MAY DAY" situation, perhaps the result of some of the experiments conducted by Dr. Stanley Millgram.

That is to say, curtailment or suppression of a signal may be achieved in many ways. To my mind, a signal source and transmission system is now vulnerable, quite vulnerable, to psychological terrorism, given that a great deal of the particulars of our lives, achievements, even psychological profiles and history, may be contained within a system of some kind, on record, so to speak. Unless that system be secured, hostile operators may get to work! (The nightmare situation is that even with a secured sytem, criminals (for that is what they must be) could be at work.

In any loosely coupled system, enough michief can be done by what may be termed "Bloody Mindedness." Of course, "Bloody Mindedness" has its uses (The Harry Palmer scenarios), but I wish to raise the argument a notch or two by giving a definition of what is meant by sociological terrorism.

As I understand it, a definition goes something like this:

> An inversion of the psychological values and beliefs of a person by individuals with the ultimate aim of the psychological destruction of the personality.

NATO ASI Series, Vol. F32
Information Systems: Failure Analysis
Edited by J. A. Wise and A. Debons

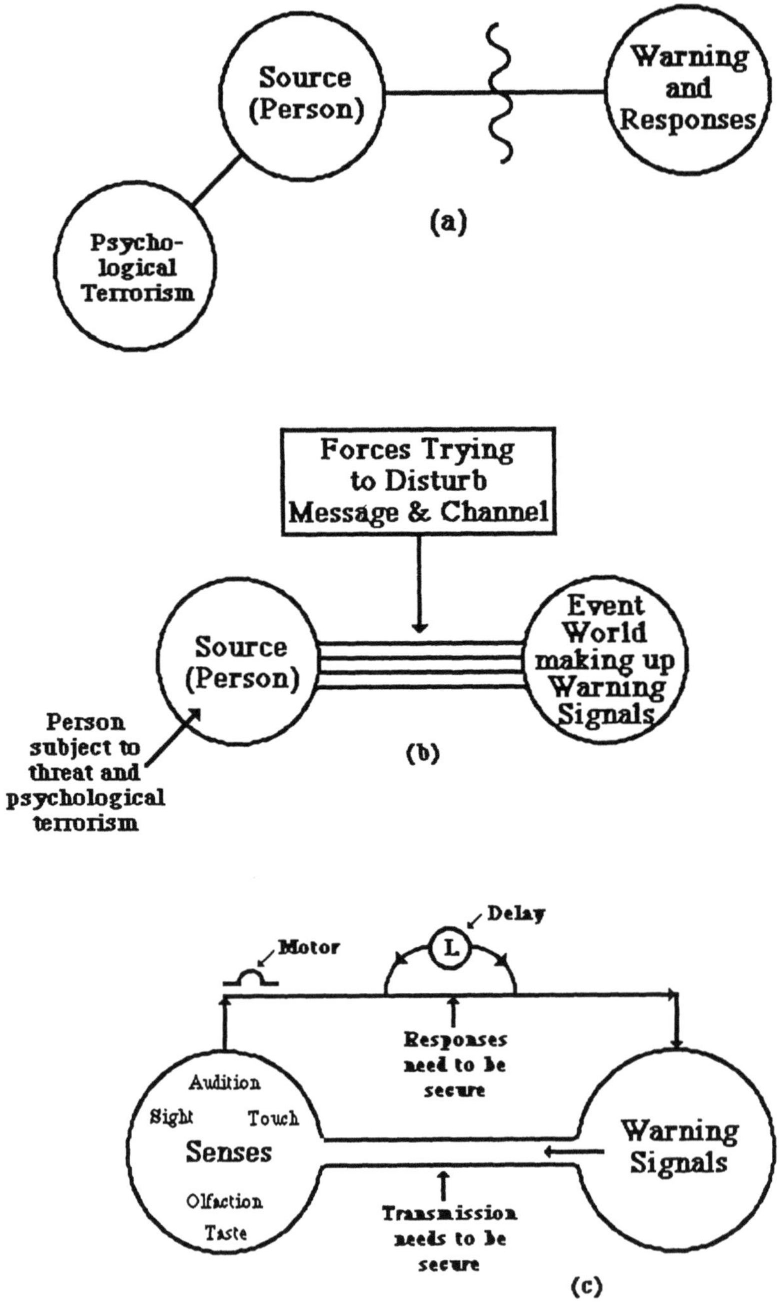

Figure 1. Simple diagrams (a, b, c) demonstrating the psychological effects of terror.

To be sure, if one encounters a system that continually inverts one's values, probably a nebulous and diffuse system, one's mind or personality could be seriously affected or destroyed if only by psychological isolation and confusion.

If industrial, commercial, and business enterprises can be adversely affected by number, accounting, or computer digit accidents or sabotage, the same process can operate upon an individual to cause his or her failure (e.g., code debit for credit in a personal banking account). The hazards of life in a physically moving environment have been particularly well researched by Professor James Reason and working hazards at sea by Surgeon Captain John William Davies, RN.

On psychological terrorism, the "Yapping Dogs" etc., including the work of Professor John Cohen , similar to the adverse effect s of physical motion, the line is subtle, for often physical and psychological variables are entwined However, it is probably vacuous to state that there is a great deal of difference between a person choosing to be in a certain physical or psychological position or state, as opposed to being put into that position. Here again I am reminded of the boss experiments conducted by Dr. Millgram. To summarise, man in psychological motion may be acted upon by adverse psychological processes -- psychological interference, etc.

This last point leads me to the important question that could be asked by an individual: "How do I know when I an under attack?" How can I relate this to individuals or groups whom I can trust to try to understand the reality of the position? Similar questions can be asked by groups or organizations (e.g., governments). (I am reminded of redundancy in a system by photographs of World War II bombers, shot to pieces but still flying!)

The question of knowing whether one is under attack, psychologically so to speak, is often only the precursor to physical attack. First apply the psychological operation, reduce resistance, and then apply physical or psychological force.

To take the argument down by a few notches, one particularly notices these days the demise of academic philosophy and arguments against it. It is obvious that only by the use of the critical mind does one come to realize that a state of psychological warfare is in existence. These may be harsh terms.

In the twentieth century we cannot afford to neglect the media and it effects, be they original or copied. Recall the hysteria of the Orson Welles broadcast (1938) created by a radio play. (Nothing New Here.)

Newspapers, books, cinema, TV, radio and theatre can produce psychological terror to affect us all to a degree. When attempts are made to remove one's individual control of one's own mind, we are often aware if it and may take some kind of avoidance action. However, if the opposing forces have done their homework, constraints can be introduced in a subtle manner so that when avoidance or evasive action is attempted, all paths are blocked!

One can be made physcially fatigued before the psychological coup-de-grace is delivered. In this repect, I am referring to my earlier point that to suppress a signal may have the disastrous effect of preventing some kind of action. (Again, to maintain a certain degree of good humor, one could call this a vacuous statement.) It appears that the argument here is centered around two psychological variables, that is, judgment and will.

Somehow we use our senses to detect or initiate a signal and our motor system to initiate some kind of action (pursuit, avoidance, etc.) to set wheels in motion.

During the workshop at this conference, the notion of a black box was introduced as was the 'spy in the cab' recorder and the logic bomb. Still considering the topic of psychological terrorism, I would like to raise the point of the psychological time bomb -- a set of psychological circumstances, operators, values to be liberated upon an individual, crew, or system at any one time, generally in a psychologically closed state. Persons involved in emergencies or confined for long periods of time are vulnerable to this kind of threat.

Here we are not responding to an overload of signals or to possibly fictional physiological hazards and their subsequent off-spring, but to the psychological security of a system, e.g., civil, military, organizational, that could be vulnerable to attack from psychological forces. During times of crisis (e.g., war), such systems are tested and have to be made secure or avoidance action must be taken to nullify the adverse effects of the attack - be it propaganda or political warfare.

In this shadowy area we want to be vigilant on two accounts. First, we may be wrong; second clinicians may see our accounts to be unreal and pathological. But when the stature of the individual cannot be in question, it is necessary to listen, to take heed.

So as a summary, at any one time the degree to which an individual, social group, or society is being terrorised depends upon those who are in a position to understand and to act as counter, taking the steam off a little, so to speak. We could be considering the area of professional malpractice, of incompetence, etc. However, I personally believe the positive act of psychological terrorism to be more prevalent, a hand at work.

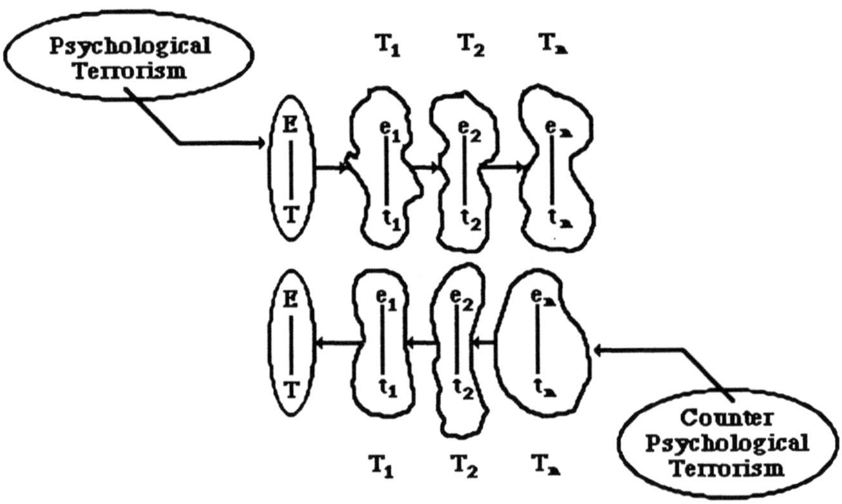

Figure 2. Effects of counter terrorism.

In a very interesting little book, Julian Critchley has related important events concerned in a number of military operations. He gives a nice example of how our perceptions may be fooled. A drawing of a cat/owl changes slowly to an owl/cat over a period of some twenty drawings. Accordingly, not until drawing 18 or so do people notice the change.

Going back (and I do not wish to give the impression of a grasshopper mind but simply some sort of cybernetic aspect to my point of view), the concept of change and time has been referred to and quoted by the director of this Workshop, Professor Anthony Debons, referring to the thesis of Dr. John Wise and to the controlled change of the ASW schedule. It is only when such changes become exaggerated or out of control (falling into the command of an idiot or adversary) that we suffer ill effects.

Despite a great deal of misinformation about the aims of industrial enterprises, we should perhaps look to the ideal of the best industrial enterprises, where only the best and best interests are allowed to prevail to counter the seeds of disaster.

Finally to survive the worst effects of this psychological nightmare, we need leadership--leadership, of course, of the right kind. There is no doubt that the cinema provides us with many mental models or scenarios, which are called to mind when the moment decides. When dealing with psychological terrorism or injustice, it is sometimes good to know that the wrongs are righted, the good guys win.

When the threshold to fear becomes too great, we need that external operator (e.g., "A" Team) to get us out of that situation. I understand that during battle the only place a soldier wants to be is anywhere away from the action.

Ideally, an information system should be allowed to exist unperturbed, that is, without noise. In a competitive, hostile, idealogically oriented world, I fear this is seldom the case or state of affairs. Security exists to maintain an information system intact, and at each vulnerable point interference will be introduced. Given freedom of information, the security of an information system, open or closed, is vital if true records are to be maintained (e.g., proceedings of British parliamentary debates, records of U.S. congress, etc.). Under conditions of psychological terrorism, the EATPUT model becomes distorted until the terror is removed. This point is illustrated below by the various schemata. In the paper presented by Yovits et al. (this volume), it is the out-put stage of the decision process which, to my mind, is most vulnerable to interference and distortion, implementing the decision. I have constructed several schemata to demonstrate the action of the psychological terrorism dimension, with its opposite counter psychological terrorism.

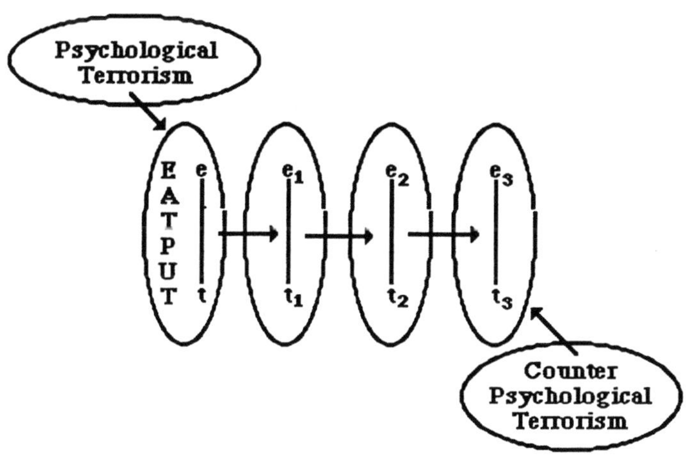

Figure 3. Changes in the information system as a function of terror.

I am pleased to say there is an element of cybernetics about the schema, one demonstrating change. Bearing in mind that I have been concerned with psychological terrorism and other forces involved in an information system, I would like to quote from a published book by a medical practitioner:

> There is a great deal of experimental work drawn from the laboratory and field studies that shows how the best ordered people can become totally disorganized when frustrated, isolated, kept without sleep, or made anxious or afraid for their lives.

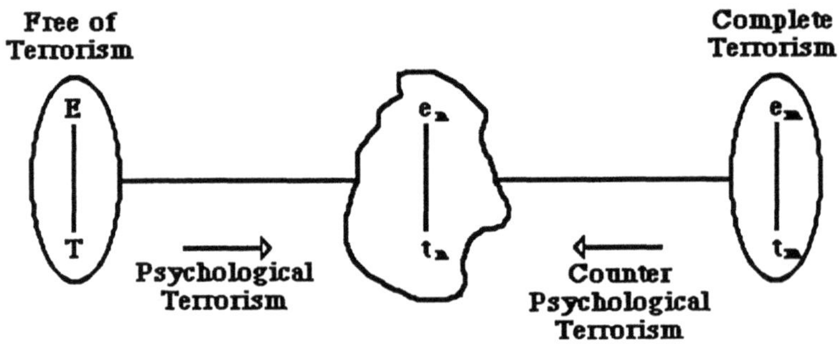

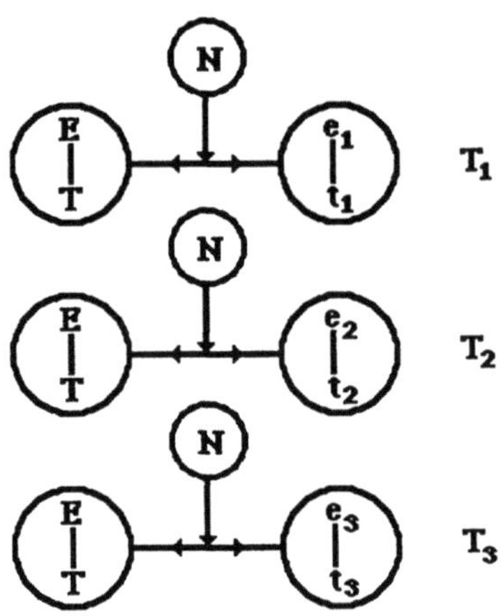

Figure 4. The effects of terror, like noise in a cybernetic system, completely destroy the operation of an information system.

TWO EXAMPLES OF PARTLY FAILING INFORMATION SYSTEMS

J. G. Kretzschmar

Nuclear Energy Research Centre
Mol, Belgium

INTRODUCTION

During the months preceding the August 1986 NATO Advanced Research Workshop on Failure Analysis of Information Systems, two specific events illustrated the weaknesses of the presently available information systems when one is really looking for information and not for data. The first example concerns the use - by a non expert - and the failure of computer-based on-line information retrieval systems. The second one relates to the absence of reliable information systems to inform in due time the authorities and the general public with respect to major socioeconomic issues such as the Chernobyl accident and its present and future impact. In both cases, it appears that the main reason for failure resides in the fact that we do actually collect, store, and manipulate billions of data without much attention for:

- The fact that information is the result of a complex process of deterministic and heuristic reasoning, using exact as well as fuzzy data from interrelated knowledge fields.
- The skills required to access the system and. if successful, to use the data.
- The knowledge and expertise already available in the brains of the real experts in the specific fields of science, technology and management.

With the ever increasing speed and storage capacity of our so-called information systems, and the continuously decreasing prices of computer and communication hardware, it looks as though the time has come to automate some of our reasoning processes and delegate the most straightforward parts of it to machinery, in order to avoid a tremendous mismatch between data collection and information extraction capacity. Knowledge-based systems or expert systems could possibly help us to bridge this gap.

Although there are several definitions of an expert system, is, we will stick for the time being to the following one given by the British Computer Society, Expert Systems Specialist Group (BCS ESSG): "An expert system is regarded as the embodiment within a computer of a knowledge-based component, from an expert skill, in such a form that the system can offer intelligent advice or take an intelligent decision about a processing function." A desirable additional characteristic, which many would consider fundamental, is the capability of the system, on demand, to justify its own line of reasoning in a manner directly intelligible to the enquirer.

On-Line Information Retrieval Services

Many different on-line information retrieval services exist, and new entries are noted regularly. They all have been developed almost independently, using different file structures, command languages and access protocols. The complexity of searching, the proliferation of databases, and the lack of standardization make the knowledge required for successful searches extensive, as will be illustrated by what follows.

Going to the appropriate library service an inexperienced user would like to find some relevant publications concerning "failure analysis of information systems." Suppose that this user - with the help of the person behind the keyboard - has access to approximately 300

NATO ASI Series, Vol. F32
Information Systems: Failure Analysis
Edited by J. A. Wise and A. Debons

databases in Dialog and some 80 in the ESA Information Retrieval Service. (Note that more than a couple of database are accessible in both retrieval services.) He starts with INIS, but for some obscure reason the link cannot be made. Switching to PASCAL, 1973 - 1986, he gets exactly four references for the combination "failure analysis" and "information." None seems to be appropriate. Turning to INSPEC, 1971 - 1986, with "failure analysis" and "information system" the following five references do appear:

- Can informatics help those responsible for maintenance?
- The life of boiler parts is a problem unsolved as yet.
- A study and manufacture of a voltage contrast system for the analysis of LSI integrated circuits.
- Trouble information organization to meet assigned reliability objectives.

Not very pleased with the result, the user combines "failure analysis" with "computer system" and 12 references are printed. They seem to cover a broad range of different and quite specific topics. None of the previously obtained five references reappears. The user feels that he is coming closer to the required end result and now tries EDF-DOC, 1972 - 1986 with the same keyword combination, but zero documents are found. So why not File 175: COMPUTER DATABASE, 1983 - 1986, Issue 14? The combination "failure + analysis + information + system" generates the following:

- CAM = Process Control?
- Case study of two information systems implementation efforts using the Kolb-Frohman Model of consultation.
- A computer expose dies.
- Why systems fail

Going on with "failure + analysis + computer + systems" he discovers another twelve references. Among them he finds again the second and fourth references of the previous search. Also uncovered is the *New York Times*, Vol. 134, No.465O1, Aug. 14, 1985, p. 1 an article by D.E. Sanger on "Carbide computer could not track gas that escaped: programming not done." Other items found include the following:

- Evaluating response time in a faulty distributed computing system.
- Where trouble strikes.
- A fault-tolerant dataflow system.
- Fault simulation becomes design verification tool.
- Statistical failure analysis of system timing.
- Simulation of computing processes with breakdown for a given program set.
- Simulation and design : failure mode behavior analysis.
- Solving social security's computer problems : comprehensive corrective action plan and better management needed.
- Assessing your needs accurately.

Reading the last title, the inexperienced user realized that it was useless to continue. Apart from forgetting the whole thing, the only other alternative was to ask for some of the most promising abstracts and eventually try to get reprints or copies of some of the articles. In this way, he would possible be able to figure out what interpretation the information retrieval system was giving to the keywords he was actually using. In short, it was clear that the on-line system was not at all transparent to this user.

An expert system, including knowledge about search strategy, would have been very helpful to this non-expert user. From his point of view, the information system failed - at least to some extent - to fulfill his expectations. It could, of course, be argued that this is a subjective interpretation. It nevertheless points to a very important and quite specific aspect of "failure analysis in information systems," namely the "human interface" in its most general meaning. Even without any technical failures, an information system can still fail. It is the end user who decides whether he or she got information at the right time, place, and speed, under the appropriate form (qualitative), and in sufficient but not excessive amount detail (quantitative).

THE CHERNOBYL CASE

On 26 April 1986 at 1:23 a.m., explosions occurred in the fourth unit of the Chernobyl nuclear power station. The explosions caused fire. The reactor building, the reactor itself, and the reactor core were extensively damaged. This resulted in important radioactive releases, having their major impact upon the immediate neighborhood but also significantly raising radioactivity levels in the air moving towards most of the Scandinavian and European countries. Through the mechanism of dry, and especially wet, deposition, crops and grassland were subsequently contaminated to such an extent that vegetables, milk and even meat had to be destroyed or at least withdrawn temporarily from the consumer market.

Various information systems played or were obviously supposed to play an important role, in the early phase of the accident and during the post accident period, both locally and at different national and even international levels.

How far or to what extent the power plant's local information system eventually failed or gave insufficient support to the human decision-making process is not known. We do know that the involved reactor of the RMBK type has a positive void coefficient of reactivity and that this is not impossible to control. In a core as large as that of the RMBK-1000, the conditions do nevertheless require very good instrumentation and computer control systems to cope effectively with localized disturbances in the power level. It has been questioned if computer systems and instrumentation in the Soviet Union have developed enough in recent years to reach the required level of performance. As was the case for the 1979 Three-Mile Island accident, some time and a lot of investigation will be required before this point and many others are clarified. Since it is too early, we will omit this topic for the time being to focus our attention on the information systems, that did fail or were unavailable.

Indeed, during the three or more weeks following the accident ,the general public was quite often confronted with an information mess rather than information at the level of its mental mode and capacities. Facts and numbers, expressed in almost continuously changing units, were coming down on Mister Nobody like the Niagara Falls. Every day new experts appeared on the TV screen and were quoted on the first page of every daily newspaper. Acceptable or safe levels of radioactive contamination in air, water, soil, or foodstuffs changed as a function of time or as a function of the specific nationality of the population at risk. The forecasting of movements of the upper air masses, and the subsequent transport, dispersion and deposition of material released some thousands of kilometers away from home, seemed a new science still to be invented. The billions of dollars spent over the past decades in R&D in the relevant fields of science and technology - including the field of information technology in its broadest sense - were obviously not able to create understandable information. Was all the information only for the academics and the technocrats? Or was finding the appropriate filter for the information glut the Number One problem to be solved?

At a somewhat different level, quite a few national and supranational authorities were facing their problems too, with ever returning questions like "What do we know with respect to this?" "Where is there an expert for Y ?" "On what evidence is Z based?" "What if X happens?" Quite a large spectrum of different information systems were actually used - from the simple technical textbooks and handbooks of rules and regulations to the information retrieval databases and the computer-assisted emergency response systems - and results were obtained despite the failure of more than one system. Confusion and delays were at least partly due to the fact that up to then the scientific and technical community, as well as the authorities, had been dealing with the eventual consequences of hypothetical reactor accidents of this size and impact. Hypothetical scenarios that turn into real ones have nevertheless their own unpredictable flavours as was illustrated once again. Information - or, shall we say data - seemed on the other hand to be abundantly available, or at least filed somewhere, accessible by some means and interpretable by someone. With a lot of effort and some headache, quite a number of people got their information at last. Information systems for tackling the problems at hand smoothly and in real time did not demonstrate their presence.

CONCLUSIONS

The previous two examples - although possibly somewhat biased and limited - illustrate that an information system is not a properly engineered system on the basis of some more or less sophisticated hardware, software, communication devices and links, data, and interfaces. Information is a product that is produced again and again by human beings, each time using what has gone before. Making this product available to other human beings, at the required moment and with the appropriate speed, is still a matter of major concern. The addition of intelligent knowledge-based systems or expert systems as subsystems of our overall information systems, could be of help in solving some of the problems and the failures that occur when human operators actually have to use - with or without stress - a complex information system.

In failure analysis of information systems the role of human factors, under various conditions and constraints, will certainly be important. Such failures cannot be tackled in the same way as failures in hardware, software, and communication or interfacing equipment.

INVESTIGATIVE METHODS FOR ANALYSIS OF INFORMATION SYSTEM FAILURE

APPLICATION OF TEAM CONCEPT/SYSTEMS APPROACH TO INVESTIGATION OF MAJOR MISHAPS

Rudolf Kapustin

Intercontinental Aviation Safety Consultants
Baltimore, Maryland, USA

PREPAREDNESS

After a major disaster involving loss of life and property damage, the authorities and other organizations responsible for investigating the cause and taking immediate steps to prevent a recurrence are often not prepared to assume these tasks. The authorities in many disciplines are caught by complete surprise when the unexpected happens and often must look to outside sources to assist, organize or conduct the investigation. The Three Mile Island and Challenger accidents are examples of this type of unpreparedness.

ORGANIZATION OF INVESTIGATION

If an investigating body must be formed when a state of emergency already exists and other immediate tasks such as damage control and emergency rescue must be addressed, the recruitment of highly experienced and qualified technical personnel to conduct the investigation is often difficult. The formation of the investigative body may be further complicated and delayed by the need to have totally dependent, unbiased personnel recruited and assigned to the various investigative tasks. Personnel employed by or closely associated with the operation, design, manufacture, or maintenance of the system that was intimately involved in the disaster could well lack the objectivity to conduct an unbiased investigation, create a true and factual record, analyze the facts, and eventually make recommendations to prevent a recurrence.

That is not to say that a well-structured investigative organization cannot exist where there is totally independent leadership and management nor that technically qualified personnel who are in fact closely associated with the design, manufacture, maintenance, or operation of the system that failed cannot be used effectively, provided the proper safeguards are in place. As is often the case, the really qualified technical experts can be found only within this group of personnel.

Once the nucleus for an independent and objective investigative organization has been formed, the next and even more complex task is to ensure that the staff of the investigative team has the capability to address each aspect of the system in which one or more segments failed to function and eventually caused the disaster.

ANATOMY OF A DISASTER: A CASE HISTORY

It is important to realize that in order for most system failures to develop into a catastrophic occurrence, several events must take place. A disaster is seldom, if ever, caused by a single event, a single action or inaction of the human element, or a single failure of a mechanical segment of a complex system. In most cases, there is an identifiable initiating factor, followed by three or more events that eventually create an irreversible course to a disaster.

On September 25, 1978, Pacific Southwest Airlines (PSA) Flight 182, a Boeing 727, while descending for a landing to San Diego's Lindbergh Field, collided with a single engine Cessna 172 making practice instrument approaches to the same airport. Both aircraft crashed into the downtown San Diego residential area. The 137 occupants of both airplanes and 7

NATO ASI Series, Vol. F32
Information Systems: Failure Analysis
Edited by J. A. Wise and A. Debons

persons on the ground were killed, 9 persons on the ground were injured, and a large portion of the downtown San Diego residential area was transformed into a major disaster area within a matter of seconds.

As in most disasters, even before the investigation began there was instantaneous speculation by both the professional community and the lay segments as to the cause of the accident. Obviously, even the most "expert" speculators were wrong. In fact, even after an exhaustive investigation, an exhaustive review of the voluminous data, and lengthy deliberations by the quasi-judicial body responsible for the formal determination of the "probable cause," there were major disagreements as to the proper assignment of responsibility for the initiation of the disaster.

Briefly, the air traffic procedures for the San Diego area permitted simultaneous operation of light aircraft (general aviation) training flights and operation of large air carrier aircraft. This in itself, while potentially hazardous under given conditions, is not necessarily a cause for an accident. In this particular case, the light aircraft was conducting simulated instrument approaches which, during practice maneuvers, are abandoned before an actual landing is made. The aircraft is then flown to a higher altitude where additional practice approaches can be initiated. The Cessna had just completed such a practice approach and was cleared by the Lindbergh Field tower to climb to a higher altitude.

In the meantime, PSA Flight 182, under the control of the Miramar Approach Control facility, had been cleared to descend through the altitude through which the Cessna was climbing. Traffic advisories given to PSA Flight 182 indicated that other light aircraft were in the area. The PSA flightcrew indicated they had the traffic in sight. It was assumed by the air traffic controller that the PSA crew did, in fact, have the correct airplane in sight and still had it in sight 66 seconds later when a computer/radar-generated "conflict alert" was activated in the Miramar Approach Control facility. Because of this assumption, the controller did not pass the conflict alert warning on to PSA Flight 182. The collision occurred 19 seconds after the activation of the conflict alert.

Obviously, the simplest means for avoiding the numerous failures of redundant systems and safeguards would have been for the controller to issue altitude restrictions to both airplanes until they had safely passed each other. Under the existing procedures, this was not required nor was it done. The system depended on a "see and be seen" concept.

The investigative team needed to be composed of specialists in air traffic control procedures, flight operations, aircraft systems, aircraft structures, powerplants, human factors, witness interviewing, aircraft performance, aircraft records, flight data recorders, and cockpit voice recorders.

The National Transportation Safety Board (NTSB) is a federally mandated quasi-judicial body charged, among other things, to "...conduct and control accident investigations involving civil aircraft..." (49 CFR 821.2). The nuclei of the NTSB teams that manage and lead investigations are pre-formed and are always in a standby/ready mode. The PSA accident occurred at 0901:47 PST, and the team was en route to San Diego within about two hours of the collision.

Based on the statutory provision of 49 CFR 831.9, parties to the field investigation were also responding to the accident and were proceeding to the accident site.

FEDERAL REGULATIONS

Relevant parts of the applicable law, 49 CFR 831, are as follows:

Authority

Title VII, Federal Aviation Act of 1978, as amended. 72 Stat. 781, as amended by 76 Stat. 921 (49 U.S.C. 1441 et seq) and the Independent Safety Board Act of 1974. Pub. L. 93-633, 88 Stat. 2166 et seq. (49 U.S.C. 1901 et seq.)...

831.2 Responsibility of Board

(a) The Board is responsible for the organization, conduct, and control of all accident investigations involving civil aircraft, or civil and military aircraft, within the United States, its territories and possessions. It is also responsible for investigation of accidents which occur outside the United States and which involve U.S. civil aircraft or civil and military aircraft, at locations determined to be not in the territory of another state (i.e., in international waters)....

831.7 Investigator-in-Charge

The designated investigator-in-charge organizes, conducts, and controls the field phase of investigation. He shall assume responsibility for the supervision and coordination of all resources and of the activities of all personnel, both Board and non-Board, involved in the onsite investigation....

831.9 Parties to the field investigation

(a) The investigator-in-charge may, on behalf of the Director, Bureau of Accident Investigation, designate parties to participate in the field investigation. Parties to the field investigation shall be limited to those persons, Government agencies, companies, and associations whose employees, functions, activities or products were involved in the accident or incident and who can provide suitable qualified technical personnel to actively assist in the field investigation.

(b) Participants in the field investigation shall be responsive to the direction of the appropriate Board representative and may be relieved from participation if they do not comply with their assigned duties or if they conduct themselves in a manner prejudicial to the investigation.

(c) No party to the field investigation designated under 831.9(a) shall be represented by any person who also represents claimants or insurers. Failure to comply with this provision shall result in loss of status as a party.

(d) Section 701(g) of the Federal Aviation Act of 1958, as amended, provides for the appropriate participation of the Administrator in Board investigations. Thus, the FAA will normally be a party to field investigations and will have the same rights and privileges and be subject to the same limitations as other parties....

In accordance with the statute, the following representatives were parties to the PSA/Cessna midair collision investigation and were en route to the San Diego area: Federal Aviation Administration (FAA), Pacific Southwest Airlines, Gibbs Flite Center (the owner of the Cessna 172), the Southwest Flightcrew and Flight Attendants Association, the Boeing Company, Cessna Aircraft Co., Professional Air Traffic Controllers Association, Pratt & Whitney, Air Line Pilots Association, and the Aircraft Owners and Pilots Association.

CONDUCT OF THE INVESTIGATION

Soon after arrival in the San Diego area, the Investigator-in-Charge (IIC) for the NTSB convened an organizational meeting at which the required working groups were established and an NTSB specialist was assigned as chairman of each group. A senior representative for each of the parties, designated as coordinator for his organization, presented qualified specialists to participate in the investigative activities of each working group under the direction of the NTSB Group Chairman. The Group Chairmen are directly responsible to the IIC and must ensure the adequate scope and technical accuracy of their individual phase of the investigation.

All factual findings are documented in the form of group notes, which are reviewed by all group members for technical accuracy. Depending on the complexity of the individual groups' investigative areas, the field work leading to final preparation of the group notes may require up to ten days -- again depending on the amount of follow-up or laboratory work required. Several months can elapse before a group's work is completed and its adequacy concurred with by the IIC.

Generally the on-site activities are complete in about two weeks, and the wreckage site can be released and cleared. In the instance cited, the hardware portion of the investigation was completed in about ten days. The off-site investigation required about three additional months.

THE SYSTEM CONCEPT

To cite specific examples of how the system concept was applied to the PSA/Cessna investigation, I will refer to several areas that were explored.

The basic certification of both the Boeing 727 and the Cessna 172 were examined insofar as cockpit visibility is concerned. The documentation of the history of both flights and flightcrew training was the responsibility of the operations group. Injuries, toxicological examination of pilots and other crew members, and the formal determination of the pilot's ability to function properly were the responsibility of the human factors group. At the time of the PSA accident, flightcrew background and activities immediately preceding the accident flight were also the responsibility of the human factors group. Since then, the NTSB has then established a human performance group for this phase of the investigation. Aircraft damage, collision angle, breakup mode, and flight control positions were established and documented by the structures group. The aircraft records group, in addition to pro forma documentation of aircraft historical data, made factual determinations as to the configuration of both aircraft relative to the installation and use of anti-collision devices. Power plant performance and integrity were the primary function of the power plant group, as was the examination of the remains of the fuel system to determine the source of the fire, its propagation, and its final effect on the outcome of the accident.

In this case, it was necessary for the air traffic control group to examine the entire concept of air traffic control in terminal areas, interfacing between adjacent and overlapping air traffic areas, air traffic controllers' training, radar system capabilities, application of the conflict alert system, national policy vs local and regional application of air traffic control procedures, and even the working environment in the Miramar Approach Control facility.

The systems group examined the remains of the hydraulically powered flight control systems and the hydraulic, electrical, and pneumatic system components to establish if and how aircraft controllability was affected by the collision.

The cockpit voice recorder group was primarily responsible for ensuring that an accurate transcript of the cockpit voice recorder was made, that voices of cockpit occupants were properly identified, and that flight deck noises were identified and evaluated for appropriate application of the flight regime. In addition, the crew members' comments were evaluated relative to the acquisition of the targets that were pointed out by air traffic controllers.

Flight data recorder readout is essentially a mechanical function monitored by the members of the flight recorder group. It establishes airplane speed, altitude, "g" forces, heading, and time. These data are eventually the key to the activities of the performance group, which makes a factual and analytical assessment of the airplane's performance parameters.

While highly diversified interests are represented on all of the working groups, experience has shown that effective supervision and management of the group chairmen by the IIC and equally close control of the group members by their group chairman virtually eliminate what could be a problem. If, however, a situation develops where pursuit of self-serving interests, rather than the development and documentation of pertinent facts, becomes obvious, the rules (14 CFR 831.9.b, c) provide ample authority for the IIC to remove the offender from the investigation. The fact that the participating parties are, in most cases, knowledgeable regarding the procedural regulations probably makes the enforcement of this part of the rule rarely, if ever, necessary.

I would not want to create the impression that the system is flawless -- it is not. For example, it would be contrary to human nature for a structural engineer employed by an aircraft manufacturer to totally ignore the fact that his employer's very existence may be at stake if a serious design defect is determined to be primarily responsible for a catastrophe that resulted in the loss of hundreds of lives. Likewise, the manager of the air traffic control facility might be less than totally objective if it were his facility controlling the aircraft involved in a midair collision. The director of operations of the airline involved in a midair collision could also be less than objective if he were the brother of the dead captain who was flying the airplane involved in the collision. Obviously, every effort must be made during the formative stages of the investigative group to avoid any situation that would compromise the objectivity of the investigation.

As stated above, the system is not flawless. Nearly eight months after the accident and the launching of an exhaustive and unbiased investigation, which generated voluminous data that was placed in the public file, the final report of the investigation, which was to contain "probable cause" and recommendations, was ready to be adopted by the full National Transportation Safety Board. (Only four members of the politically appointed five-member Board were in office at the time.) In April 1979, after lengthy deliberation at the required "Sunshine" public session, the majority (three) of the members adopted the following probable cause statement:

> The National Transportation Safety Board determines that the probable cause of the accident was the failure of the flightcrew of Flight 182 to comply with the provisions of a maintain-visual-separation clearance, including the requirement to inform the controller when they no longer had the other aircraft in sight.
>
> Contributing to the accident were the air traffic control procedures in effect, which authorized the controllers to use visual separation procedures to separate two aircraft on potentially conflicting tracks when the capability was available to provide either lateral or vertical radar separation to either aircraft.

The above probable cause was adopted despite lengthy and factually supported arguments by the Investigator-in-Charge and a highly qualified member of the Board that it is virtually impossible, even under the most ideal conditions, to use visual separation procedures effectively, as the pilots of PSA Flight 182 were expected to use to maintain safe separation from other airplanes in the vicinity.

The April 23, 1979, issue of the *Aviation Daily*, one of the principal air transport industry publications, reported on the deliberations:

> Procedures allowing controllers to issue visual separation instructions when other separation methods are available were cited by the National Transportation Safety Board as contributing to the midair collision of a Pacific Southwest Airlines 727-200 and a Cessna 172 over San Diego last Sept. 25. Last week, NTSB formally adopted as a probable cause of the accident that killed 144 persons, the "failure of the flight crew of (PSA) flight 182 to comply with provisions of a maintain-visual-separation clearance, including the requirement to inform the controller when they no longer had the other aircraft in sight" (DAILY, April 20).
>
> Two principals in the investigation, NTSB Senior Investigator Rudolf Kapustin and San Diego hearing officer and Member Francis McAdams, wanted to cite the visual separation procedures as the main probable cause, but their suggestions were rejected by the other three safety board members. McAdams cast the sole dissenting vote for adoption of the probable cause and report, and said he would be issuing a dissenting opinion for inclusion in the report.

Dissent by Board Member Francis McAdams stated the following probable cause:

> ... the failure of the flightcrew of Flight 182 to maintain visual separation and to advise the controller when visual contact was lost, and the air traffic control procedures in effect which authorized the controllers to use visual separation procedures in a terminal area environment when the capability was available to provide either lateral or vertical radar separation to either aircraft. Contributing to the accident were the following:
>
> 1. The failure of the air traffic control system to establish procedures for the most effective use of the conflict alert system at the San Diego approach control facility.
> 2. The failure of the controller to restrict PSA 182 to a 4,000-foot altitude until clear of the Montgomery Field airport traffic area.
> 3. The improper resolution by the controller of the conflict alert.
> 4. The procedure whereby two separate air traffic control facilities were controlling traffic in the same airspace.
> 5. The failure of the controller to advise PSA 182 of the direction of movement of the Cessna.
> 6. The failure of the Cessna to maintain the assigned heading.
> 7. The possible misidentification of the Cessna by PSA 182 due to the presence of a third unknown aircraft in the area.

The IIC proposed a similar probable cause but cited the inadequacy of existing air traffic control procedures as a primary factor -- and all others as contributory.

CONCLUSIONS

1. In a complex investigation that has potential political, social, and economic impact, the key to success is the investigative body's complete independence from self-serving interests and from outside influences on this quasi-judicial body that is in a position to affect the final adjudication of the matter. While the laws may be designed to ensure the independence of this body, true independence can be judged only by the final outcome of the matter under consideration and the effectiveness of the recommendations that result from the investigation.

2. While no methodology is flawless and guaranteed to produce perfect results consistently, the "team concept/system approach" can be applied to the investigation of any catastrophic or

non-catastrophic event or system failure. The nucleus of the investigation team and a comprehensive investigation plan must be in place and in a state of readiness at all times, and may be specifically tailored to the system that may be subject to failure and subsequent investigation.

RECOMMENDATION

Using as a model the Public Law pertaining to the investigation of transportation accidents, the formulation of safety improvement recommendations, and independent oversight of agencies responsible for regulating the transportation systems in the United States, the following is suggested for incorporation in a future NATO document:

> Proper conduct of the responsibilities of this investigative body requires vigorous investigation of accidents, incidents, and other system failures regulated by other agencies of NATO. It demands continual review, appraisal, and assessment of the operating practices and regulations of all such agencies, and calls for the making of conclusions and recommendations that may be critical of or adverse to any such agency or officials. No NATO agency can properly perform such functions unless it is totally independent of any other department, bureau, commission, agency, or component of NATO.

REFERENCES

Aviation Daily. (1979). April 23.

National Transportation Safety Board. (1979). *NTSB Accident Report: PSA Flight 182, San Diego, 25 September 1979*. Washington, D. C.: Author. April.

National Transportation Safety Board. (1985). Title 49 CFR-Transportation - Chapter VIII - National Transportation Safety Board Part 83. (Revised 20 March 1985).

U. S. Public Law. Independent Safety Board Act of 1974. U. S. Public Law 97-74- November 3, 1981. 95 stat. 1065.

U. S. Public Law. Tittle III Independent Safety Board Act of 1974. U. S. Public Law 93-633, 49 use 1654 Section 301-302.

THE INVESTIGATIVE TECHNIQUES USED BY THE CHALLENGER COMMISSION TO ADDRESS INFORMATION SYSTEM FAILURES AS THEY RELATED TO THE SPACE SHUTTLE ACCIDENT

Randy R. Kehrli

Presidential Commission on the Space Shuttle Challenger
Washington, D.C., USA

Soon after the Presidential Commission on the Space Shuttle Challenger Accident was appointed (February 6, 1986), Commission hearings revealed that the decision making process which led to the launch of the Space Shuttle Challenger on January 28, 1986 was "flawed" (Report of the Presidential Commission on the Space Shuttle Accident, 1986,Chap. V, pp.82-119; hereafter "Report"). Moreover, the Commission eventually concluded that the information system used by the National Aeronautics and Space Administration (NASA) to track problems had also failed. The purpose of this paper is to describe the techniques utilized by the Commission to investigate information system failures as possible contributing causes to the accident.

ESTABLISHMENT OF THE PRESIDENTIAL COMMISSION

President Reagan established a Commission to investigate the Challenger tragedy on February 3, 1986. The Commission was charged with two responsibilities: a) to review the circumstances surrounding the accident in order to establish the probable cause or causes; and b) to develop recommendations for corrective or other action based upon the Commission's findings and determination (Executive Order 12546). The executive order required that the investigation be completed in 120 days.

The personnel composing the Commission included a Chairman and twelve Commissioners, an Executive Director, an Executive Secretary, fifteen investigative staff members, thirteen administrative staffers, eight individuals providing writing support, over one hundred people working in a support capacity charged with document control, and six independent test observers. The Commission's investigation began on February 6, 1986, and culminated in the presentation of the Report to the President on June 6, 1986, and publication of four volumes of appendices on August 5, 1986.

The Commissioners included experts in engineering, aeronautics, physics, journalism, law, current and former NASA astronauts, and representatives from the military and aerospace industries. The investigative staff included professionally trained aviation accident investigators, astronauts, Department of Defense personnel, FBI agents, attorneys, engineers, metallurgists, and human factors experts. All were selected on the basis of related experience and/or special qualifications that lent valuable support to the investigation.

COMMISSION INVESTIGATIVE STRUCTURE, PLAN, AND TECHNIQUES

Originally, the Commission began its investigation by acting in an oversight capacity in relation to NASA's own accident investigation. However, during a hearing on February 10, 1986, the Commission initially learned of the troubled history of the Solid Rocket Motor joint and 0-ring seals, the Shuttle hardware that malfunctioned and caused the Challenger accident (Report, 1986, Appendix A, p. 206). Moreover, the Commission discovered the first indication that NASA's contractor, Morton Thiokol, initially recommended against launch because of concerns regarding low temperature effects on the performance of the 0-ring seals.

NATO ASI Series, Vol. F32
Information Systems: Failure Analysis
Edited by J. A. Wise and A. Debons

On February 15, 1986, the Commission changed its role from that of oversight to that of independent active investigation and analysis of data presented by NASA and its contractors, with the following statement:

> In recent days, the Commission has been investigating all aspects of the decision-making process leading up to the launch of the Challenger and has found that the process may have been flawed. The President has been so advised. Dr. William Graham, Acting Administrator of NASA, has been asked not to include on the internal investigating teams at NASA persons involved in that process. The Commission will, of course, continue its investigation and will make a full report to the President within 120 days (Report 1986, Appendix A, p. 206).

In order to accomplish its objectives, the Commission continued to hold public and closed hearings, where over 70 witnesses gave sworn testimony amounting to 2,800 pages of hearing transcript. The Commission itself also divided into four investigative panels:

1. Development and Production, responsible for investigating the acquisition, test, and evaluation processes for the Space Shuttle elements (the panel investigators).

2. Pre-Launch Activities, responsible for assessing the Shuttle system processing, launch-readiness process, and pre-launch security (consisting of five Commissioners and two staff investigators).

3. Mission Planning and Operations, responsible for investigating mission planning and operations, schedule pressures, and crew safety areas (consisting of four Commissioners and one staff investigator).

4. Accident Analysis, charged with analyzing the accident data and developing both an anomaly tree and accident scenarios (consisting of five Commissioners and two staff investigators) (Report, 1986, Appendix A, p. 207).

In addition to the panel structure described above, a general investigative staff was formed, consisting of four investigators under the direction of two staff attorneys, as well as a human factors expert. The general investigative staff supported the work of the Commission panels, and was charged with documenting fully the factual background of various areas of the Commission's interest, including telecon between NASA and Thiokol officials the night before launch; the history of the Space Shuttle's Solid Rocket Motor joint design and O-ring problems; NASA safety reliability and quality assurance functions; and the assembly of the Solid Rocket Booster for STS 51-L (the Challenger mission). Subsequent investigative efforts by the general investigative staff focused on the effectiveness of NASA's organizational structure, particularly the Shuttle program structure, and on allegations that there had been external pressure on NASA to launch on January 28.

When the Commission's investigative units began to function, it soon appeared that NASA's information system had failed on two levels, and both were eventually determined to be contributing causes of the accident:

1. There was a failure to communicate the initial concern of Shuttle contractors about the negative effect of the cold temperatures on the launch of STS 51-L to NASA personnel responsible for the launch decision (Report, 1986, Chap. V, pp.82 -119).

2. There had been a long history of problems with the O-ring seals that was not adequately communicated to upper level NASA management, resulting in the problem not being adequately addressed and resolved at an early stage (Report, 1986, Chap. VI, pp. 120 - 151).

In order to investigate the causes of the accident, including these information failures, the Commission and its staff travelled throughout the United States, conducting on-site inspections of contractor facilities and all NASA operation and management centers. Thirty-five formal panel investigations were conducted in which NASA and/or contractors made presentations to panel members. The launch site at the Kennedy Space Center in Florida, the flight control center at the Johnson Space Center in Texas, the Marshall Space Flight Center in Alabama (responsible for the Shuttle Solid Rocket Booster), and the various contractor manufacturing and production facilities were repeatedly inspected and visited by Commissioners and staff members. The Space Shuttle wreckage recovered from the Atlantic was inspected, photographed, and analyzed by Commissioners and staff experts as well as by NASA personnel.

During the investigation, over 160 interviews were conducted, usually on a one-on-one basis, where statements were given by witnesses and recorded by a stenographer. Over 12,000 pages of transcript were obtained via the panel presentations and these individual interviews (Report, 1986, Appendix A, p. 208). The transcripts were normally available to the Commission 24-48 hours after the panel discussion or interview.

The Commission obviously was interested in examining the flow of information within NASA prior to, during, and after the accident. Almost 6,300 documents, totalling 122,000 pages or more, and hundreds of photographs were obtained from NASA, its contractors, or other sources, and were made a part of the commission's permanent data base and archives (Report, 1986, Appendix A, p. 208). Each request for information made by the commission was made to a central office at NASA established to coordinate such requests. Moreover, such requests were formalized by outgoing letters from the Commission and confirming memoranda from NASA, both identifying what document or information was requested and what was produced in response to the request.

Obviously, a vast amount of material was made available to the Commission and was analyzed in a short period of time. To assist in that effort, the Commission engaged a computer document control and search capability contracted from the United States Justice Department. This automated system afforded the Commissioners and staff investigators the opportunity to identify, search, retrieve, and reproduce any document in the Commission's custody. The document control system permitted investigator searches by subject, date, author, title of document, source of document (either original source or Commission source), or the computer identification number assigned to the document. A second computer capability made it possible to locate and search all of the hearing testimony, Commission panel discussions, and interview statements. These two systems proved to be efficient tools in managing and analyzing the information obtained during the investigation. A more detailed description of the "Inquire" and "Juris" computer systems can be found in the Commission's Report to the President (Appendix B, pp. 214 - 218).

APPLICATION OF THE EATPUT MODEL TO THE CHALLENGER COMMISSION'S INVESTIGATION

The Challenger Commission's Report to the President concluded that NASA's information system failed in several respects, which contributed to the cause of the accident. As noted previously, the Report identifies two problems in this regard -- the failure to communicate the contractor's concerns about cold temperatures to the launch decision makers, and the history of inadequate detection, tracking, and resolution of the faulty 0-ring seal design. Although the Commission did not plan its investigation under the EATPUT model of information system analysis, the model can be applied to the NASA information systems used to 1) make the launch decision, or review pre-flight readiness, and 2) track and identify the history of Shuttle flight problems. The table below is an attempt to apply the EATPUT model in the two instances listed above.

Table 1. Two Aspects of NASA Information System Analyzed according to EATPUT

EATPUT COMPONENT	LAUNCH DECISION	PROBLEM TRACKING
EVENT WORLD	Any flight or mission specific concern related to launch	Any design or performance problem or anomaly
ACQUISITION	People, satellites, sensors, computers, contractor and program personnel	Engineering test data, computer sensors recording in-flight data, post-flight inspections
TRANSMISSION	Telephone, messages, telefax, meetings, communications networks via teleconferences	Telephone, teleconference, letter, telefax, computer tracking system
PROCESSING	Computers, letters, reports, Flight Readiness Reviews, Mission Management Team	Computers, problem reports, Problem Assessment Reports, Flight Readiness Reviews meetings
UTILIZATION	Data considered by Shuttle Program personnel and launch decision makers	Data reviewed by contractor and program engineers/ managers, NASA management
TRANSFER	Launch/No Launch	Action Items to resolve problem, launch constraints

With regard to the Challenger 51-L mission, the EATPUT model can be applied to aspects of the contributing causes of the accident -- specifically, the launch under cold temperature conditions and the history of 0-ring seal problems in prior flights. The information in the table below is taken from findings of the Commission report.

Table 2. Two Contributing Causes of Challenger Accident Analyzed According to EATPUT

EATPUT COMPONENT	LAUNCH DECISION DECRIPTION	O-RING PROBLEM TRACKING
EVENT EVENT	Cold temperatures of 0-rings (29-31RF joint temperature)	History of 0-ring seal erosion/anomaly in prior flights
ACQUISITION	Infrared sensor, thermometer personnel	Engineering contractor tests, post-flight inspection
TRANSMISSION	Teleconferences between Thiokol and NASA Level III managers	Erosion experience tracked to Thiokol to NASA via mail, computer, and tele-conference
PROCESSING	Thiokol report and recommendation to NASA LEvel III; none to higher level	Problem addressed at lower level

	launch decision makers	Flight Readiness Review, reported in Problem Reports
UTILIZATION	No consideration of temperature effects on O-Rings by launch decision makers	Problem reviewed by by Level III Level management, erroneously closed out on computer tracking system
TRANSFER	Launch	Launch constraint attached but waived; Action Item issued but design changes not made, and problem not eradicated prior to launch

Finally, I have constructed the table below to show which investigative techniques were used by the Commission and its staff to detect and analyze the various event states identified in the two tables above.

Table 3. Commission Investigative Technique
(In Chronological Order of Use per Event State)

EATPUT COMPONENT	COLD TEMPERATURE	0-RING HISTORY
EVENT WORLD	Hearing Testimony Document Request	Hearing Testimony Document Request Interviews Inquire System
ACQUISITION	Hearing Testimony Document Request Interviews	Document Request Interviews Inquire and Juris Hearing Testimony
TRANSMISSION	Hearing Testimony Document Request Inquire Interviews	Interviews Document Request Hearing Testimony
PROCESSING	Hearing Testimony Document Request Inquire Interviews	Interviews Document Request Hearing Testimony
UTILIZATION	Hearing Testimony	Interviews Document Request Hearing Testimony
TRANSFER	Hearing Testimony	Interviews Hearing Testimony

As can be seen from the Table 3, the failure of the information system as it dealt with the cold temperature data was initially discovered and, for the most part, detailed in testimony before full Commission hearings. On the other hand, the 0-ring anomaly experience was discovered via panel presentations, staff interviews, and document requests; the document search system was then used to detail the history of problems; and finally testimony was taken during a Commission hearing. Both techniques gave satisfactory results, although the latter method appeared to facilitate a more thorough and efficient consideration of the facts and questioning during the Commission hearing.

CONCLUSION

This paper has described, in a general way, the investigative techniques and methodology utilized by the Presidential Commission on the Space Shuttle Challenger Accident. Two aspects of the information system failures found to contribute to the accident by the Commission were listed and placed into the EATPUT model. The applicable techniques used by the Commission to investigate the various EATPUT event states of NASA's information system were also identified.

The Commission's success in making a definitive determination of the causes of the Challenger tragedy in a short time period are due, in large part, to the expertise and "hands on the problem" attitude of the individual Commissioners and staffers. The three techniques of investigative hearings, panel presentations and inquiry, and individual recorded interviews all were invaluable in fulfilling the Commission's mandate. Also, the extensive and efficient computer systems used to identify, search, and control the evidence obtained by the Commission were invaluable in successfully completing the investigation.

REFERENCES

Report of the Presidential Commission on the Space Shuttle Challenger Accident. (1986). Chapter V, pp. 82-119, June 6.

Executive Order 12546. (1986). February 3, p. 206.

ACCIDENT ANALYSIS AND INFORMATION SYSTEM FAILURE ANALYSIS

M.Trung Ho

Institut National de Recherche et de Securite
Paris, France

INTRODUCTION

The objective of accident analysis is to highlight the factors that bear on the probability rate of accident occurrence, to study the way in which accidents take place, and ultimately, to develop preventive measures to reduce their probability. The methods used for this analysis could be applied to different kinds of unwelcome occurrences: major accidents, minor accidents, near accidents, technical incidents, material damages, etc.

When we speak of "accident analysis," we generally refer to an *a posteriori* analysis carried out after the unwanted occurrence has taken place. An *a priori* analysis, which takes place before the unwelcome occurrence arises, is known as "system safety analysis."

It is generally admitted that the most efficient investigation methods are those based on events analysis - that is to say, analysis of the events that have led to the accident and of the logical connections which exist among them. It is also generally acknowledged that investigation methods based on an incomplete model or vision of the accident phenomenon (for example, the conception of the accident as resulting from a violation of rules or of the accident as resulting from a dangerous act and/or from dangerous conditions) do not always lead to interesting conclusions.

For fifteen years, the Institut National de Recherche et de Securite (INRS) has been working to develop and put into practice in industry a method for analyzing occupational accidents. It is a method based on events analysis and allows investigators to start from the final unwelcome event and trace events back toward incidents or failures that occurred in the system, in particular toward the failures of the part of the system that deals with information (the information sub-system).

A BRIEF LOOK AT THE INRS METHOD

It is not the purpose of this paper to give a detailed description of the content of the method since it is described by Meric et al. (1976), Leplat (1978), Monteau (1983), and Ho (1983). A simple outline with an example of application will suffice. The INRS method was born of the need to replace the incomplete and reductive vision of an accident mentioned above by a broader one, that can be summed up in two statements:

- Even if one considers only events immediately prior to it, an injury may be the result of several combined facts.

- It is not enough to consider simply those events that occur immediately prior to an accident; there may be other causes, more remote in time or in the organization, which must be considered.

This broader vision of an accident, which goes beyond mere technical or individual factors to take into account organizational factors when necessary, is admirably served by the fault tree method. This method is basically a very open means of studying events and presenting them in

NATO ASI Series, Vol. F32
Information Systems: Failure Analysis
Edited by J. A. Wise and A. Debons

an efficient graphic form, conveying a great amount of information clearly and concisely.

Broadly speaking the procedure proposed by the INRS for analysing accidents can be broken down into the three stages described below.

Stage 1: Draw up an accident report.

- Gather information, taking care to include only the facts, avoiding value judgments and interpretations (such as "clumsiness", "carelessness", etc.).
- Look for the connections which exist between the facts, and organize them in a logic diagram starting from the injury (or the final event) and systematically asking the following questions:
    - What was needed for (X) to happen?
    - Was (Y) needed for (X) to happen?
    - Was (Y) alone sufficient for (X) to happen?

Three types of link are used:

- Simple chains: one fact, one antecedent

- Disjunctions: several facts, one antecedent

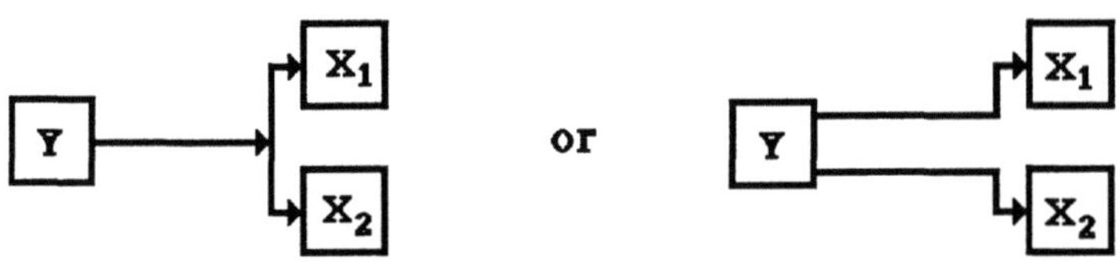

- Conjunctions: one fact, several antecedents

The diagram obtained is called an antecedent tree. The act of drawing up the antecedent tree can lead the investigator to search for additional information and to pursue the analysis often very far back from the accident.

Stage 2: Exploit the accident report.

- Draw up a list of possible curative measures capable of neutralizing the different antecedents without excluding any measure a priori.

- Decide upon appropriate preventive measures, bearing in mind the following efficiency criteria:

  The stability of the measure in time
  The non-transference of the risk
  The scope of the measure (suitability for general application)
  The cost of the measure for the operator and the firm.

- Prepare to follow up measures by deciding who will implement them and who will be responsible for enforcing them and supervising their efficacy.

- At this second stage, the investigator may introduce certain interpretations that would permit a better understanding of the accident process, but these interpretations are well distinguished from the fact-gathering of the first stage.

Stage 3: Management of accident reports.

The third stage takes us away from the actual analysis of an accident to consider the more general problem of safety management in the firm. Such management includes several activities:

- Early detection of risks based on potential factors of accidents detected in past case studies.

- Design and application of preventive (and not simply curative) measures.

- Follow-up of measures as a whole (curative and preventive).

- Periodical recapitulation of results.

DETECTING FAILURES OF AN INFORMATION SYSTEM BY ANALYZING ACCIDENTS OR INCIDENTS

In this section, I shall examine a sample accident by drawing its antecedent tree. This diagram will be used for searching for failures of the information system and for possible preventive measures.

Report of the Accident

The accident occurred in a quarry. Mr. A had to weld a steel plate onto a hopper feeding materials to a belt conveyor. He climbed up on the belt conveyor, stood on its lower part, and carried out the welding. Mr. B took advantage of the fact that the conveyor was stopped to adjust the upper drum. B was called to the telephone. When he came back, he wanted to check the adjustment he had made and, having forgotten the presence of his fellow worker, started the conveyor. A fell down.

Use of the Antecedent Tree

With this accident we have a typical example of the situation called co-activity. In such a situation one can find two or more people (or teams) carrying out uncoordinated activities, at the same time, at the same place, or in the same system.

The antecedent tree of the accident is presented in Figure 1. Boxes on the left represent A's activities; boxes on the right show B's activities.

When A began his task, he probably had in mind the image of a limited system, the components of which are the belt conveyor and himself. Since he has stopped the conveyor, there should be no danger in standing on it. He did not think of the possibility of intrusion of another activity by another person. He had not signalled his presence by, at least, hanging on the control panel a notice saying something like "Zero energy stage. Man working on the conveyor." Such a measure is not generally considered reliable enough, and it is often recommended to lock the switch and keep the key. But even if A had placed only a notice, B probably would not have switched on the conveyor.

We can say here that vital information was not transmitted (transmission failure). When B took advantage of the conveyor's being stopped to adjust the upper drum, the system was altered: it became more complex with, as components, the belt conveyor and two men working on it. This new system, larger and more dangerous, was not supported by an information sub-system. B knew that A was on the conveyor, but he contented himself with keeping this information in his memory. Since there was a certain distance between the two work stations, this information sub-system was incomplete and very weak. A did not know that B was adjusting the upper drum and could start the conveyor. B should have at least informed A of his projected operation on the conveyor and placed a notice signalling the presence of A and B. (A more reliable system could be constituted by two padlocks locking the switch, the key of the first padlock being kept by A and the second's by B.) Without these minimum precautions, a telephone call, acting as an interference or disturbance, and probably combined with other factors, had been sufficient to erase from B's memory the presence of his fellow worker.

We will probably never know what other factors contributed to B's memory lapse, but it is often suggested in human failures studies that a significant percentage of failures are due to the fact that a person's intellectual capacities are so taken up with immediate goals, partial criteria, or an unadapted model of the reality that he is unable to see other essential points or to be aware of the need to modify his mental model. This phenomenon could be called inadequate mental representation. In this case, the immediate goal of "Check the adjustment," and perhaps along with the content of the telephone call, had probably contributed to prevent B from having an adequate representation of the situation.

The foregoing analysis of a rather common accident allows us to define several interesting failure modes of an information system:

- Non-emission of a vital information.
- Harmful effect of a disturbance.
- Failure of the information sub-system to grow in efficiency and reliability while the whole system is growing in size and complexity.
- Inadequate mental representation that prevents a person from making a good decision.

CONCLUSIONS

If a system is to survive, it should have an information sub system. Analysis of accidents, near-accidents, or technical incidents can provide interesting, if not vital, information on system failures, particularly failures of the information system.

For carrying out useful analysis, the method selected should not be based on an incomplete model of the accident phenomenon. The model should be an open one that allows the investigator to collect all the pertinent facts it is possible to gather. It should help him to organize the facts and to provide a sound description of the accident process. Methods based on events analysis seem to meet these criteria.

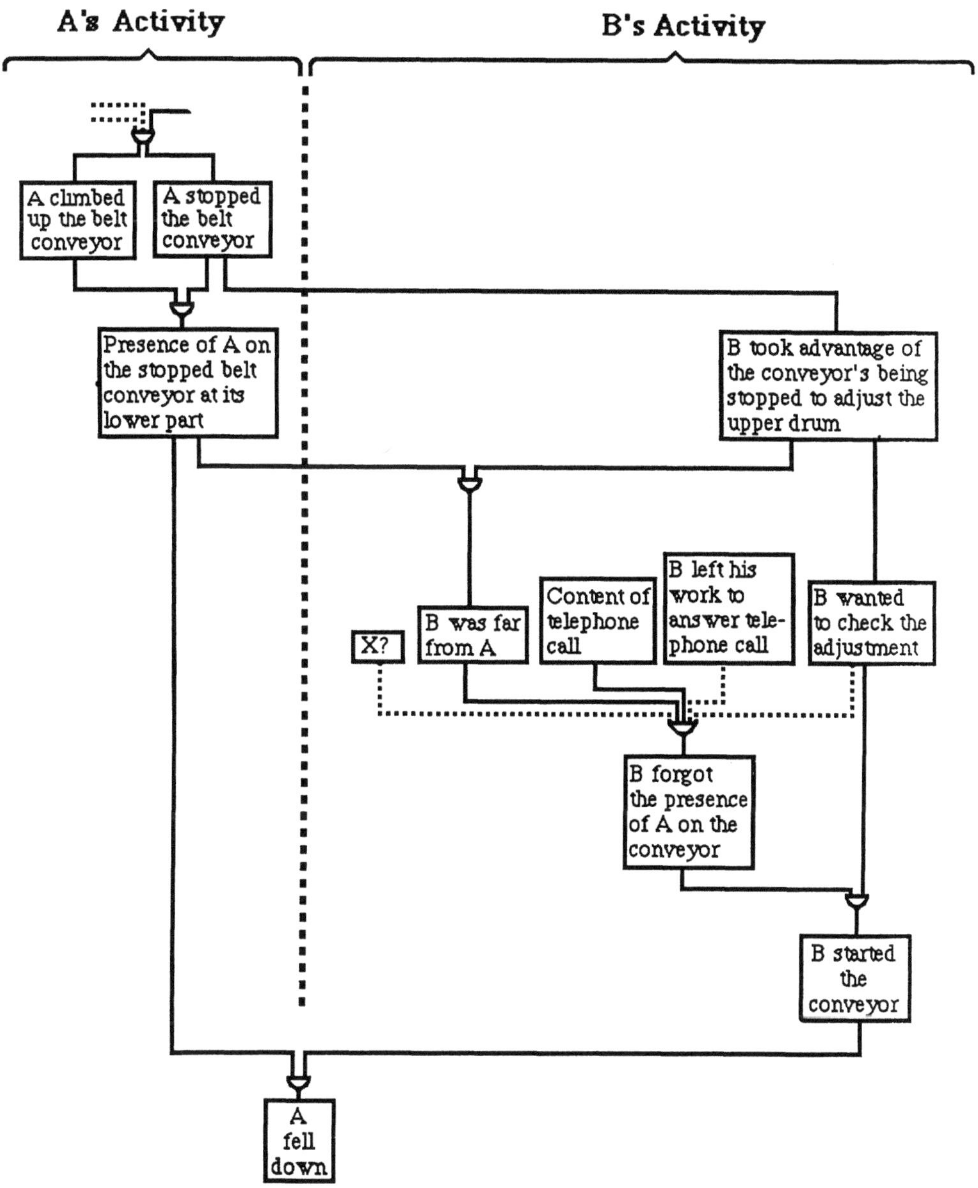

Figure. 1 Antecedent tree of an accident.

REFERENCES

Ho, M.T. (1983). System analysis (safety): methods and analysis. In *Encyclopedia of Occupational Health and Safety*. Geneva; International Labour Office, pp. 2133-2136.

Leplat, J. (1978). Accident analysis. *Journal of Occupational Accidents*. 1, pp. 331-340.

Meric, M., M. Monteau, and J. Szekely. (1976). *Safety Management Techniques. Analysis of Occupational Accidents and Use of the Notion of Potential Factors of Accident in Risk Prevention.* Report No. 243/RE, Institut National de Recherche et de Securite, Paris.

Monteau, M. (1983). Accident analysis. In *Encyclopedia of Occupational Health and Safety*. Geneva; International Labour Office, pp. 13-16.

Saari, J. (1984). Accidents and disturbances in the flow of information. J*ournal of Occupational Accidents*, 6, pp. 91-105.

FACTORS IN THE INVESTIGATION OF HUMAN ERROR IN ACCIDENT CAUSATION

Michael L. Barnett

College of Maritime Studies
Warsash, Southampton, U. K.

INTRODUCTION

In determining the procedures to be used in an accident investigation, the first two factors to be considered are a) the type and purpose of the investigation, and b) as a consequence, the role of the individual investigator.

Investigations into accidents are usually either for the purposes of an adversarial inquiry in which conflicting parties may be seeking compensation, or an inquisitorial inquiry for reasons of improved safety. Generally, the avowed main purpose of an inquisitorial accident inquiry is to establish the cause of the accident in order to prevent similar disasters in the future, with the secondary and subsidiary aim of establishing who, if anybody, is to blame.

Preferably, an investigator should be independent of the organization he is investigating, and charged only with establishing the cause of the accident. In actual practice, however, this might not always be the case. Therefore, as well as taking account of the purposes of the investigation itself, the individual investigator should also be aware that there may be certain motives for offering or not offering pieces of information to different investigators, not only by the participants in the accident but also by other parties involved in obtaining information during the investigation.

For example, the essential information considered necessary to determine the cause of an accident, the best techniques and procedures to obtain that information, and the cooperation and responses from survivors of the accident might be very different if the investigator is a policeman, a systems engineer, or a psychologist.

The role of the investigator and his technical background will therefore determine, to a certain extent, not only the sort of information being sought, and thus the method of obtaining it, but also after the analysis of the data, his interpretation of the cause of the accident.

IDENTIFICATION OF ACCIDENT CAUSES

Whether the aim of the investigation is culpability or accident prevention, there is a tendency to ascribe to an accident a single major cause, although it is recognized by researchers, for example Norman (1980), that accidents rarely occur as a result of a single cause but from the interaction of a number of contributory factors. It has also been suggested by Spurgion and Young (1981) that such factors can strictly only be considered contributory and not causal, because their occurrence leads to accidents in some cases but not others.

It appears, then, that an accident occurs as a result of a unique and subtle combination of many contributory factors, which seem to conspire to create an accident. Some of these factors may seem trivial in themselves and may therefore pass unrecorded or unnoticed, but even if it were possible to determine precisely all the circumstances surrounding an accident - and this is obviously not always possible - the cause is still open to interpretation.

The classic story to illustrate this problem is told by Chapanis (1959) and is repeated here in full because it so clearly demonstrates one of the fundamental paradoxes of accident investigation:

NATO ASI Series, Vol. F32
Information Systems: Failure Analysis
Edited by J. A. Wise and A. Debons

> A man has a protracted argument with his wife. He stamps out of the house to the nearest bar and drinks four highballs. He then decides to go for a ride. It is night-time, there is a skim of snow on the ground, and the tires of the car are smooth. In rounding a poorly banked curve at excessive speed, the right front tire blows out. The car leaves the road and is demolished. What was the cause of the accident? The argument? Drinking? Speed? The weather? The smooth tires? The blow-out? The poorly designed highway? It is impossible to say, for if we had changed any of these factors, perhaps the accident would not have happened. We have no way of assigning a "cause" even though we may have a complete description of the circumstances leading up to it. The policeman, minister, psychiatrist, and highway safety engineer would each find a different cause in this event.

The complexity of the different interacting systems in a modern technological industry like aviation, for example, requires that a multi-disciplinary team of experts be assembled for a thorough and detailed investigation in order to achieve a clear understanding of the circumstances of an accident. However, for the reasons given above, the team concept does not guarantee that at the conclusion of the investigation, there will be unanimous agreement as to the cause of the accident. (See Kapustin 1986.)

In summary, because of the many and varied contributory factors that can play a significant part in the events preceding an accident, the causes of a particular accident are legitimately open to interpretation. An individual investigator's background and likely interpretation of events will influence the specific procedures adopted to gain information and the type of information sought.

HUMAN ERRORS

One interpretation of accident causation is to regard all accidents as the result of a mistake by someone, somewhere, at some time or another. This analysis traces the cause of any accident to an error on somebody's part, whether he be the actual operator of the system, a maintenance engineer, or even the original system designer. As Singleton (1979) points out, the extent to which mistakes on the part of the designer are considered either human errors or technical failures is very much a matter of the current fashionable convention.

Singleton goes on to explain that one of the problems of any description of human error is that there are several different ways of classifying errors, i.e., there is no single universally agreed upon taxonomy of human error. One method is to describe human errors in terms of an information processing model of the particular system, organization, or technological industry under consideration.

The marine industry, with which the author is most familiar, provides a typical example of several interacting man-machine systems in operation. The safe control of bulk liquid cargoes is a particular technological system, akin to process plant control ashore, where human activities like monitoring, fault detection and diagnosis, problem solving, and decision making are all involved.

Using the EATPUT Model of an information system, it might be possible to describe the different types of error that could occur in the handling of a hazardous bulk liquid cargo aboard ship, including errors in the acquisition of information about the status of the cargo, errors in the transmission of that information, errors in the interpretation of information and in the utilization component of the system, and finally errors in the transfer component of the system, such as the operation of a wrong valve.

The problem facing the accident investigator is that establishing the type of error that has occurred in a particular incident does not help in determining the underlying cause of the error. A number of different causes can produce the same type of error. The operation of a wrong valve, for example, could be due to many things: the bad display of information to the operator; negligence; ignorance of the system on the part of the operator through either incompetence, poor training, or inexperience; a calculated risk taken by the operator, perhaps influenced by outside pressures; the effects of drink, drugs, or fatigue on the operator; or, as Reason and Mycielska (1982) have shown, an unfortunate slip or mental lapse.

It can also be seen that proposed measures to improve the safety of the system can take many forms, such as improved hardware, improved software including new procedures, improved training, improved selection of personnel, or improved working conditions.

The problem of establishing the underlying cause of an accident has been recognized in the aviation industry by researchers like Feggetter (1982), who has attempted to devise methods for the investigation of the human factors aspects of aircraft accidents, and to go beyond a surface description of the errors to a deeper understanding of the underlying cause of an accident.

TECHNIQUES FOR INVESTIGATING HUMAN FACTORS IN ACCIDENTS

Although some researchers, for example, Humphrey (1981), have suggested that special, short questionnaires might be useful in obtaining certain data after an accident, questionnaires suffer from a number of disadvantages. Perhaps the most significant is that because the questions are predetermined, there is no completely satisfactory design for exploring the more relevant or interesting areas in sufficient depth. By their very nature, accidents are the result of a complex set of circumstances, and the questionnaire approach is likely to be too inflexible to capture the subtleties and nuances of the events that have occurred.

Interviews also have their disadvantages, as shall be seen, so Feggetter has suggested that the ideal balance is a structured interview incorporating a checklist, conducted by a skilled interviewer. Feggetter emphasizes the importance of the interviewer being a human factors specialist trained in interviewing techniques. While accepting that special training in these skills is required, it should also be borne in mind that the interviewer will require considerable knowledge of the technical aspects of the operation, particularly a sympathetic understanding of the practical problems encountered by the operator.

It is a matter of debate whether the optimal interviewer is a human factors expert well versed in technical matters or a technical expert with a good understanding of human factors. It may well be that the interviewee will feel more comfortable with a professional peer and thus be more likely to talk openly, but whatever the status of the interviewer, it is important that the interview be conducted in confidence and without recrimination, and that this intent be made clear to the interviewee.

Some of the problems of interviews are the same as for any method of investigation. It is likely, for example, that the evidence from participants is biased, either for legal or personal reasons. In addition, recollections of an accident can be confused and incomplete, and the accounts of genuinely independent witnesses can vary widely since each person perceives the events leading up to an accident in his own way. To overcome the problem of the accurate recollection of past events, it is generally held that interviews with participants should take place as soon as possible after the accident.

The problem of the wide variation in individuals' perceptions of events leads to a more theoretical point about the relevance of the information recollected, or what may be called the problem of hindsight. Because accident investigations inevitably take place after the actual accident, the total environment or context in which the events prior to the accident took place is lost forever and cannot normally be retrieved by subsequent investigators because of the intervention of the accident itself.

Since the accident must lie in the time sequence between the events leading to it and their recall, participants tend to remember their pre-accident experiences in terms of those recollected events that appear to have had a direct bearing on the final accident. Consequently, investigators can explain the human factors aspects of an accident only in terms of these remembered events, i.e., in terms of chains of supposed causes and effects.

Sheehy (1981) has suggested that such inferences may not be justified, and that since witnesses will attempt to provide a concise, logical, well-structured account of the events prior to the accident, there is a danger that they will naturally omit all the half-forgotten, seemingly unrelated and unimportant thoughts, actions, and pre-occupations that, in reality, might be highly significant in an explanation of the accident. It is presumably to overcome this problem that Feggetter has suggested that accident investigators should consider using hypnosis or "truth drugs," as well as more conventional interview techniques.

In the normal interview situation, however, the interviewer will need to build as complete a picture as possible of the situation prior to the accident, and this may involve exploring the interviewee's background, training, motivation, etc. Naturally, this requires skill and experience on the part of the interviewer, and at this stage, he should avoid introducing his own bias through preconceived theories of the likely cause.

The inevitable incompleteness of the information relating to the human operators in the system makes subsequent analysis difficult. Techniques like fault-tree analysis are very effective in pinpointing technical failures because it is possible to model the system accurately. To analyze the thoughts and actions of the human operators, however, can be far more difficult due to the lack of complete and appropriate models of their processes.

CONCLUSION

Accident investigations are an important and necessary part of the safety strategy in a technological system, but it should be remembered that if the purpose of the study of accidents is to improve the safety of the system, then the sorts of caused factors involving human operators that are generally derived do not necessarily suggest easy or immediate remedies. For example, remedies to reduce accidents caused by a factor as general as "error of judgment" still present a complex training and assessment problem.

If the aim is safety improvement, then, as Singleton (1979) suggests, other methods are required to complement accident investigation, such as forms of work analysis like task and skill analysis.

REFERENCES

Chapanis, A. (1959). *Research Techniques in Human Engineering*. Baltimore: John Hopkins Press, p. 94.

Feggetter, A. J. (1982). A method for investigating human factor aspects of aircraft accidents and incidents. *Ergonomics*, 25(11), pp. 1065-1075.

Humphreys, M. (1981). The collection and analysis of in-depth road crash data. *Ergonomics*, 24 (6), pp. 423-435.

Kapustin, R. (1986). This Volume.

Norman, D. A. (1980). *Errors in Human Performance*. San Diego: California University.

Reason, J. and K. Mycielska. (1982). *Absent-minded? The Psychology of Mental Lapses and Everyday Errors.* Englewood Cliffs, NJ: Prentice-Hall.

Sheehy, N. P. (1981). The interview in accident investigation: methodological pitfalls. *Ergonomics*, 24(6), pp. 437-446.

Singleton, W. T. (1979). The human factor in oil pollution. In Wardley-Smith, J. (ed.). *The Prevention of Oil Pollution.* London: Graham and Trotman, pp. 233-252.

Spurgeon, P. and K. M. Young. (1980). The role of work analysis in the study of shipping accidents and the implementation of computer-based collision avoidance systems. In Oborne, D. J. and J. A. Levis (eds.). *Human Factors in Transport Research. Volume 1.* London: Academic Press, pp. 49-57.

THE INVESTIGATION OF INFORMATION FAILURES IN ORGANIZATIONS

Donald Mason

College of Librarianship, Wales
Llanbadarn Bawr, Aberystwyh, U. K.

In a perfect information system, the information would always flow freely through an organization and would be utilized effectively. Unfortunately, we live in an imperfect world and have to take account of a range of barriers to information flow, some of which deflect or alter the flow, while others just soak it up and become, in effect, information sinks (Unesco, 1977).

Information cannot, therefore, just be allowed to flow, it has to be pumped; and the well-designed information system will do this and provide an efficient and effective service. Efficient in the sense of operating smoothly at an economical cost, and effective in that it meets the objectives and standards laid down for it in terms of precision, penetration, response time, reliability, etc.

However, in the general consideration of the analysis of information systems failure, we have to take into account the fact that much information in an organization is of relatively little value, and that the value of information is not necessarily related directly to the cost associated with it (Christie, 1981, p. 56). For the analysis to be cost-effective, the cost involved in carrying it out must be related to the financial and other losses arising from the failure. There is a difference in dramatic effect between an aircraft crash and the collapse of an industrial company, although the latter may result in as much human anguish as the former.

In looking at methods for analysis of information failures, it may therefore be necessary to consider a range of technique, each suitable for the investigation of failures of different orders of magnitude. This paper is based on the practical experience of the author in investigating information failures in industrial companies, all of which have been costly but none of which has led to the closure of the company concerned.

The Debons EATPUT model is based on the assumption that certain components are necessary to any information system. The model can therefore be used not only to conceptualize the system itself but also as a framework for investigation into failure. EATPUT is an acronym for

Event World:	the event as it occurs, its nature and perception.
Acquisition:	the sensor that picks up the event.
Transmission:	the means by which the information flows through the system.
Processing:	the method by which information is manipulated for use.
Utilization:	systems output in usable form.
Transfer:	communication allowing feedback with the environment.

For individual information systems, it is also necessary to specify the objective, and to consider the effect of the system on decisions arising from it. An outline of failure analysis would therefore consider whether an event had been detected or not, and if so, whether an event had been noted accurately; whether there had been a sensor failure; whether there had been a breakdown or corruption of information in the transmission; whether processing rules were incorrect or not sufficiently refined for significance to be noted; and whether the output had been presented in the right form and to the right person. Also, attention would be paid to

NATO ASI Series, Vol. F32
Information Systems: Failure Analysis
Edited by J. A. Wise and A. Debons

the objectives and to the utilization of the information in the decision-making process in order to check that the objectives were relevant to the required usage.

In the investigation of information failures in organizations, a systems approach similar to that used for systems analysis and design can be effective. The first step is the identification of the problem, and the investigator has to ensure that the outward and visible symptoms are not mistaken for the genuine problem. Data can be gathered, analyzed by various techniques and conclusions reached on the reasons for failure. It is also usual for the investigator to be asked to make recommendations on the improvement of the information system to prevent future failure.

The information system within which the failure occurred cannot be studied in isolation. Its position within the information environment of the organization and its interrelation with other systems have to be considered. The organization's policy in relation to information systems and the formal and informal channels of communication within the organization all have to be taken into account in arriving at an understanding of how, and with what level of effectiveness, the information systems operate. The investigator also has to bear in mind the value of information in relation to costs and in relation to needs, and seek to identify those critical nodes in the information network at which failures could lead, or have led, to heavy financial loss (Christie, 1981, p. 57)

The particular stage of development in the life cycle of the information system is also relevant (Duffy and Assad, 1980). The introduction of new information technology into a system can lead to a range of problems, including loss or corruption of information; non-availability of information; unauthorized and poor communication; disorganized indexing and retrieval systems; inappropriate time utilization by staff; unspecified responsibility for technology introduction; and shortage of knowledgeable staff for system evaluation (European Industrial Research Management Association, 1985). These are problems which management has to overcome if the systems are to work effectively. Often failures occur because insufficient thought has been given to the necessary re-design of systems, apart from those problems that arise, on the one hand, from over-confidence in new gadgets, and on the other, from the cynicism of the latter-day Luddites.

A GUIDE FOR FAILURE ANALYSIS OF INFORMATION SYSTEMS

The following approach to information systems failure analysis poses a range of questions that will enable the investigator to obtain an overview of the information system and its place within the information environment or the organization, together with an understanding of how the system does or does not work. The questions should identify the cause or causes of failure and indicate possible remedies.

1. *What stage in organizational development has the organization reached?*

Studies of organizational development have shown that companies tend to develop in a series of stages. The first stage is based on the creativity of the founders, who are usually technically or entrepreneurially oriented. When the company has developed to a size where more management control is needed, a capable business manager is appointed. As the company increases in size, management control becomes more difficult, and decentralization with delegation takes place, followed later by the development of coordinated activities, with company-wide programs of control and review instituted for line managers and with technical functions, such as data and information processing, centralized. Each of these stages usually has a different approach to the supply of information needed for decision making.

2. *What is the information policy of the organization?*

Information policy is often related to the stage of development. It is rarely clearly stated in a policy document and has to be deduced by the investigator from observation of information flows and from the attitudes of senior managers (which may be contradictory). What one is looking for here, both for the company as a whole and for individual departments and managers, is whether the company ethos encourages or discourages the flow and use of information. The failure of unwelcome information to flow upwards in an organization is a well known phenomenon. A complaint on the workshop floor becomes a compliment by the time it reaches top management.

3. *What is the objective of the information system under investigation, and how does it relate to other information systems?*

The objective may be clearly stated or may need to be deduced from examination of the use of the information. The purpose of the enquiry is to define the true objective so that it may be kept in mind at all the later stages of the investigation as a check that the system is designed and operating in a way that will meet the objective. Also, the investigation should determine that he objective matches and forms part of any higher information system to which the system under investigation belongs, and whether the objective clashes with those of interrelated systems at the same level in the hierarchy.

4. *Does the information system have any sub-systems, and if so, what are their objectives?*

Again, the purpose here is to check for compatibility among the objectives of the sub-systems, and to ensure that they form part of the overall objective.

5. *What is the environment from which the information is collected?*

For some systems, the environment can be worldwide. For example, a company will wish to collect strategic information on the politics, stability, and finances of various governments, on international and regional development plans, on the activities and plans of competitors, suppliers, customers, etc.

6. *How is the information collected and input to the system?*

The methods of collection and input need to be assessed in terms of validity, relevance, and exhaustiveness. The form of input should be such that the later manipulation and correlation of information can be effectively carried out.

7. *How is the information stored in the system?*

The investigator needs to discover what is recorded and in what form; what information is static; what is changed by being over-written; and whether there is a historical record for comparative purposes. This last point can be important because if information is lost from the system, it can be difficult to discover how it was actually used in making a decision at some previous point in time.

8. *By what means does the information flow through the system?*

This is a check on the transmission system (electronic, clerical, or oral) to see whether the information flow is facilitated or blocked by the system, and also whether the information is corrupted in any way.

9. *What is the effectiveness of the retrieval system?*

The retrieval system may work on the basis of producing regular printouts in response to built-in criteria of selection and correlation; or it may operate on the basis of provision of information on demand. The check here is to see whether retrieval produces information that is relevant and timely, and whether it will produce all the information that is required. It is necessary, however, to distinguish between system relevance and user relevance. If the system has been properly designed, it will produce all the information relevant to the "wording" of the enquiry. Whether the enquirer judges the information to be relevant to his purposes is much more subjective. The system may supply information of which he is already aware, or may supply it before his mental processes can accept it as being relevant. (My mind is made up; do not confuse me with the facts.)

10. *In what form is the information output?*

Here the concern is with the way information is presented to the user. This may vary from raw information for the user to assimilate and draw his own conclusions, to a report in which the information has been correlated, important factors highlighted, conclusions drawn, and recommendations made. It can include text, illustrations, statistical data in diagrammatic form, etc. The presentation should be in a form that meets the objective of the system and is useful to the recipient in effective decision making. Information overload should be avoided.

11. *What use is made of the information?*

Although in following the system through as if it were a logical sequence this question comes last, it will usually be asked at the same time as question 3 because of the essential interrelation between usage and objectives. Investigation will show whether the information needs and usage have gradually changed over a period of time, whereas the system is still operating according to the original, but no longer relevant, objective.

The application of the critical analysis and evaluation techniques at appropriate stages of the investigation will help to elucidate the situation. The questions to be asked are as follows:

- What is this system (or sub-system) being operated? Need it be operated at all?
- What is being done? Is it the right action at this stage? Would some other action be more effective?
- Who is doing it? Is it the right person? Could it be done better by someone else, or by a machine?
- Where is it being done? Is it the right place? Would another place be better?
- When is it being done? Is it the right time? Would another time be better?

This technique may appear to be simple-minded and perhaps old-fashioned, but nevertheless it does enable the investigator to get to the core of the problem (Mason, 1978, p. 62).

Two final points need to be made. First, prevention being better than cure, the critical nodes in the information system should be identified and information audits carried out at regular periods to check that the system is functioning correctly, that the information collected is relevant, and that the conclusions drawn from the output are valid. Secondly, that for any investigation of an information system failure to be effective, the investigator should be a person who comes from outside the organization concerned.

REFERENCES

Christie, B. (1981). *Face to File Communication: A Psychological Approach to Information Systems*. New York: Wiley. p. 56.

Duffy, N. M. and M. G. Assad. (1980). *Information Management: An Executive Approach.* Oxford University Press. pp. 58-60.

European Industrial Research Management Association. (1985). *The Impact of New Information Technologies on R & D and on I & D Management*, p. 52.

Mason, D. (1978). *Information Management.* Peter Peregrinus, pp. 55-69.

Unesco. (1977). *Handbook for Information Sytems and Services*, pp.11-14.

INFORMATION SYSTEM FAILURES DURING HAZARDOUS MATERIAL SPILLS

William F. Stubler

Westinghouse R&D Center
Pittsburgh, Pennsylvania, USA

Darlene K. Susa Anderson

Koppers Company, Inc.
Pittsburgh, Pennsylvania, USA

BACKGROUND

When a hazardous material spill occurs, individuals, organizations, and physical resources must be coordinated to achieve an effective response. For example, the chemical manufacturer must often provide accurate and timely data about the chemical to prevent personal injury and property damage. Because information is critical for the effective application of human and physical resources, an information system must channel critical data and support decision-making processes.

An information system failure may be defined as a failure of the system to respond to changes in its environment and achieve intended objectives. Information system analyses and design methodologies provide the tools necessary to analyze information failures, overcome deficiencies, and achieve effective information systems.

EATPUT INFORMATION SYSTEM MODEL

An information system may be described as an environment, consisting of people, technology, and procedures, which augments human decision-making and problem-solving capabilities. Debons (1980) described the EATPUT information system model, which is based on analogies with organisms and the functions they must perform to survive in their environments. The EATPUT model has six components:

Event World The environmental (spatial and temporal) data within which the system is encapsulated.

Acquisition The mechanisms for sensing the event world and capturing data about events of interest.

Transmission The modes of signal transmission used to carry data from the environment to the information system and to carry data between the components of the information system.

Processing The manipulation of acquired data to put it in a form useful to the system or the user.

Utilization The contemplative, interpretive, and deliberative function which converts processed data into information and knowledge prior to execution of a response.

Transfer The function in which the system or user takes action based on the information or knowledge gained and causes some change in the event world.

The EATPUT model provides a framework for understanding how an information system should function in its environment to accomplish its specific objectives. It may be used to:

NATO ASI Series, Vol. F32
Information Systems: Failure Analysis
Edited by J. A. Wise and A. Debons

- To design new information systems.
- To uncover conditions where existing information systems may fail.
- To analyze information system failures after they occur.

In the following case study, the EATPUT model is used to identify conditions where failures may occur in the type of information system used by many chemical companies.

CASE STUDY

Background

The liability concerns of companies that manufacture and transport hazardous materials have been well publicized in recent years. The 1984 accident in Bhopal, India, tragically illustrated the danger of hazardous material releases. Increased public awareness and regulatory pressure from government agencies, such as the U.S. Department of Transportation (DOT), the Environmental Protection Agency (EPA), and the Occupational Health and Safety Administration (OSHA) have focused attention on the need for safe handling of these materials and better emergency response to minimize losses when incidents do occur. The ability to obtain needed data and transfer them into correct and timely actions is essential to successful emergency response.

An analysis was performed to define the requirements for an ideal information system that could support decision making activities of chemical manufacturers during transportation-related spills of hazardous materials. An existing information system, typical of those used by chemical manufacturers, was analyzed and compared with the ideal information system. Discrepancies that may result in information system failures were identified.

Analysis

System Objective

The objective of the information system is to minimize the chemical company's liability through the prevention of personal injury and property damage and the distribution of accurate and timely data. The decisions to be supported include the following:

- Identifying and evaluating hazards (including identification of the spilled product and contributing hazard factors in the spill environment)

- Determining the best responses to the situation (spill containment, material disposal, procurement of auxiliary equipment, and communication of data about the spill to involved agencies and the media).

Information System Boundaries

The information system addressed in this report consists of several components:

- Personnel from a chemical manufacturing/distribution company including the Hazardous Material Response Coordinator, management, and technical specialists (people with special expertise who provide guidance in the event of a product spill)

- Chemical Transportation Emergency Center (CHEMTREC), an organization established by the Chemical Manufacturers Association to provide technical assistance in the event of a transportation- related hazardous material spill

- Person at the spill site (e.g., fireman, common carrier employee, or a bystander)

- Local, state, and federal agencies and authorities that become involved when a spill occurs.

The primary user of the information system is the chemical manufacturer's Hazardous Material Response Coordinator. This person interfaces with emergency site personnel, government agencies, CHEMTREC, local authorities, and company technical specialists to identify the hazards associated with the spill and recommend a course of action.

The structure of this information system is similar to information systems used by many chemical companies. The ideal information system discussed in this report refers to the way the information system should function if it is to achieve the objective and support the decisions outlined above. The typical information system refers to the way this information system actually functioned for the chemical company studied.

Event World

The Event World of a transportation-related hazardous material spill may be organized into three categories, (1) factors that define the spill, (2) complicating factors, and (3) liability concerns.

Factors that define the spill include the following:

- Type of chemical involved
- Mode of transportation (truck, rail, boat or aircraft)
- Quantity involved
- Rate of spill
- Specific hazards to people, property, and the environment

Several complicating factors tend to increase the risks involved or the difficulty of clean-up. These include:

- Proximity to the manufacturer
- Physical access to the spill
- Factors that inhibit containment such as the weather (wind and precipitation) and terrain (streams, hills, etc.)
- Time factors which may affect the accessibility of people.

Liability concerns include:

- Involved parties at the spill site whose safety and health must be protected emergency personnel, bystanders, neighboring populations, etc.)
- Involved parties away from the spill site who must be informed of the event (CHEMTREC, company personnel, and various authorities)
- Federal, state, and local regulations and company policies which must be satisfied in the case of a spill
- Media personnel who require timely and accurate data if the incident is to be reported fairly and accurately.

The ideal information system should satisfy the following criteria:

- 24-hour availability from spill sites.
- Recommendations for immediate actions to be provided immediately upon request and follow-up actions to be provided within 10 minutes of initial request.
- Adherence to all government and company policies.

In the typical information system, the Hazardous Material Response Coordinator cannot begin his role until he has been notified of the incident. Despite a notification plan, which allows for alternate coordinators to be contacted, delays have occurred due to the unavailability of coordinators during non-business hours (e.g., holidays, vacations, travel time to work, etc.). One historical analysis has shown that 55% of spill-type incidents occurred during non-business hours, 4:00 P.M. to 8:00 A.M.

Acquisition

When a spill occurs, specific data must be obtained such as the name of the chemical(s) involved and DOT hazardous material classification(s). Other variables that must be estimated include: quantities involved, rate of spill, proximity to populations, injuries and property damage, and environmental factors that may complicate the spill. After the cleanup has begun, data must be acquired on an on-going basis to monitor progress.

The Acquisition component of the typical information system consists of a person located at the spill site, such as a fireman, common carrier employee, or bystander. This person must agree to become a part of the information system and act as a data collector for the Hazardous Material Response Coordinator.

Since this person may not belong to the chemical organization, he or she may have little or no knowledge of chemistry and training in hazardous material identification or clean-up procedures. This may affect his or her ability to acquire and communicate accurate and complete data in a timely manner. For example, inaccurate estimates of the rate of spill or proximity to local populations may cause the Hazardous Material Response Coordinator to recommend inappropriate actions. Failure to communicate all available data on a product label has caused unnecessary delays as the Hazardous Material Response Coordinator tried to obtain these data by other means. Misinterpretation of product label warnings, coupled with a failure to promptly contact CHEMTREC or the Hazardous Material Response Coordinator, has resulted in over zealous actions by local authorities at the site and adverse publicity for the chemical company.

Transmission

The information system must be able to send data from accident sites anywhere in the country to the Hazardous Material Response Coordinator at the manufacturer's headquarters and CHEMTREC in Washington, D.C. Transmission should be immediate because delays may impair the effectiveness of the emergency response. Due to the volume of data related to a spill and the likelihood of multiple incidents, provisions must be made for prioritizing the most critical data. Also, the reliability and availability of the Transmission component must be ensured by redundant channels, periodic calibration, outage monitoring, and other strategies.

The telephone is the transmission mode typically used by chemical companies because it is readily available and satisfies most of these requirements. However, the telephone has some inherent deficiencies that limit its effectiveness. For example, incidents have occurred in which messages could not be transmitted because telephone lines were busy due to other parties discussing the incident. Also, ambient noise or telephone line interference may impair the clarity of the verbal signal. Other modes of transmission may satisfy the transmission requirements and eliminate some of the deficiencies inherent in verbal presentation of data.

Processing

The role of the Processing component is to convert acquired data into a useful form and file and retrieve it for the Utilization component. When a hazardous material spill occurs, data acquired from the site and extensive product and hazard response data on file must be processed with minimal delays and presented in a readily usable form to the Hazardous Materials Response Coordinator (the Utilization component).

Processing for the typical information system consists largely of manually sorting paper files containing descriptions of products and hazards. Due to the volume of data to be sorted, the Hazardous Material Response Coordinator's response may be slower than desired and he or she may not have time to retrieve all relevant data.

Utilization

The Utilization component makes use of processed data for decision-making. Decisions include selection of a course of action, evaluation of progress, and modification of response strategies.

In the typical information system, much of the application of available data (e.g., analyzing product data) is performed by a human -- the Hazardous Material Response Coordinator. Opportunities exist for enhancing this component with decision aids (e.g., expert systems) as they become available through state-of-the-art technology. For a single spill scenario, several courses of action may be appropriate. Decision aids may help the Hazardous Material Response Coordinator weigh alternatives and select the optimum response and also ensure consistent responses for similar scenarios.

Transfer

The Transfer component should facilitate the implementation and execution of decisions between the Hazardous Material Response Coordinator and the many people and agencies involved in the spill response. The spill response may cross many organizational boundaries (chemical company, CHEMTREC, federal, state, and local authorities, fire department, etc.). The site emergency personnel, if they are not company personnel, are not obligated to follow the course of action recommended by the Hazardous Material Response Coordinator. Therefore, documentation of events and decisions is necessary to protect the interests of the chemical company. Documentation is also necessary so that experiences gained from past spills can be fed back into the Event World Analysis.

Chemical companies typically rely on telephone conversations to carry out the functions of the Transfer component. However, involved parties may not be at the telephone when calls arrive due to other response activities. Also, telephones may not be present or free for use during the incident. Strategies to overcome these deficiencies should be investigated. Other modes of Transfer, which present data visually, should be investigated to improve the user's comprehension of the data and improve the likelihood that the Hazardous Material Response Coordinator's recommendations will be acted upon.

CONCLUSION

As a result of comparisons between the ideal and typical information systems, the following discrepancies were found:

- Planning for accidents is virtually impossible due to the many scenarios that may result from the multitude of chemical products and possible contributing factors.

- Acquisition of data about the incident relies on people at the spill site who may have limited knowledge of chemistry and training in hazardous material identification and clean-up procedures.

- Transmission relies heavily upon use of the telephone, which is limited by the availability of telephone lines, ambient noise, and line interference.

- Processing and utilization of hazardous material data is largely carried out by the human component of the information system -- the Hazardous Material Response Coordinator -- and is subject to delays.
- Problems with cooperation and coordination of activities may occur where the information system crosses organization boundaries.
- The presentation of data to users is primarily verbal and may not be as effective as other modes, which present data visually.

These discrepancies represent opportunities for improvement. Information system analyses and human factors principles of design may play an important role in the development of better information systems for emergency response to hazardous material spills. Specific areas of activity may include the following:

- Fault analyses to assess potential accident scenarios
- Procedure development and training to facilitate response
- Development of alternate data acquisition, transmission, and display modes to enhance the Hazardous Material Response Coordinator's access to and comprehension of accident data
- Development of enhanced methods for analyzing product hazard data and developing courses of action (e.g., expert systems)
- Evaluation of organizational interfaces to enhance coordination and cooperation of involved parties.

CURRENT ACTIVITIES

The following summarizes activities currently undertaken by the chemical industry that address some information system deficiencies associated with hazardous material spills.

In 1985, portions of the OSHA Hazard Communication Standard came into effect requiring manufacturers in the standard industrial classification (SIC) codes 20 through 39 to supply material safety data sheets (MSDS) to customers (Chemical Manufacturers Association, 1986). These sheets must include descriptions of the chemical's physical properties, reactivity data, health effects, and recommended personal protection equipment, and procedures for first aid, fire fighting, storage and handling, and spill and waste disposal. As of May 1986, the standard requires manufacturing companies to provide safety training to their own employees working with hazardous materials. An American National Standards Institute (ANSI) committee is developing a standard for electronic transfer of data associated with the material safety data sheets.

In September 1986 the Chemical Manufacturers Association (CMA), a trade organization, introduced a new service for transmitting computerized data on commercial chemicals to emergency response teams handling accidents involving hazardous materials (Bureau of National Affairs, 1986). The service, known as Hazard Information Transmission (HIT), uses a computer program to transmit data on chemicals, their hazards, and mitigation techniques. Emergency response teams must have access to a telephone, modem, and personal computer printer or similar device to receive transmissions. Data on 1,700 chemicals produced and shipped in the United States in large quantities currently is contained in the data base. Eventually data on more than 4,000 chemicals may be included.

CMA is also sponsoring a Community Awareness and Emergency Response (CAER) program to coordinate public and private sector efforts to establish or renew integrated community emergency response plans and communications networks (Chemical Manufacturers Association, 1986). Some chemical manufacturers have developed their own emergency response teams to be available at the site to provide guidance and/or clean up the spill.

APPLICATION TO OTHER INFORMATION SYSTEM FAILURES

The EATPUT model proved to be a valuable tool for analyzing the information system used by chemical companies. It permitted a complex system to be broken down into functional parts and analyzed, while providing an understanding of how the parts should function as a whole to achieve the system's objective. The functional emphasis of the model concentrates on what the components should accomplish without getting bogged down in the details of the people and equipment used to accomplish it.

Problem formulation is key to the application of this model. As Debons (1986) noted,

> In no sense can the componential functioning of information systems be considered as linear or sequential. Each component (or subcomponent) of a system possesses its own acquiring, processing, using, and transfer mechanisms.

Thus, it is important that the analyst establish from the start the level of detail to be pursued in the analysis. Also, the frame of reference used in an analysis may sometimes cause terminology to appear confusing. For example, one person's act of transfer (notifying the Hazardous Material Response Coordinator of a chemical spill) becomes another person's acquisition (Hazardous Material Response Coordinator acquires data about the spill). Finally, the time element associated with information failures should be explicity stated to maintain a clear understanding of which part of the failure scenario is being addressed.

The authors recommend that a rigorous and systematic approach be taken to organize the problem prior to analyzing functional requirements of the components. The analysis should be conducted at two levels: first, an overall level to understand the scope of the problem and, next, a detailed level to understand the functional requirements of specific elements of the man-machine system during selected scenarios. The following methodology is proposed for identifying potential failures of existing information systems:

1. Conduct a high level review of the information system components placing major emphasis on the system objectives, system boundaries, and event would (including scenarios of interest). Requirements for the acquisition, transmission, processing, utilization, and transfer components should be stated in general categories at this time to obtain an overview of the problem.

2. Select scenarios from the event world analysis for further analysis (e.g., initial detection of a chemical spill).

3. Identify each link where transfer occurs between elements of the information system (e.g., on-site fireman to CHEMTREC, CHEMTREC to the chemical company's Hazardous Material Response Coordinator, and Hazardous Material Response Coordinator to the on-site fireman).

4. For each link, analyze the Event World of the scenario and determine how the acquisition, transmission, processing, utilization, and transfer components should function to accomplish the system objectives.

5. Contrast this performance with way these components actually function to identify functional deficiencies in the information system.

6. Examine the physical elements of the information system to determine what changes (equipment, personnel, procedures, training, etc.) are necessary to correct the functional deficiencies.

When analyzing an information system failure that has already occurred, the analyst should define how the components of an information system should have functioned during the failure scenario to achieve system objectives. The analyst should compare this with the best understanding of how each component did function. Discrepancies between ideal and actual performance illustrate deficiencies in the design of the information system. Alternatively, the analyst may use the accident scenario as a starting point to hypothesize failures for each information system component. By determining the human or machine elements involved, the analyst can create a profile of likely failures that may have contributed to the overall information system failure.

The methodology discussed above may be applied to any information system to analyze actual or potential failures. It clarifies the intent of the information system, the environment it must function in, the decisions to be supported, and the functions required to support these decisions, so that appropriate decisions may be made about equipment and personnel. Methodologies that do not consider these factors are likely to overlook the true scope of the problem or suffer from the type III error pitfall of analyzing the wrong problem.

ACKNOWLEDGMENTS

The authors wish to thank Robert Krugh, Alfred Rebersak, Arlene Shaner, and Louise Studniary for their participation in the information system analysis presented in this paper.

This project was conducted for the University of Pittsburgh and not for Westinghouse Electric Corporation or Koppers Company, Inc.

BIBLIOGRAPHY

Bureau of National Affairs. (1986). Emergency access to chemical data available. *Job Safety & Health*, September, 230.

Chemical Manufacturers Association. (1986, May/June). Special issue: community safety. *ChemEcology*. (Available from CMA, 2501 M Street, NW, Washington, DC 20037).

Cilliers, I. (1985). Problems in information science. *Information Age*, July, 7-3.

Debons, A. (1980). Foundations of information science. In O. Harbo and L. Kajberg (Eds.), *Theory and Application of Information Research. Proceedings of the Second International Research Forum on Information Science, 3-6 August 1977, Royal School of Librarianship, Copenhagen* (pp 75- 81). London: Mansell.

Debons, A. and Robertson, C. (In preparation, 1986). *Informatology: Foundations of Information Science.*

Dulaney, T. (1978). Hazardous materials: How to cope. *Distribution Worldwide*, November, 24-40.

National Safety Council. (1978). *Accident Prevention Manual for Industrial Operations* (8th ed.) F. E. McElroy (Ed.). Washington, DC.

Pijawka, K. D., and Radwan, A. E. (1985). The transportation of hazardous materials: risk assessment and hazard management. *Dangerous Properties of Industrial Materials Report* September/October, 2-11.

CONTROL OF INFORMATION GENERATED DURING FAILURE ANALYSIS OF INFORMATION SYSTEMS

J. Howard Petrie

Consultant in Information Systems
Liverpool, U. K.

INTRODUCTION

The amount of information generated by the investigation into the failure of an information system will depend on the size of the operation. In the event of the failure of a complex system, involving large numbers of people as well as plant and equipment, the investigation may necessitate the use of a large team of personnel.

The discovery of the reasons for the failure of the system may depend on the concatenation of a number of pieces of evidence that may seem to be unrelated. A large investigation will be likely to generate a considerable amount of information, making the task of analysis more difficult.

Some of the types of information generated from an investigation will include the following:

- Interviews with relevant personnel (staff, witnesses, victims).
- Observations by investigator.
- Documents containing information describing the various aspects of the system (e.g., manuals, management documents, technical reports). This documentation will make up the internal information files of the information system being investigated.
- Computer data bases that are part of the information system being investigated.
- Press statements.
- Reports from failure investigators to superiors.

OFFICE SYSTEMS

The information acquired and generated by the investigation team needs to be stored systematically in order to allow controlled access to the information, and to provide the basis for sophisticated retrieval functions, which are described below. The processing of the information, and particularly the creation of documents, will be best carried out by an office automation or management system. There is a wide range of such systems on the market, and their use makes possible the fast and efficient preparation of documents.

Some office systems allow documents to be retrieved by providing ways of searching for information contained in them. For instance, it may be possible to retrieve a particular document by entering its date of production, report number, author, title, or a combination of these elements. For large investigations, however, there will be a strong case for setting up the investigation information system in conjunction with sophisticated information retrieval software. The use of such software will allow the team to create data bases and to retrieve and utilize the information generated by the investigation in the fullest possible way. It is even possible to process the information so that every word, number, or other data element can be used as a search element.

NATO ASI Series, Vol. F32
Information Systems: Failure Analysis
Edited by J. A. Wise and A. Debons

The description of such a system is given below. An outline of the flow of work from the office management system to an information retrieval system is shown in Figure 1. An example of the use of information retrieval systems for an investigation can be seen in the Challenger investigation (see Kerhli, 1986, Report of the Presidential Commission on the Space Shuttle Challenger Accident, 1986).

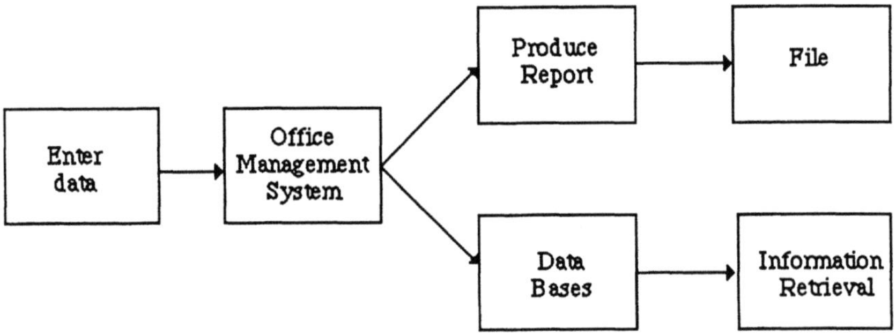

Figure 1. Information flow diagram for information retrieval

BUILDING THE INVESTIGATION INFORMATION RETRIEVAL SYSTEM

Acquisition

The relevant information will be acquired by the investigation team. Once the acquisition process is underway, it will be necessary to organize and prepare the data for the construction of one or more investigation data bases. For this activity, it will be useful to devise a number of categories of information, derived from the event world of the system being investigated. An analysis of the acquisition process will produce a definitive set of these categories, e.g., interviews, system documents, reports.

The interviews with witnesses can illustrate how to build a data base. A set of interviews might have both structured and unstructured components. The former might include a number of standard questions together with responses; the latter might include unprepared questions and responses, together with other information freely provided.

The construction of a data base for the interviews (see Figure 2) begins with the devising of a number of data elements, which will make up a record corresponding to an individual interview. Examples are name, address, date, questions, answers, comments. The form of each element should be described. It might be "free text," in which case any data can be freely entered without any control. However, it is likely that control will need to be exercised on how data are entered into some fields. (A field is the name given to the place in the record.) In an interview data base, such controlled data elements might be the date a record was created and name of interviewer (e.g., surname, forename, middle initial).

Data Elements
- Interviewer
- Date
- Time

- Interviewee
- Address
- Position within system
- etc.

Structured Questions
- Question 1
- Answer 1
- Question 2
- Answer 2
- etc.

Unstructured Questions
- Question A
- Answer A
- Question B
- Answer B
- etc.

Comments
- Interviewer's comments
- Interviewee's comments

Figure 2. Some data elements for an interviews data base.

If data are always stored in the same way, retrieval will be simplified. A search for, say, a date will retrieve all records made on that date if it has been entered correctly in each record. For a small investigation, the information generated by the team is likely to be less, and hence it may be possible to make do with a single data base covering all aspects of the information collected (see Figure 3). In this case, the format of the data base must accommodate all types of data elements. In addition, a further element (type of information or document) will be needed. The interview records would, in this case, form part of a general data base.

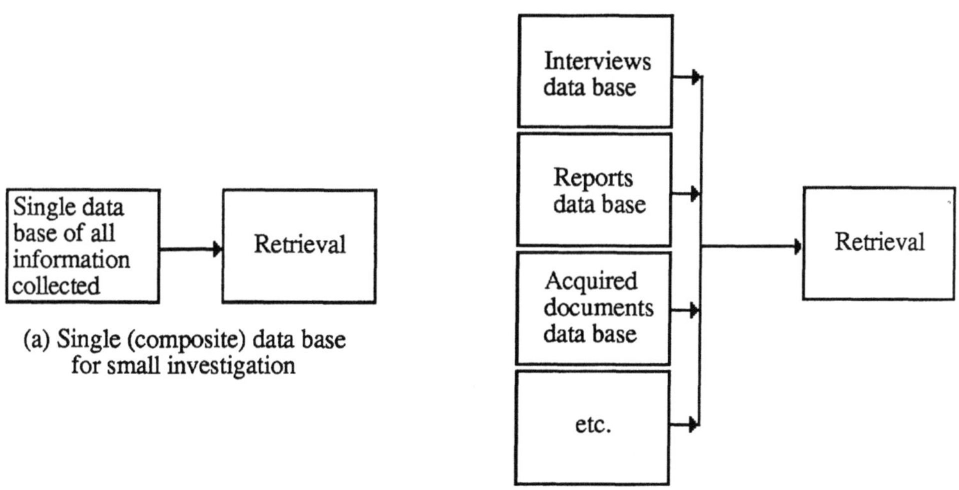

Figure 3. Differences in data base design for small investigations (a) and large investigations (b).

The decision of which data bases to set up will depend on the circumstances of the investigation. The design of the data base should not be taken lightly since the efficiency of the investigation management process will very likely be dependent on this data base.

Processing of the Data Base for Retrieval

Information retrieval systems contain some powerful features that are potentially valuable for the investigation of information system failures. The data base designer will need to decide which of these features is required. It is possible to designate each data element that is to be searchable by specifying that it be entered into an index. Such an index functions somewhat like an index at the back of a book but is more powerful and flexible. The software will need to specify how to index each data element. For instance, each "free text" field may need to have each word individually placed in the index. The names of multiple authors will need to be separated, and the forenames will need to be kept next to the surnames in the index. The chosen software will have a standard way of executing these operations.

Retrieval of Information

The retrieval of the data generated or acquired during an information system failure analysis will be unpredictable. Some of the retrieval features which can provide the required retrieval flexibility are the following:

- Search on individual words or data elements (e.g., find all interviews carried out by interviewer x).
- Range searching (e.g., find all interviews carried out between 10:00 and 12:00 on a particular day).
- Searching for partial words (e.g., find all interviews that mentioned the partial word "comput" to retrieve interviews containing the words "computer," "computing," etc.).
- Use of Boolean logic (e.g., find all interviews that gave the same answer to question 1 and the same answer to question 4).

These are only some of the possible search features in information retrieval software that would exploit the investigation data base in order to help reach a satisfactory conclusion to the investigation. The particular features available will depend on the software chosen and on the power of the computer hardware.

CHOOSING A SYSTEM

Office Management and Retrieval

The production of reports and other documents will best be undertaken using a word processing/office management system. If the system is carefully chosen and designed, it will be possible to use the office management system to produce the proper reports and to prepare the input for the data base(s). The same data will be usable for each of these purposes. It may be possible to choose an integrated office management system that has an inbuilt data base system and retrieval features. Such a system would be sufficient for small investigations.

Computer Hardware

The selection of hardware will depend on individual circumstances. Hardware may already be available, in which case it can be used if suitable software can be found.

The size of the data base(s) will need to be taken into account. For a small operation, a microcomputer may be sufficient. Quite a powerful single-user machine can be acquired for less than $5000. The largest amount of software is available for IBM-compatible machines. Hard, or Winchester, disks are advisable for storing reasonable amounts of data. If a more powerful solution is required, a minicomputer or access to a large (mainframe) computer is advisable.

Software

Both information retrieval and office management software are available for a wide range of machines. The project team should seek advice on this matter if it is not available within the team itself. For an IBM or IBM-compatible microcomputer, one of the standard word processing system and data base/file management packages may be sufficient. For a minicomputer, such as the Digital Equipment range, proprietary information retrieval software is available, e.g., BASIS from Battelle Laboratories, or BRS from BRS. A range of products is available. It might be possible to use a service bureau for a short-term project, which may be more cost-effective for an investigation.

CONCLUSIONS

The organization of the information collected or generated from an information system failure analysis will be extremely important to the success of the investigation. The use of an office management and information retrieval system will contribute greatly to the progress of the investigation. Proven hardware, software, and information on how to use such systems are widely available.

REFERENCES

Keherli, R. R. (1986). The investigation techniques used by the Challenger Commission to address information system failures as they related to the space shuttle accident. *This volume.*

Report of the Presidential Commission on the Space Shuttle Challenger Accident. (1986). Washington D.C: U. S. Government Printing Office, 6 June.

HUMAN ISSUES
IN
FAILURE ANALYSIS

MANAGEMENT STRATEGIES AND INFORMATION FAILURE

Ron Westrum

Eastern Michigan University
Ypsilanti, Michigan, USA

PART I: SOME KEY CONCEPTS

INTRODUCTION

Success in large-scale projects demands useful information flow. The quality of information available strongly affects the quality of the decisions systems can make. When key items of information are missing, faulty and even disastrous decisions may result. Testimony before the Rogers Commission on events surrounding the destruction of the space shuttle Challenger underline the damage that missing information can cause. Yet major information failures are seldom accidental. Rather, they reflect basic strategies used by organizations to process the events around and within them. The organization's structure and its philosophy of management are both vital determinants of how information in it will be treated.

This paper will examine two very different management strategies for information processing. I will argue that adherence to one of these strategies is very much more likely to cause information failures than the other. Then I will suggest some of the organizational features likely to provide more adequate treatment of information.

A key concept in understanding information failure is organizational culture. Recent management books such as *In Search of Excellence* (Peters and Waterman, 1982) have made popular the idea of "corporate culture" (Ouchi and Wilkins, 1985). Here we will examine only one aspect of corporate culture, an organization's philosophy about rationality. We could loosely define rationality as those actions and procedures that organizations feel are likely to lead to effective and efficient performance. An organization feels it is rational when it implements these procedures, a kind of internal "scientific method" for achieving positive results. Feelings associated with these methods are likely to be strong and are not easily changed. Individuals who achieve results by other than the approved methods in the organization are likely to be punished for their development methods rather than rewarded for their positive results (Westrum and Samaha, 1984).

Another key concept is the idea of organizational thought. My own research strongly suggests that thought is a generic property of intelligent systems on several scales: the individual, the small group, the organization, the inter-organizational field (e.g., science), and the society. I have found this concept extremely helpful in understanding information failures, and I suggest we use it in examining the problems on which we will focus here. Our major concern in pursuing this inquiry will be to improve the quality of thinking by organizations. As we will see, information failures are only one manifestation of a thinking strategy that is also likely to have other negative aspects.

CALCULATIVE AND GENERATIVE RATIONALITIES

Here I am going to outline two general organizational forms of rationality, which I call calculative and generative respectively (Westrum, 1985). These are admittedly ideal types, and every organization is likely to have some calculative, some generative aspects. These types are valuable, however, because they emphasize certain organizational traits that are very important for information failure.

NATO ASI Series, Vol. F32
Information Systems: Failure Analysis
Edited by J. A. Wise and A. Debons

Calculative rationality is a strategy that uses the past to compute the future. A calculative system uses the knowledge it has acquired to predict what is going to happen in future experience. The tables of life expectancy prepared by insurance actuaries are a good example of such calculative rationality. Furthermore, this strategy uses past experience to compute the best course of action. Thus, past experience and experiment provide a seemingly infallible guide to the best way to do things. This experience is confided into doctrines and is expressed through the organization's structure, which is set up to react to events occurring in the expected way (Miles and Snow, 1978).

Such a strategy is likely to concentrate the responsibility for creative thoughts in a special part of the system. This specialization is possible because 1) it is believed possible to encapsulate such expertise; in a special part of the organization; 2) certain individuals have wider training and expertise, and 3) the volume of rethinking is not expected to be large. The system, then, can be divided into central and peripheral parts, with the role of the latter being largely action - rather than thought-oriented.

The performance of such a system has certain properties, which the sociologist Max Weber saw as the hallmarks of bureaucracy: speed, certainty, and predictability (Weber, 1978). The ideal system of this kind, Weber felt, would act like an automation: for every input, the output would exactly be predictable. Frederick Taylor's "scientific management" popularized this system for American manufacturing. V.I. Lenin adapted this idea for the U.S.S.R. as a major tenet of Bolshevik strategy for party control and for soviet industry (Selznick, 1952).

Calculative rationality seeks to make the organization into a rule-driven machine. This machine is extremely efficient under a certain set of environmental conditions, i.e., minimal change. To the extent that the organization faces an unchanging environment, this strategy will provide a reasonably good performance. It may have high internal costs in terms of stress for organization members who dislike being treated as cogs in the machine, but these people will often be "cooled out" with high monetary rewards or perks in kind. Self-selection for recruitment will lower the amount of stress as the organization attracts those of a similar personality to current members.

Organizations like this, however, can be disastrous when coping with changing environments and high technology, because the very features that make them predictable also make them highly vulnerable.

THE VON NEUMANN BOTTLENECK

The cause of this vulnerability is the organization's mechanisms for thought. The structure of the calculative organization is such that all major thinking is expected to be done by the central part of the organization. This is fine if all the information and ideas happen to be there, but they usually aren't. Furthermore, the recognition of ideas in the periphery is inherently threatening to the center, since it suggests that there ought to be structural change, perhaps reducing the center's role. Thus, the system is able to think only those thoughts that will leave its structure unchanged --- nothing, in other words, very significant or very threatening to current conceptions. As time goes on, the system's channels of information tend to become clogged, as more and more evidence builds up on the periphery that things need to change and more of the center's resources are spent protecting its status rather than engaging in creative problem-solving.

In many respects this situation reminds me of what is called the "von Neumann bottleneck" in computer science (Douglas, 1983). Traditional computer architecture makes all operations pass through a central processor, a design first codified by John von Neumann. With a computer's rapidity of operations, the delay this causes is relatively minor. It only becomes serious for very complicated problems requiring billions of calculations. For these applications, more sophisticated architectures, "parallel processing," etc. have been devised.

For computers, the only real problem posed by this bottleneck is speed, since computers make very few errors. The information is in the system, it stays in the system, and eventually it gets processed.

Not so with its human analogue, the organization. Here a centralized thinking responsibility may mean that important information 1) is not transmitted, 2) is not utilized, and 3) may not even be generated in the first place. Here the bottleneck is more serious, since it is likely to cause severe breakdown of information generation and flow. At this point it may be useful to consider some small scale examples.

Bottlenecks in the Hospital

A traditional problem in hospitals is getting information to flow from nurses to physicians. Nurses frequently have medical competence or tactical observations which are essential to correct action with patients (e.g., the level of drug dosage). But since physicians have higher status and tend to think of themselves as omniscient, nurses can seldom give advice directly to physicians. Instead, they must make it appear that the physician is advising them, while actually the information is flowing the other way. When this doctor-nurse game is understood by both parties, information flows smoothly. When one party will not co-operate in its tacit conventions, however, hostilities result and patient care suffers (Stein, 1967).

Bottlenecks on the Flight Deck

A serious barrier to effective communications on the airplane flight deck is differences in status between pilots and other members of the deck crew, such as co-pilots and engineers. Pilots often strike a "macho" pose and, like physicians, see themselves as omniscient. Accordingly, they do not encourage communications about problems and difficulties from other members of the crew. The net result is that important information does not flow freely, resulting in errors and mishaps. The status of the pilot is protected but the safety of the aircraft is not (Foushee, 1982).

Barriers in the Boardroom

In his classic study *Organization and Innovation*, Chris Argyris found that top executives in technical organizations tended to be low on interpersonal effectiveness in board meetings. The result was considerably reduced information flow and lessened creativity for the organization. Argyris's (1965) findings were expanded by Irving Janis (1972) in his book *Victims of Groupthink*, who made many striking observations, e.g., about the processes that keep disagreements tacit and keep top decision-making groups from reassessing mistaken strategies.

In each one of these cases, we see a basic conflict between preserving the system's centralized structure and effective performance. Information on the periphery of the system cannot be used because it would threaten this structure. Each one of these contexts represents a small-scale model of what is likely to happen to larger organizations using a purely calculative strategy.

CALCULATIVE STRATEGIES AND LARGE ORGANIZATIONS

Calculative strategies for large organizations are most common and seem to work best in environments where tasks are stable and market conditions are relatively static. This "mechanical" style of management, then, is appropriate for situations where rapid change is not a major factor in the organization's environment, and where the technology is not highly complicated and involves little research. Such organizations usually have rigid hierarchies, with clear separation between line and staff, and strong definitions of responsibilities for individual departments. These organizations tend to be more bureaucratic in character and attract more traditional managers and workers.

This style is also typical of government bureaucracies, where spelling out responsibilities is often a matter of law. In bureaucracies there is a strong emphasis on written communication to provide records as to who did what, in case something goes wrong. Going "by the book" is a reflex with organizations, which often excuse low performance with reference to legal or administrative requirements.

Such organizational forms, however, cope very badly with complex technological systems and a rapidly shifting tasks. Information flow in such organizations is highly routinized and regulated. Such channels are not flexible enough to cope with tasks for which there is no obvious algorithm. We can state as a fundamental hypothesis that high technology in particular requires a very different kind of management philosophy, one more oriented to

1. The free flow of information, often orally transmitted;
2. Overlapping and highly coordinated responsibilities, with good inter-team communication;
3. Relatively low emphasis on hierarchy as opposed to the technical and organizational requirements of the task.

This is usually referred to as "organic management" (Burns and Stlker, 1961) to distinguish it from mechanical (bureaucratic) management. Its more complicated forms are called "matrix management" and include a large number of ad hoc project teams. This kind of management style is strongly generative of new ideas, internal critiques, and is usually oriented to future performance rather than past fault-finding. This kind of generative rationality is well matched with the requirements of rapid change and high technological complexity (Emery and Twist, 1965).

Now comes the interesting question. Suppose we take a high-technology project and give it to an organization with a strongly mechanical style of management. Here the task requirements and the organizational structure are grossly mismatched. We can anticipate a number of negative consequences:

1. Creativity will be under-utilized by the organization. Novel ideas ideas and approaches will be suppressed by the organization's center because they interfere with the hierarchy's status. They will be a strong emphasis on doing things "the company way" as opposed to ways that might work better.

An example of such blockage is provided by Elting Morison's (1966) classic essay *Gunfire at Sea* about the efforts of Percy Scott and William Sims to bring modern gunnery to the British and American navies. Scott's innovations, although offering incredible improvements over traditional practice, were largely ignored because Scott was a line officer. Only through clever use of publicity and the press was Scott able to force the admiralty to listen to him. Sims got the American Navy to pay attention only by going directly to the President.

Another example is William McLean's invention of the Sidewinder missile (Fialka, 1985). Development of this extremely efficient and inexpensive weapon took place contrary to explicit Defense Department policy, which favored a more expensive radar-controlled version over McLean's infrared one. The project was actually hidden from military brass until it was ready to operate, at which time it proved wildly successful, with a reliability so high that working missiles could be assembled with parts found in junk yards.

A third example is the problems surrounding the development and deployment of the M-16 automatic rifle in the United States Army. Because the U.S. Army ordnance establishment was oriented largely to weapons production and disliked the idea of a weapon invented without its initiation, American military forces in Vietnam were given an inferior rifle and inadequate supplies and instructions to maintain it. Development actually degraded the weapon's performance. A congressional investigation was needed to sort out this tangled tale of bureaucratic fumbling (Fallows, 1981).

2. Information flow about problems will be faulty. It will encounter serious bottlenecks, and will not be transmitted in timely, useful form. Often external resources will have to be used to get a hearing for problems and difficulties.

The Quebec Bridge disaster (Tarkov, 1986) provides an object lesson in information failure. In 1907 the partially-built Quebec bridge collapsed, dumping 19,000 tons of steel on the banks and in the waters of the St. Lawrence river, and killing 75 workmen. Examination of the causes of the accident showed that danger signs had repeatedly been ignored, efforts to get independent review of designs had been scotched, and last-minute warning signals were not transmitted. The major engineer in charge, Theodore Cooper, had run the entire project from his office in New York, and successfully deflected attempts to check or undermine his authority.

The aircraft carrier may or may not be an effective strategic weapon, given modern missile capabilities. What appears certain is that negative findings on carrier survivability are being suppressed. Both computer studies and war games are frequently rigged so that threats to carriers are minimized. Professor Thomas Etzold (quoted in Greve, 1982) of the Naval War College stated that

> In more than five years of experience as an umpire and adviser in high-level naval war games, I have witnessed the unwillingness of senior naval officers to permit carriers to be sunk, even when taken under overwhelming attack....[At times, Etzold said], a guy simply says 'It's my game and I don't want the carrier put out of action,' which is a good representation of the larger mental problem in the Navy.

3. When failure occurs, furthermore, rather than self-examination, there will be systematic dishonesty about the nature of the failures. Self-examination would call into question the prerogatives of the hierarchy. Instead, effort will be placed on silencing and punishing whistle-blowers in the organization who are pointing out the people and procedures responsible for failures. This contributes to problem #2, since it discourages others from coming forth.

Very serious problems have plagued the nuclear reactors of the Tennessee Valley Authority. Hundreds of employees have reported such problems as faulty welds in high-pressure tubing and improper installation of cable systems controlling the reactor core. Rather than encouraging employees to come forward about such problems, however, management has reportedly harassed and intimidated the employees who do. An outside firm has had to be called in to hear employee complaints (O'Connor and Wright, 1986).

In the wake of the shuttle Challenger explosion, the contractor, Morton Thiokol, punished two of its engineers who gave the Rogers Commission important testimony about conditions which led to the failure. Both had also personally opposed launching. Allen McDonald and Roger Boisjoly, both of whom had held important positions at the contractor, were given less important jobs, which deprived them of access to information about the event that might have a material bearing on responsibilities for the disaster (Sanger, 1986). Also certain documents bearing on the event were apparently shredded by NASA officials at the Marshall Space Flight Center. A spokesman said that the shredded *Weekly Engineering Reports* were not a system of records, because the the information contained in them is preliminary and may later prove to be incorrect. For that reason, they are routinely disposed of (UPI Reports, 1986).

Calculative organizations, therefore, are likely to do badly on tasks that require rapid information flow, innovation, and critique. The reason for such failures is that the use of ideas from the organization's periphery threatens the status of the organization's center. Rather than tolerate new or critical ideas, whatever their value for the mission, they will be sacrificed to protect the organization's structure and statuses. This observation strongly suggests that when organizations engage in genuine thinking, their structure is likely to change; to think is therefore to risk structural change. Organizations afraid of such change will block the new ideas, a process that Donald Schon (1971) has called "dynamic conservatism." On the other hand, organizations using generative rationality, by creating a periphery empowered to think and capable of thought are willing to grow and change. As Robert Hayes (1986) has said, "The role of top management is not to spot and solve problems as much as to create an organization that can spot and solve its own problems."

CHARACTERISTICS OF GENERATIVE ORGANIZATIONS

Let us consider for a moment what we would expect the characteristics of a generative organization to be. First of all, there would be an enlargement of the responsibility for thinking. A generative organization would encourage all its members to observe, to critique, and to generate new ideas. Each person in the organization would be given the role of thinker, observer, and critic, to the extent that he or she wished to adopt it (Feaver, 1982). Without useful ideas or observations, there is no information to flow. A first step toward such enlargement would be transfer of decision-making authority to lower levels.

In 1949, J. C. Worthy (1984) carried out a study of Sears and Roebuck stores in the middle size range, and found that there were two predominant patterns in regard to supervision. In pattern A, there was no intervening layer between the store manager and the individual department managers. In pattern B, there was such an intervening layer. Now most business people would tend to see pattern B as allowing more organizational control since closer tabs could be kept on department heads. Actually, however, the "A" stores did better during the recession of 1949, probably because the department heads in such stores had learned to think and act for themselves, rather than waiting for initiatives from above. Nor was this the only difference. Managers at the "A" stores were more likely to be promoted to higher posts at Sears; employee morale was higher at the "A" stores, and so on.

Similarly, Zald and Denton showed that the YMCA managed to survive a major change in its environment because of decentralized decision-making authority. J. C. Penney currently allows a greater degree of autonomy to local buyers in an effort to keep the stores tuned to local fashions (Gilman, 1986). Current training for airline cockpit interaction emphasizes the ability of all flight deck personnel for "inquiry, advocacy, and critique" (Feaver, 1082). This empowerment of the periphery in terms of thinking and acting is an essential element of generative rationality.

Second, we would want information in this organization to flow freely between organizational subsystems. I think of a memo sent by Admiral Sir Percy Scott when he became Inspector of Gunnery Practice for the Royal Navy (about 1903). This is part of the memo:

> Heretofore it has been the custom to boycott all ideas from sea-going ships I am trying to change that and make use of the talent of the Navy please distribute enclosed to any officers who have ideas, and explain that the boycott is off as far as Xlent is concerned (Padfield, 1966, p.136).

I want particularly to call attention to the idea of "use of the talent of the Navy." One of the salient features of a generative approach is making efficient use of the entire cognitive resources of the system. John Leitch, an early pioneer in organizational development, wrote in 1919 about something very similar:

> The making of men involves the developing of the brain service of the whole human element and then concentrating this force along a specific line of action and toward a definite goal (Leitch, 1919).

One of the values of the free flow of information lies with what it allows the system to know about itself. Harold Wilensky (1967) has coined the term "organizational intelligence" to describe how organizations get information important for their survival. He does not distinguish, however, between what the organization knows about the world outside and what it knows about itself. We might call this extelligence versus intratelligence. A generative system does not, for the most part, keep very close track of what is going on within it, except in regard to information that is essential for getting the task done. What it must get information about, however, is ideas and observations that would allow it to improve its performance. (I realize that this is an extremely approximation and would appreciate for ideas as to how it might be refined.)

A key difference between calculative and generative rationality is how the two strategies tend to use information. For calculative organizations, information is largely used to develop closer controls on the periphery. For this reason, information flow about problems is inherently "political" in character and is therefore likely to be restricted. Lowered risk-taking and innovation are a common result. For generative organizations, however, information tends to take on more the character of a contribution to the organization. People are more willing to transmit critical information because they are more concerned about how the organization is doing than how they (or their supervisors) will look.

An anecdote from Leitch (1919) will make this point clear. In the course of implementing an "industrial democracy" plan in a foundry, the workers discovered they needed an inspector skilled in process control. They suggested someone who was an expert molder, and this person was then promptly appointed to the new post. The attitude that followed is worth our attention:

> It might be that the gate was too large or too small; but whatever the cause, Harry found it and the men, recognizing that he knew what he was talking about, were glad to have his advice. When they were puzzled on a mold, they began to get Harry's approval before the pouring began. They realized that it helped dividends to avoid the waste of poor castings and *they dropped the too common attitude of letting pride forbid them to ask questions* (emphasis added) (p. 128).

An obvious corollary is that one characteristic of a generative system is the ability to identify a good idea and make use of it, regardless of where the idea originates in the system. The kind of thing I have in mind is an anecdote told by Douglas MacGregor (1958) about Charles Kettering:

> I remember an experience Kettering had when a ship was being built and one of his first diesel engines was being installed in it. As he was wandering around the shipyard just a few weeks before the launching, he noticed one of the workmen standing at the stern of this vessel looking at it. Ket walked up to him, got into a conversation with him, hauled up a nail keg and sat down for a while. He had noticed the man staring intently at the stern of the ship and finally asked: "What are you looking at?" The fellow said: "That propeller." Ket said: "What about it?" He said: "It's too big." Ket said: "How much too big?" "Oh," he said, "at least five inches; maybe five and a half."
>
> I think most people would have dismissed such a comment at this point. This was an ordinary workman; he was not a member of management; he was not an architect--just an ordinary guy. But Ket did not drop the subject. He called up the architects and asked them to check the plans and find out the diameter of the propeller on this ship. They did and after a little while came back with an answer. Then he said, "Will you send somebody out to measure it?" And of course they did. Somewhat later he got a phone call, very

> shamefacedly, from the architectural firm. They said they did not know who had made the mistake but the diameter of that propeller was five and a quarter inches too great! Under the conditions of design of this ship, this was a serious error. This kind of belief in the intelligence of his fellow men was as characteristic of Joe Scanlon and Kurt Lewin as it is of Kettering. (p.10)

Writings (Leslie, 1983) about Kettering's character suggest (contrary to MacGregor) that this was a departure from his usual reflexes. He was not called "Boss Kettering" by accident. I think of another telling anecdote which emerged in the wake of the shuttle disaster. In a letter to the New York Times, Don Friedkin (1986) wrote of an acquaintance who had consulted for NASA:

> He spoke in unending praise of the [former] free flow of information at NASA, without which, in his view, we would not have reached the moon within years of the actual achievement. All efforts were to have a rocket go all the way to the moon and then take off from the lunar surface on its journey back to earth, which posed an almost impossible liftoff problem.
>
> An engineer far down the pecking order came up with the idea of circling the moon and using a lunar module to reach it, greatly minimizing the lunar liftoff problems. My acquaintance made the point time and again that in the normal communication structure this lowly engineer's idea would never have reached top management.

If the system requires restructuring as a result of the idea, furthermore, the organization is willing to let this happen. Thus a third characteristic is a certain flexibility of organizational structure. What Mary Parker Follett (1941) called the "Law of the Situation" governs action instead of the status or power needs of organizational subsystems. On a short- term basis, this may mean what one observer called "flickering authority," where leadership may shift during different phases of a task as different kinds of expertise are required (Withingham, 1969). This handing-off of authority is unusual except in high performance groups, where the welfare of the team is considered more important than individual egos (Vaill, 1978). Consonant with this emphasis is impersonal evaluative feedback after team performances to identify the sources of error rather than fixing blame (See Jordan, 1968).

In the long-term, not only roles but structures may shift. Thus new ad hoc groups and lines of communication can form to respond to current needs. Generative institutions are likely to be responsive to their environments in a creative way, not simply "rounding up the usual suspects," but striking out boldly in new directions. This occurs in the development of new scientific specialties organized around novel interests (Menard, 1971) and in technological "hothouse" environments such as Silicon Valley(Saxeniam, 1985). Within established organizations, such developments may occur as entrepreneurship (Pinchot, 1985). The "Skunk Works" at Lockheed is an example of such an internal development (Johnson, 1985). It has been responsible for a number of important and novel aircraft, often produced in record time with minimal budgets and personnel.

SOME GENERATIVE PRINCIPLES

It may be worthwhile at this point to sum up some of the ideas which are threaded through the previous discussion. Here is a rough summary of generative principles:

1. Encourage system-wide awareness for all members of the system No one can be expected to help solve the system's problems if they do not understand what those problems are. An empowered periphery must be one aware of overall goals and approaches. (I think that it is no accident that Kelly Johnson (1985) got everybody who worked on a project out to see its first flight. This shows

the kind of involvement I have in mind. From a calculative point of view, this kind of sensitivity would make sense.)

2. Encourage creative and critical thought for all organization members. Although some members of the organization will contribute disproportionately, it is vital to realize that some important ideas may come from unlikely sources.

3. Link the parts of the system whose work is interdependent. The members of a task system must understand each other's work if they are to cooperate in solving the system's problem--not just their own. It is not enough to identify with the system as a whole. Without seeing integration as an important task, organization members will perform their contributions often in blissful ignorance of what the rest of the organization requires.

4. Scan the system's parts for relevant solutions or contributions. Use the best solutions regardless of their origins. Every organization should examine the ability of its intratelligence system to do this. It may be useful to develop formal exercises to generate alternatives. The fruits of these exercises should be formally transmitted and acknowledged.

5. Reward communications and activities that show a desire to contribute to the entire system's thought processes. Although today's contribution may not be the answer sought, tomorrow's contribution will never come unless today's is recognized. "Good try" is always superior to "No good."

6. Avoid over-structuring. Most of the organization's resources should be used in coping with problems, not in building up the private domains of its leaders. It is a natural tendency for parts of systems to entrench themselves. It is equally certain that resisting this tendency is necessary to maintain generativity.

7. Examine mistakes honestly. Generative systems characteristically deal with mistakes as system problems rather than as person problems. While genuine negligence should be punished, oversights and inadequacies are human. The important issue is to identify the source of the the mistake, not punish the person who made it. The ability of the system to repair its problems is strongly related to the willingness of people in it to open themselves to criticism. This willingness is greatest when criticism is dispassionate and impersonal.

PART II: INFORMATION SYSTEM FAILURE INVESTIGATIONS IN COMPLEX ORGANIZATIONS

INTRODUCTION

The question initially posed for this NATO Advanced Research Workshop reads in part: "Can a procedure, such as that used to investigate aircraft accidents, be established to investigate information system failures?" This question itself calls for considerable comment, since it sidesteps a key issue. Aircraft accident investigations may be more or less sophisticated, but they have a definite practical purpose: to improve the functioning of the aircraft, its passengers, and crew. The adequacy of the investigation must be gauged in relation to this purpose; one cannot separate the value of the investigation from its ultimate end. The relation between the investigation and performance improvement needs to be studied as well as the conduct of the investigation itself. Sophisticated investigations of marine transport accidents take place, but do not result in improvements comparable to those in aviation (Perrow, 1984). The reason they do not lead to improvements is that the inter-organizational learning system which exists for marine transport is far less adequate than that for aviation. If we are going to investigate the failures of information systems in complex organizations, the major question is how such investigation will lead to improvements in organization functioning.

This part contains three sections. In the first ,we consider briefly the nature of aircraft accident investigations. In the second, we discuss, again briefly, the development of organization science. This section is important since investigations of failure assume basic knowledge of organizations. In the third ,I suggest how we might improve the learning system which tracks and analyzes organization performance. We need to consider this learning system, since it will be the recipient of information failure analyses.

AIRCRAFT ACCIDENT INVESTIGATION

The investigation of aircraft accidents has run parallel to the development of aviation itself (Brown, 1966). As the technology has become more sophisticated, so have the investigations. Newton, 1964). Air crash investigation has now become a small branch of applied science or engineering (Barley, 1969). Furthermore, with larger and faster aircraft, the stakes in an individual accident have gone up. In the 1977 Tenerife collision some 577 persons were killed when two 747's collided on the ground (Carroll et al., 1977). This number might have been multiplied had the accident taken place over an urban area. As mechanical sophistication has improved, attention has turned to detailed examinations of the human factors aspects of accidents ((Beaty, 1980). This includes examination of cockpit interaction patterns and their impact on performance (Cooper, 1979). Investigation of aircrashes is frequently carried out in parallel by government investigative bodies, the aviation industry, and pilot unions (Perrow, 1984). As a result, investigations are probably more thorough than in any other sphere except space technology.

What is supremely important to the effect of these investigations on performance improvement, however, is the inter-organizational learning system. Suggestions for improvements are generally implemented because 1) there is a legal requirement to do so, and 2) because failure to implement can lead to grave and immediate consequences for manufacturers and operators. These range from declines in ticket sales, refusal of pilots to fly, to fines and withdrawal of operating licenses. Airline accidents in particular take place in the public eye and have high saliency to legislators and their constituencies. A highly skilled inquiring system is coupled with strong social sanctions. In other fields involving "high technology," the inquiring systems may be adequate, but the implementation is not. In marine accidents, for instance, pressure to fix problems is simply not as great as with aviation. Fewer people get killed in an accident, so public pressure is less. In nuclear power, public stakes are just as high but the technological system is much more complicated. Another reason for carelessness in nuclear power is that catastrophic accidents like Chernobyl are rare and most consequences from accidents are delayed by years. Humans learn much more quickly when feedback is immediate.

A number of other factors make aircrash investigation valuable. Standardization allows problems with a single aircraft to translate into structural improvements for others of the same type. Radar, telecommunications, and flight recorders provide an accurate record of pre-crash events. The establishment of a legal right to investigate by government bodies allows access to key data in a timely way. Intense public interest encourages investigation. Above all, the investigation takes place immediately. These helpful factors should be kept in mind when we consider the investigation of information system failures in complex organizations, since in organizations they are usually absent.

ORGANIZATION SCIENCE

As with airplanes, human beings have tried to increase the effectiveness of the organizations they use. (I beg the pardon of the foreign colleages for the parochialism of this analysis, which is largely restricted to the United States. I shall be happy to have my omissions corrected.) The field of organization design is roughly analogous to aeronautics in its general aims: to make organizations work better. But progress in organization science relating to organizational effectiveness has been much slower than in aeronautics. And implementation has a dismal record. The reasons for this situation are beyond the scope of this paper, but have much to do with the general social structure of modern societies, in which organizations play a major part (Westrum and Samaha, 1985). Organization science is disorganized. There are no *Feynman Lectures*, no really outstanding textbooks; the ostensibly comprehensive *Handbook of Organizational Design* (Nystrum and Starbuck, 1982) doesn't tell one how to design organizations, no competent history of the field (or introduction to it) exists, and so forth. Lack of standard terminology is part and parcel of the problem, but also reflects the varied practices of organizations themselves.

While studies or organizational activities (such as war) go back into ancient times, structural study of the organization itself in a scientific manner is relatively new and has largely taken place only since World War II (Westrum and Samaha, 1985). Myths regarding organization science are abundant. (The negative assessment of the field I offer here is necessarily subjective, and might not be agreed to by others. I suspect that this situation reflects both a failure of intellectual leadership on the part of the field's top scholars and also the promotion practices of American social science, which reward analysis rather than synthesis.)

The place of "scientific management" in organization science is one such myth. Established by Frederick Taylor in the early years of the century, scientific management was at best scientific only in relation to its studies in ergonomics (Friedmann, 1955). Structural changes proposed by Taylor (Nelson, 1980) were based on intuition and were either reflective of ordinary good sense or actually harmful (e.g., the functional foreman system). Nor was Taylor's strategy necessarily the best contemporary one for operational problems (Hirschborne, 1985). At the same time that Taylor was trying to reform workers by a "divide and conquer" strategy, others were using group methods intuitively to increase factory production and increase the performance of naval units (Leitch, 1919). Neither the factory successes nor their naval counterparts were developed into general models for management theory, however. (I plan to examine this failure and its consequence in a paper being prepared for *Administrative Science Quarterly*.) In the factory case, this was the result of the association of the innovations with anti-union sentiment, and in the naval case because the authors never thought to develop a general management theory, even though they might well have done so.

Scientific study of the effectiveness of supervision patterns did not begin until World War II. Study of the effectiveness of organizational designs themselves began with Worthy's (1984) studies of Sears and Roebuck, though many earlier studies dealt with aspects of design, notably military studies. After 1955, scientific studies of organizational design really "got going," but the results were more apparent in books and scientific journals than in actual practice. The preeminent leader in this work, the Tavistock Institute in London, discovered a great deal and was able to improve the organizations with which it worked but had only a modest influence on organizations generally (Trist, 1981).

While researchers may know how high-tech installations ought to be run, getting practitioners to follow these guidelines is very difficult. This failure to implement is having a serious impact on nuclear power plants and other sensitive installations. That some airlines are now implementing these practices for cockpit resource management is a hopeful sign but only a beginning. Nor is there general agreement among management experts on what should be done. The safety of a great many vital but fragile systems meanwhile remains in peril (Lovias and Lovias, 1982).

A first step in developing a sound investigation system for information system failures in this context would be a useful synthetic compendium of organization studies, especially those concerned with information system performance. Part of the effort would be an attempt to search out "best practice" cases, much like the popular *In Search of Excellence* (Peters and Waterman, 1982).

HOW INFORMATION SYSTEMS FAIL: FORMS OF FAILURE

As an exercise, we might look at ways in which information systems do not work, the forms of failure. Among them would be the following:

1. Failure to coordinate actions between units.
   Example: Grenada invasion radio compatibility problems

2. Failure to inform about problems
   Example: DC-10 door blowout problem (Shields, 1979).

3. Failure to inform about innovations, including inadequate advocacy.
   Example: Lime juice and scurvy in British Navy (Mosteller, 1981)

4. Management failure to inquire as to conditions and changes; desire not to know bad news.
   Examples: Bays of Pigs debacle
   Challenger 0-ring situation
   Loss of U.S.S. Pueblo

5. Failure of nodal subsystems to combine information in a useful way.
   Example: Pearl Harbor intelligence failure
   Prelude to Three Mile Island accident (Welchman, 1982)

6. Failure to record transmitted information about problems
   Example: Quality problems at Tennessee Valley Authority nuclear reactors

INVESTIGATING FAILURES IN INFORMATION SYSTEMS

My discussion here will be concerned solely with information system failures in large, complex organizations. Although most of my past research has dealt with systems on an even larger scale, I will not make any attempt here to deal with problems of organizational fields or societal monitoring systems (Westrum, 1982). Even with this limitation, however, the field in question is enormous. Let me make a few remarks on methods in use. Currently, a number of extremely loose procedures exist for the investigation of information system failures. Four such procedores are noted below.

Government Hearings and Boards of Inquiry

Here the setting is judicial rather than scientific, but there are certain advantages. Witnesses can be subpoenaed, forcing them to some degree of cooperation. Their testimony is frequently given under oath, or other threat of punishment for lying. Proceedings are public, and lacunae are a matter of record. This may establish a motive for later action by the responsible bodies.

Hearings, unfortunately, also provide occasions for investigators to show off or to punish those they do not like. The function of hearings is often to place blame rather than to develop better procedures: the focus may be on the perpetrators of the information system failure rather than its systemic causes. As a result of government hearings, we know much more about the failure of military and other governmental organizations than we know about the operations and failures of private firms. Government hearings do not always, however, result in a useful catalogue of lessons learned, and in any case there is no compulsion to read this catalogue when it does exist.

Journalistic Investigation

Such investigations can be useful if they are intensive. Generally, the longer journalists or free-lance are given to investigate something, the better and more valuable these efforts will be. Long articles are more competent than short ones; books are better yet. However, the financial economy of these persons frequently demands a rapid turn-around time. Leisurely investigations are the privilege of elite writers whose previous writings allow a comfortable bank balance; or of those who have a regular reporting job and must write books after working hours. As a social scientist, I have often been astonished at the persistence and audacity of a close friend who is a prize-winning journalist, but at the same time appalled at the lack of expertise and background in the relevant specialty.

Although journalists often do better investigative work than social scientists, they frequently lack an appropriate theoretical framework to interpret what they find, and they are not usually motivated to develop one. A closer collaboration between journalists and social scientists might be beneficial to both.

Social Science Investigations of Information System Failures

This type of investigation could potentially have the greatest value because the investigators would know best what to look for and how to interpret it. But past efforts of this kind have serious problems associated with them. One problem is timeliness. Unless the social scientist gets to the scene early, evidence and witnesses may vanish. Forgetting, selective memory, covering up, etc., increase as time goes on. Getting the grant to do the work may take time.

More important may be severe limitations on access. Many information failures involve organizations that do not want prying outsiders sifting internal information or getting access to dirty laundry Furthermore, social scientists often display an amazing naivete regarding the operations of organizations, especially when complex politics is involved.

Internal Investigations by Private Organizations

This genre is often carried out by consultants like Donald Mason, whose paper appears elsewhere in this volume. Melville Dalton's (1959) classic study *Men Who Manage* documents an extremely serious information system failure in a large petroleum refinery. The result of this intratelligence failure was an inability to develop sound internal policies of maintenance. The kind of difficulties described show the importance of investigating the soundness of internal information flows. Such private efforts, however, seldom enter the literature, thus leaving few public tracks and doing little to improve institutional wisdom and managerial theory. While such private efforts are often quite sophisticated relative to investigations by social scientists, it would be highly desirable to have the wisdom gained preserved in some publicly accessible form.

TOWARD BETTER PROCEDURES

The insufficiency of current investigation procedures is a fault not only of analysis but also of synthesis. The analysis may be extremely well done, but its lessons are not applied. The inter-organizational learning system is not learning and implementing past experience successfully. Unless the learning system is improved, better investigations will advance understanding but not performance (Katz, 1966). For instance, the lessons of combined action of air, armor, infantry, and artillery units learned during the World War II African campaign were still conspicuously unlearned in the invasion of Grenada (Cushman, 1986). The lesson of the faulty risk analyses leading to the losses of the EC-121 and the U.S.S. Pueblo (Strach, 1971) could have helped avoid the loss of the Challenger (McKean, 1986). Fitzgerald's (1972) exposes on bad practices in weapons acquisition should have led to massive reforms, but didn't. (It is notable that the pattern here is exactly what one would expect in a calculatively rational system: the specific facts are fudged into accord with the "party line.")

SPECIFIC PROCEDURES

1) Investigations should be done soon after the event. What happens when investigation is delayed is that the trail gets cold. Even the genesis of major decisions may be difficult to discover if one waits too long (e.g., Kraft, 1980).

2) Documentary evidence as to communications and decisions should be gathered and organized as rapidly as possible. Since memory is fallible, this material should be put in order and properly reviewed by qualified persons at a very early stage of the investigation. Gaps in documentary evidence should be filled by interviews. Memories degrade rapidly with time, so these interviews should be done with minimal delay, and preferably by a qualified oral historian. Files should be set up and organized properly.

3) The investigating team personnel should include the following:

   a) A forensic psychologist specializing in testimony or an oral historian (Trankell, 1972).
   b) A historian skilled in documentary analysis (e. g., Brown an Graff, 1977).
   c) Practitioners skilled in the domain in question.
   d) A management theorist. One should not forget that the major reason for the investigation is better management practices. Unless one of the members of the investigating team is skilled in translating the specific findings into more general reçommendations, the investigation's major purpose may not be served. (Argyris et al., 1985).
   e) A representative of higher authority. In the military, this might be an aide to a high-ranking general officer. Elsewhere it is not clear who should fulfill this function. This person's duty would be to see that the recommendations of person (d) get implemented.

CONCLUSION

In this paper I have discussed some important concepts and examples of the relation between management philosophy and organizational outcome, with special consideration given to the involvement of turbulent environments and complex technologies. My basic argument is that certain situations demand a generatively rational strategy. I would argue further that investigations of failure also require a learning system equipped with this philosophy if organizations are going to profit from them. Without such a learning system, there is no way that the lessons of the past will be applied to the design of the future.

This argument is naturally a first approximation to truth rather than a definitive statement. I feel secure in speculating and transmitting my speculations because one of the characteristics of generative strategy is to act before all the information is in. Yet another characteristic is that information processing does not stop when further information comes in. I am certain that considerable adjustment will take place as time goes on, in regard to both major and minor premises of this argument Thus, it is essential that each of these points now be subjected to creative scrutiny if progress is to be made.

REFERENCES

Argyris, Chris. (1965). *Organization and Innovation.* Homewood IL: Dorsey Press. (See also Interpersonal Barriers to Decision-Making. *Harvard Business Review* volume On Management. New York: Harper and Row, 1975, and Irving Janis, *Victims of Groupthink* . Boston: Houghton Mifflin, 1972.)

Argyris, Chris, Robert Putnam, and Diana McLain Smith . (1985). *Action Science: Concepts,*

Methods, and Skills for Research and Intervention. San Francisco: Jossey-Bass.

Barley, S. (1969). *Aircrash Detective: An International Report on the Quest for Air Safety.* London: Hamish Hamilton. (Very good popular treatment by a comprehensive journalist. As a layman, I found this a useful introduction to the subject. For a more sophisticated introduction by a former Chief Inspector, see William H. Tench (1985), *Safety Is No Accident.* London: Collins.)

Barzun, Jacques and Henry Graff. (1977). *The Modern Researcher.* New York: Harcourt, Brace, Jovanovich.

Beaty, D. (1969). The Human Factor in Aircraft Accidents. New York: Stein and Day. (Also see Jane E. Brody. (1980). 'Human Factor' in Plane Crashes Is Facing More Intense Scrutiny. *New York Times*, June 23, pp.1, 18.)

Blakeslee, Sandra. (1986). Feynman's Findings: They Fooled Themselves. *New York Times*, June 11, p. 16.

Brown, Sir Vernon. (1966). The Story of Air Accident Investigation by a Onetime Inspector. *Journal of the Royal Aeronautical Society*, 70, pp. 243-247.

Burns, Tom and G.M. Stalker. (1961). *The Management of Innovation.* London: Tavistock.

Carroll, V., Loren Jenkins, and Janet Huck. (1977). Collision Course. *Newsweek*, April 7, pp. 48-55. (For human factors aspects of this accident, see J. Reason and K.Mycielska. (1982). *Absent Minded? The Psychology of Mental Lapses and Everyday Errors.* Englewood Cliffs: Prentice-Hall, pp. 210-217.)

Cockpit Resource Management (1986). *A Training Course for Group Problem-Solving.* CRM (P.O. Box 1985, Austin, Texas 78767) originated as a joint venture between United Airlines and Scientific Methods, Inc. (the Blake/Mouton "managerial grid" organization). It grew out of a collaboration between the two organizations to improve the ability of United's flight deck personnel to cope with operational problems. CRM trains flight deck personnel to interact in a more generative way, so that the entire cognitive resources of flight deck personnel are brought to bear on problems. According to the organization's publicity, some 5,000 persons have been trained using their methods. The organization has recently begun consultation with the nuclear power industry, according to William Taggart of CRM. See also Also see Douglas B. Feaver. (1982) , pp. 1,6,7. and Walter J. Boyne. *Phantom in Combat.* Washington: Smithsonian Institution, 1985, p. 72.

Cooper, George E., Maurice D. White, and John K. Lauber. (1979). Resource management on the Flight Deck. *Proceedings of a NASA/Industry Workshop* . Moffett Field, CA: Ames Research Center.

Dalton, Melville. (1959). *Men Who Manage.* New York: John Wiley.

Douglas, John H. New computer architectures tackle bottleneck. *High Technology*, June, pp. 71-78.

Emery, Fred and Eric Trist. The Causal Texture of Organizational Environments. *Human Relations*, 18, pp. 21-32.

Easterbrook, Gregg. (1984). Why Divad wouldn't die. *Washington Monthly*, November.

Fallows, James. (1981). *National Defense.* New York: Random House, pp. 76- 95. (Also see Edward C. Ezell. *The Great Rifle Controversy.* Harrisburg, PA: Stackpole Books, pp. 162-228.

Feaver, D. B. (1982). Pilots learn to handle crises - and themselves. Washington Post, September 12, pp. 1,6,7. (Also see: G. E. Cooper, M. D. White, and J. K. Lauber. *Resourcement Management on the Flight Deck.* Proceedings of a NASA/Industry Workshop. Moffett Field, CA: NASA/Ames Research Center, 1979. Write: Chief, Aeronautical Human Factors Research Office, Mail Stop 239-21, NASA-Ames Research Center, Moffett Field, CA 94035.)

Fialka, John J. (1985). Weapon of choice. *Wall Street Journal,* February 15, pp.1, 30. (I am currently working on a case study of this innovation and its social context.)

Fitzgerald, A. Ernest. (1972). *The High Priests of Waste.* New York: Norton.

Follett, M. P. (1941). The giving of orders. In H. C. Metcalf and L. Urwick (eds). *Dynamic Administration: The Collected Papers of Mark Parker Follet.* New York: Harper and Row, pp. 50-70. (Like John Leitch, Follet was a management thinker largely ahead of her time.)

Gilman, Hank. (1986). J. C. Penney decentralizes its purchasing. *Wall Street Journal,* May 8, p. 6.

Foushee, H. Clayton. (1982). The role of communications, socio- psychological, and personality factors in the maintenance of crew coordination. *Aviation, Space, and Environmental Medicine,* 53(11), pp. 1062-1066.

Friedkin, D. S. (1986). Letter, New York Times, March 9. (The engineer in question was John C. Houbolt, at that time Assistant Chief of the Dynamic Loads Division at Langley Research Center. Friedkin's letter gives only the barest clue to Houbolt's long and arduous fight to make himself heard in NASA. But it is a credit to the organization that he could make himself heard, in spite of initial rejection. See Courtney Brooks, James Grimwood and Loyd Swenson. (1979). *Chariots for Apollo: A History of Manned Lunar Spacecraft.* Washington: National Aeronautics and Space Administration, pp. 61-86.)

Friedmann, G. (1955). *Industrial Society.* Glencoe, IL: Free Press.

Greve, Frank. (1982). Navy duping itself, U.S., on strengths, experts say. *Detroit Free Press,* May 31, pp. 1,2.

Hayes, R. H. (1986). Why strategic planning goes awry. *New York Times,* April 20, Business Section, p. 2.

Hirschhorn, Larry. (1985). *Beyond Mechanization: Work and Technology in a Post-Industrial Age.* Cambridge, MA: MIT Press. (Chapter 7)

Janis, I. (1972). *Victims of Groupthink.* Boston: Houghton Mifflin.

Johnson, Clarence "Kelly" with Maggie Smith. (1985). *Kelly - More Than My Share of It All.* Washington: Smithsonian Institution Press. (See particularly chapter 16)

Jordan, N. (1968). Training information-processing decision-making man-machine systems. in Jordan, N. *Themes in Speculative Psychology* . London: Tavistock, pp. 164-191. (This paper relates some aspects of the Systems Research Laboratory experiments on radar operator team training.)

Katz, Amrom. (1966). *Some Notes on the History of Aerial Reconnaissance (Part I).* P-3310 Santa Monica: RAND Corporation, p. 4.

Kraft, Joseph. (1980). Annals of industry: The downsizing decision. *New Yorker*, May 5.

Leitch, John. (1919). *Man to Man: The Story of Industrial Democracy*. New York: B.C. Forbes, p. 154. (See also Leitch, John. (1919). The background of industrial democracy. *Annals of the American Academy of Political and Social Science*, 85, September, pp. 203-213.)

Leslie, S. (1983). *Boss Kettering: Wizard of General Motors*. New York: Columbia University Press.

Lovins, Amory and Susan Lovins. (1982). *Brittle Power*. Andover, MA: Brick House.

MacGregor, Douglas. (1958). The Significance of Scanlon's Contribution. In F.G. Lesieur (ed). *The Scanlon Plan: A Frontier in Labor-Management Cooperation*. Cambridge: MIT Press, p. 10.

McKean, K. (1986). They fly in the face of danger. *Discover*, April, pp.48-58.

Menard, H. (1971). *Science: Growth and Change*. Cambridge: Harvard University Press.

Miles, Raymond E. and Charles C. Snow.. *Organizational Strategy, Structure, and Process*. New York: McGraw-Hill. (See pp. 31-48).

Morison, Elting. (1966). Gunfire at sea. in E. Morison. *Men, Machines, and Modern Times*. Cambridge: MIT Press. (Also see Peter Padfield. *Aim Straight: A Biography of Admiral Sir Percy Scott*. London: Hodder and Stoughton, 1966.)

Mosteller, Frederick. (1981). Innovation and evaluation. *Science*, 211, pp. 881-886.

Newton, E. (1964). The investigation of aircraft accidents with particular reference to those caused by mechanical failure: their diagnosis, and the lessons learnt from them. *Journal of the Royal Aeronautical Society*, 68, pp. 156-164.

Nelson, Daniel. (1980). *Frederick W. Taylor and the Rise of Scientific Management*. Madison: University of Wisconsin Press.

Nystrom, Paul and William Starbuck (eds). (1982). *The Handbook of Organizational Design*. Oxford: Oxford University Press.

O'Connor, Colleen and Lynda Wright. (1986). Troubled times for the TVA. *Newsweek*, January 27, pp. 23-24.

Ouchi, William G. and Alan L. Wilkins. (1985). Corporate culture. *Annual Reviews of Sociology*. pp. 457-483

Padfield, P. (1966). *Aim Straight: A Bibliography of Admiral Sir Percy Scott*. London: Hodder and Stroughton. p. 136.

Perrow, C. (1964). *Normal Accidents: Living With High-Risk Technologies*. New York: Basic Books.

Peters, Thomas and Robert Waterman. (1982). *In Search of Excellence: Lessons from America's Best-Run Companies*. New York: Warner Books.

Pinchot, G. III. (1985). *Entrepreneuring: Why You Don't Have to Leave the Corporation to Become an Entrepreneur*. New York: Harper and Row.

Sanger, D. (1986). Engineers say they were punished for their testimony about shuttle. *New York Times*, May 11 p. 17. (Both were soon after reinstated, in the wake of negative publicity.)

Saxenian, A. (1985). Silicon valley and route 128: regional prototypes or historical exceptions. In M. Castells, (ed). *High Technology, Space, and Society.* Beverly Hills: Sage, pp. 81-105. (See also Dirk Hanson. (1982). *The New Achemists: Silicon Valley and the Microelectronics Revolution.* Boston: Little, Brown.)

Schon, D. A. (1971). *Beyond the Stable State.* New York: W. W. Norton, pp. 31-60.

Selznick, Philip. (1952). *The Organizational Weapon: A Study of Bolshevik Strategy and Tactics.* New York: McGraw-Hill.

Shiels, Merrill et al. (1979). How safe? *Newsweek*, June 11, pp. 34-43.

Smith, R. Jeffrey. (1985). Pentagon decision-making comes under fire. *Science*, 227 (4 January), pp. 32,33.

Stein, L. I. (1967). The doctor-nurse game. *Archives of General Psychiatry*, 16, pp. 699-703. (Similar problems surround the medical use of high-technology devices like computed tomography scanners. See Stephen R. Barley, Technology as an occasion for structuring: evidence from observations of CT scanners and the social order of radiology departments. *Administrative Science Quarterly*, 31(1), March 1986, pp. 78-108.)

Strauch, R. (1971). *The Operational Assessment of Risk: A Case Study of the Pueblo Mission.* Rand R-691-PR, Santa Monica: RAND Corporation, March.

Tarkov, John. (1966). A disaster in the making. *American Heritage of Invention and Technology*, 1(3), pp. 10-17.

Thompson, Mark. (1986). NASA inflated shuttle safety. *Detroit Free Press*, June 1, pp. 1,8A.

Thompson, Paul. (1978). *The Voice of the Past: Oral History.* New York: Oxford University Press.

Trankell, Arne. (1972). *The Reliability of Evidence.* Stockholm: Beckmans.

Trist, E. (1981). *The Evolution of Socio-Technical Systems: A Conceptual Framework and an Action Research Program.* An Occasional Paper of the Ontario Quality of Working Life Center, Toronto: Ontario Ministry of Labour.

UPI. (1986). NASA investigates alleged document shredding. *UPI Reports*, May 16.

Vaill, Peter. (1978). Toward a behavioral description of high performance systems. in McCall, M. and M. Lombardo (eds). *Leadership: Where Else Can We Go?* Durham, NC: Duke University Press.

Weber, Max. (1978). Bureaucracy. in H. Gerth and C.W. Mills. *From Max Weber: Essays in Sociology.* Oxford: Oxford University Press, pp. 196-244.

Welchman, Gordon. (1982). *The Hut Six Story: Breaking the Enigma Codes.* Harmondsworth: Penguin.

Westrum, Ron. (1982). Social intelligence about hidden events: its implications for scientific research and social policy. *Knowledge: Creation, Diffusion, Utilization,* 3(3), pp. 381-400.

Westrum, Ron. (1985). Calculative and generative rationality. *Social Psychology of Science.* Newsletter. March. (It is gratifying to see that several colleagues have used them since in articles and interviews. The inspiration for the concept was an article by Joseph C. Pitt. Conceptual change and conceptual tension. *Methodology and Science,* Vol. 14, No. 2 (1981), pp. 132-138 Pitt speaks of "reactionary" and "progressive" rationalities.)

Westrum, Ron and Khalil Samaha. (1985). *Complex Organizations: Growth, Struggle, and Change.* Englewood Cliffs: Prentice-Hall.

Wilensky, H. (1967). *Organizational Intelligence.* New York: Basic Books.

Withington, F. G. (1969). *The Real Computer: Its Influence, Uses, and Effects.* Reading, MA: Addison-Wesley, pp. 199-203. (Withington speculates about the use of the computer in a generative way, allowing access to the information in the system to a wider group.)

Worthy, James C. (1984). *Shaping an American Institution: Robert E. Wood and Sears Roebuck.* Urbana, IL: University of Illinois Press. Also see his unpublished book manuscript. Available from the Worthy, who is at the Graduate School of Management, Northwestern University, Evanston, Illinois 60201. Professor Worthy was kind enough to let me see this manuscript, which describes some of his classic studies for Sears. Worthy has shown in a recent book that the overall tendency of Robert Wood (Chief Executive Officer at Sears 1929-1954) was to decentralize decision-making to the maximum extent possible.

INVESTIGATING SOURCES OF ERROR IN THE MANAGEMENT OF CRISES: THEORETICAL ASSUMPTIONS AND A METHODOLOGICAL APPROACH

Irving L. Janis

University of California, Berkeley
Berkeley, California, USA

THEORETICAL ASSUMPTIONS

It is generally recognized that the methods used to collect and evaluate data on major human events, such as policy fiascos and disasters, are heavily influenced by the investigator's theoretical assumptions about what causal sequences are likely to account for such events. Accordingly, I begin this paper with my main theoretical assumptions about sources of error in decision making and then turn to methodological questions pertaining to empirical research on failures of information systems.

What types of failures on the part of top-level policy makers in government and industry are likely to have the most disastrous consequences? Under what conditions are such failures most likely to occur? What are the mediating steps that lead to the worst kinds of failures? These are the main questions to which my theoretical assumptions attempt to give preliminary answers. The answers are intended primarily to serve as guidelines for subsequent research on errors in policy making, including both basic research and applied research oriented toward developing and testing new procedures designed to improve the effectiveness of problem solving in governments and in other large organizations.

Herbert A. Simon's Analysis of "Good" Policy Making

I take as my starting point Herbert A. Simon's (1956, 1977) assumption that human beings sometimes approach decision making in a "rational way," and sometimes not. In his highly influential book *Administrative Behavior*, Simon's descriptive account of how good executives make their decisions calls attention to several major steps in effective problem solving that they carry out. One involves listing alternative courses of action or choices; another is to examine the major consequences that can be expected to follow from each of the choices; yet another is to carry out a comparative evaluation of the various consequences. Simon adds, however, that an objectively rational approach cannot be carried out fully because of the limited knowledge and capabilities of all human beings.

Actual behavior of policy makers, Simon points out, will always fall short in several ways of an ideal image or model of complete, objective rationality:

(1) Rationality requires a complete knowledge and anticipation of the consequences that will follow on each choice. In fact, knowledge of consequences is always fragmentary.

(2) Since these consequences lie in the future, imagination must supply the lack of experienced feeling in attaching value to them. But values can be only imperfectly anticipated.

(3) Rationality requires a choice among all possible alternative behaviors. In actual behavior, only a very few of all these possible alternatives ever come to mind (Simon, 1976, p. 81).

NATO ASI Series, Vol. F32
Information Systems: Failure Analysis
Edited by J. A. Wise and A. Debons

In practice, then, according to Simon's account, good policy makers who are striving for high quality decisions engage in analytic problem solving, but they do so in a way that only crudely approximates the requirements of a normative rational model. Their procedures involve working to the best of their limited abilities, within the confines of available organizational resources, to exercise all the caution they can to avoid mistakes in the essential tasks of information search, deliberation, and planning.

Like many other experts in administrative sciences, Simon refers to analytic problem solving as a "rational" approach to decision making. But it seems best not to use that label because of its ambiguities. The term "not rational" is especially misleading because it connotes "irrational," "wildly unrealistic," "crazy," and other extreme epithets, which are not appropriate for conceptualizing the shortcut approaches that so often are used, which I shall describe shortly. Consequently, I avoid using the terms "rational" and "non-rational" except when quoting other writers. Instead, I use terms like "sound" versus "unsound" or "high quality" versus "low quality" when evaluating the procedures used by decision makers to arrive at their choices.

The essential tasks involved in analytic or reflective problem solving are shown in Figure 1, which summarizes the main components described by Simon and other social scientists who have studied effective policy making. The four steps represented in this schematic diagram, it will be noticed, require a decision maker to do a great deal of work and to use extensively the informational resources of his or her organization.

Symptoms of Defective Decision Making

Executives in policy-making roles seldom use an analytic problem-solving approach, as (Simon 1976) and many other social scientists have emphasized. But only when all four steps of vigilant problem solving are fully carried out can the decision maker be expected to avoid the symptoms of defective problem solving listed in the fourth column of Figure 1. I assume that these symptoms are indicators of failure to meet the elementary criteria of sound decision making (Janis and Mann, 1977). When none of the steps shown in the Figure are carried out adequately by a chief executive, a board of directors, or an executive committee, one can expect to observe all seven symptoms of defective decision-making.

The presence or absence of these symptoms can be used to judge the quality of decision making, not only of top evel policy makers but also of middle level and lower level administrators who are pursuing the goals and interests of their own sub-unit. The seven symptoms can also be applied to assess the quality of decision making of executives who arrive at strategic choices by adopting a "logical incrementalism" approach, making a series of tentative or partial decisions pertinent to only one subsystem within a large organization. Those step-by-step decisions will not necessarily prove to be of poor quality, as manifested by a large number of symptoms of defective decision making. As Quinn (1980, p. 17) observed in his study of managers of multi- billion dollar corporations, "In the hands of a skilled manager, such incrementalism was not muddling."

According to my first main assumption, then, the quality of the procedures used by decision makers in arriving at a given policy decision can be assessed by examining whatever observational reports and records are available to find out whether or not each of the seven symptoms was manifested at the time the final choice was made. For example, to evaluate the quality of decision making by national leaders during a crisis provoked by a destabilizing action taken by an adversary nation (such as military invasion of an allied third country), the investigator could examine records of a policy making group's discussions and memoranda to see whether they show manifestations of any of the following symptoms:

Gross omissions in survey of objectives: The group never explicitly discusses objectives or gives them such brief or cursory consideration that the participants fail to take into account some of the major goals or values implicated by the choice.

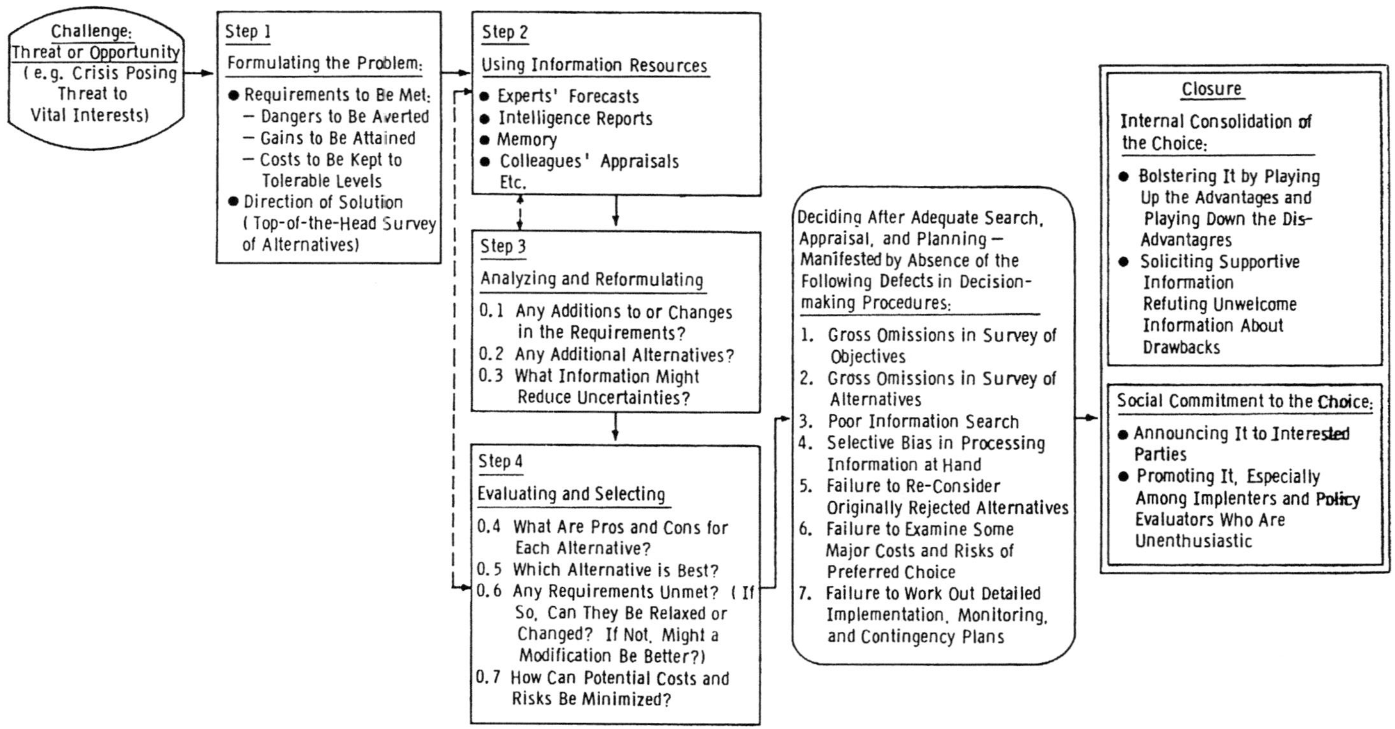

Figure 1. Main steps characterizing a reflective problem-solving approach to decision-making. (From Janis, In preparation)

Gross omissions in survey of alternatives: The group does not consider any alternative policies during its deliberations except the one that is chosen. If alternative policies are mentioned, they are immediately excluded or dropped without discussion and the group shows clear concentration on and preference for a single alternative from the outset.

Poor information search: The group does not seek for available information necessary for critically evaluating the pros and cons of the preferred course of action and other alternatives. Or the group seeks for information in such a perfunctory and incomplete manner that it fails to obtain important pieces of information that would have been available if requested from experts or other appropriate persons inside or outside their organization.

Selective bias in processing information at hand: The group generally accepts new information from experts, the mass media, and outside critics to which the members are exposed only when it supports the initially preferred alternative. The members generally ignore or refute non-supporting information.

Failure to reconsider originally rejected alternatives: The group fails to reexamine the consequences of any previously considered alternative or reexamines the formerly rejected alternatives in a biased manner, discounting favorable information and giving much greater weight to information about the negative consequences.

Failure to examine some major costs and risks of the preferred choice: The group completely fails to consider the negative consequences of the preferred alternative or examines them in such an incomplete way that its members overlook important negative consequences even though information about those consequences is available.

Failure to work out detailed implementation, monitoring, and contingency plans: The group completely ignores possible problems in implementation and does not develop any monitoring or contingency plans. Or the group discusses implementation, monitoring, and contingency plans in such an incomplete manner that they remain completely vague.

In assessing whether any one of the above seven symptoms is absent, it is necessary to examine the available evidence in the decision-making process carefully in order to detect instances where the apparent absence of the symptom is continual rather than a genuine manifestation of vigilant problem solving. Most high-level policy makers know about the criteria for sound decision making, and sometimes deliberately contrive activities to make it look as though they had carried out the essential tasks of information search, appraisal, and planning. It is well known, for example, that President Lyndon B. Johnson made some of his decisions to escalate the Vietnam war without paying any attention to the available intelligence reports and the arguments for alternative courses of action but managed to rig the decision-making activities of his policy-making committee in such a way as to make it look as though he and his advisors had done so. He was able to elicit a considerable amount of informational input and advice that supported his preconceived plan of action and to elicit sufficient debate about alternatives to make the record look good by giving the floor to in-house "doves" who seemed to be functioning as devil's advocates but who had become completely "domesticated" (Janis, 1982).

In addition to gross failures, there are partial failures that can be identified. For example, with regard to the third symptom, the group may intensively search for some relevant information but fail to obtain one important piece of information that would have been available if they had requested it from relevant experts in their organization.

Both obvious and subtle indicators can be used for rating partial failures. For example, in order to rate the policy makers on the seventh symptom, an investigator might first look for obvious indicators by examining available records to see if specific plans were discussed for implementing the decision and for dealing with salient setbacks or contingencies. If no gross omissions are observed, the investigator might look further for more subtle evidence by examining the records to see whether the policy makers actually made feasible arrangements to monitor how well the decision was being implemented and how well it was succeeding in sufficient detail as to find out quickly whether somewhat concealed obstacles were being encountered that might require implementing a contingency plan.

Judging from my own experience and that of my research collaborators, only a very small percentage of policy decisions can be expected to be rated overall as high quality decisions on the basis of receiving no ratings of gross failures, indicating an absence of all seven symptoms. Nevertheless, as will become apparent later when I discuss some of my own research data, a very substantial percentage of policy decisions can be expected to warrant being rated overall as decisions of fairly good quality -- manifesting no gross failures on six of the seven symptoms.

The reason that I have been emphasizing the symptoms of defective problem solving is stated in my next major assumption, which is that the poorer the quality of the decision making procedures used in arriving at a policy decision, as manifested by symptoms indicative of failures to carry out essential steps of reflective problem solving, the greater the likelihood of avoidable disasters and other unfavorable outcomes of the decision. In other words, the symptoms of defective decision making (listed in the fourth column of Figure 1) are predictive of unfavorable outcomes. This assumption, although it is also made by many other social scientists, is much more controversial than the other assumptions for reasons that will be discussed when we examine the available evidence bearing on it.

Making Decisions by the Seat of Your Pants

Often, as I have already indicated, policy makers do not use a reflective problem-solving approach. How do they usually arrive at their policy decisions? If you ask them, they are likely to tell you that long ago they learned to disregard much of what the textbooks say about the desirability of a systematic "rational" approach and instead they do it mostly by the seat of their pants. Many policy makers I have interviewed -- in government, business, education, and public welfare -- explain that they are appallingly busy keeping their offices running and putting out all sorts of little as well as big fires that are smoldering and raging concurrently. So when it comes to making a decision about a policy issue, they feel that very little of their time and energy can be devoted to searching for the pertinent information and deliberating about the pros and cons of alternative courses of action. When Robert McNamara was Secretary of Defense, Washington bureaucrats quipped that the reason he looked so good as a policy advisor at White House meetings, in comparison with the President's other cabinet members, was that the long drive from the Pentagon gave him eight extra minutes to do his homework in the back of the limousinc (Coopcr, 1970).

What do policy makers mean when they say that in their circumstances you have to make most decisions by "the seat of your pants"? As far as I can tell, they mean that instead of approaching a crisis or any other challenge requiring a decision as a problem to be worked on in a systematic manner, with careful information search and deliberation about alternatives, they rely primarily on a few simple decision rules that enable them to arrive quickly at a solution that seems to be satisfactory.

One such rule is to accept the first alternative you think of that is "good enough" to meet the minimal requirements, without bothering to compare it with other viable alternatives. This strategy has been described by Herbert Simon (1957, 1976), who refers to it as "satisfying." Simon points out that this approach fits the needs for simplifying complicated decision problems that exceed the information processing capabilities of human beings, all of whom are creatures of "bounded" or "limited rationality" (1976, p. xxxiii). A few other rules of thumb (or "heuristics," as they are labeled by cognitive psychologists) are likely to be combined with satisfying. For example, the availability heuristic is often used to hit upon an analogy that suggests a solution. Availability is influenced by whether the analogy evokes a vivid image and other factors that determine spontaneous salience, as described by Tversky and Kahneman

(1973), Nisbett and Ross (1980), and Taylor (1982).

It is well known, for example, that President Lyndon B. Johnson made some of his decisions to escalate the Vietnam war without paying any attention to the available intelligence reports and the arguments for alternative courses of action but managed to rig the decision-making activities of his policy-making committee in such a way as to make it look as though he and his advisors had done so. He was able to elicit a considerable amount of informational input and advice that supported his preconceived plan of action and to elicit sufficient debate about altlernatives to make the record look good by giving the floor to in-house "doves" who seemed to be functioning as devil's advocates but who had become completely "domesticated." (Janis, 1982).

Other social psychologists (Green and Conolley, 1974; Raven, 1974; Raven and Rubin, 1977) have noted similar symptoms of groupthink in the way Nixon and his inner circle handled the Watergate coverup during the early 1970's. Drawing on their work and my own detailed analyses of the unedited Nixon tapes, I have carried out an intensive case study of the Watergate coverup and have used it to elaborate on the theory of groupthink (Janis, 1982a).

The decision rules that I have just mentioned are certainly not the only ones that policy makers rely upon when they adopt a seat-of-the-pants strategy. In my case study investigations, I find a number of additional ones -- which I shall speak about shortly. All such decision rules have a number of features in common. In one way or another, each of them provides a shortcut that helps to speed up decision making, sometimes resulting in satisfactory solutions to the problems at hand. But according to the key assumptions I have just discussed, those simple rules can be shortcuts to disaster. All too often, it seems to me, policy makers save time and effort at the expense of being stuck with an ill-conceived decision entailing catastrophic consequences that could have been avoided. By and large, relying on a few simple decision rules might generally work out fairly well when an executive is making routine choices or dealing with minor crises that do not entail the threat of any serious losses. But a seat-of-the-pants approach is likely to result in serious losses when used to make strategic policy decisions, if fundamental interests of the organization or nation are at stake.

Going through the essential steps of problem solving in a way that meets the criteria for high quality decision making does not by any means preclude the use of simple decision rules. The difference between a seat-of-the-pants approach (like one that relies primarily on "satisfying") and the reflective problem-solving approach is not that simple rules are used only in the former and never in the latter. Rather, the difference lies in the degree to which decision makers rely upon the decision rules they use. A seat-of-the-pants approach is dominated by them, whereas the vigilant problem-solving approach is not. In the latter approach, the quick and dirty solutions or judgments suggested by the decision rules are regarded merely as preliminary ideas to be looked into further, and they help to guide the search for relevant information. The policy makers remain open to new information and are ready to change if the information does not bear out the expectations based on whatever decision rules that had used at the outset.

Three Types of Decision Rules: Cognitive, Affective, and Affiliative

So far, the decision rules I have been talking about are ones that are referred to as cognitive rules of thumb or heuristics, which simplify the intellectual tasks imposed by the complicated problems that beset executives who make policy decisions. When a dangerous crisis or unique opportunity arises, they must seek for solutions that will avert threats to important values, that will not exceed the level of costs that are tolerable to the organization, that will not continue to create emotional tension and cause them to lose sleep at night, that will not adversely affect their status, power, compensation, and esteem within the organization, and that will be accepted by others who are expected to implement the policy decision. Executives cope with these difficult requirements or constraints by using a number of different types of decision rules.

In addition to cognitive heuristics, there are also what I call affective decision rules, consisting of allowing one's emotional reactions evoked by the dilemma to play a guiding role in deciding what to do. Actually, a fair amount is known about how, when, and why affective rules are used -- from experimental research on the psychology of emotions and from case studies in clinical psychology.

Most often, the affective rules are instigated by the temporary but powerful constraints imposed by negative emotions, such as anger or fear. For example, when decision makers become angry as a result of being frustrated by a crisis created by seemingly unwarranted aggressive action taken by an opponent, they are likely to include in their list of requirements for a satisfactory solution the added constraint that the course of action must be some form of punishment that inflicts retaliatory suffering. The affective rule, which I refer to as the "retaliation imperative," would be something like this if it were verbalized: "When you are thwarted, injured, or humiliated, don't let the bastards get away with it; do something to punish them in retaliation."

Although statesmen seldom acknowledge that their policy decisions are influenced by passion, they are far from immune. I have encountered several impressive examples that illustrate how a national leader's anger can contribute to ill-conceived aggressive actions that inflict retaliatory punishment on other nations whose leaders are suddenly perceived as intolerably frustrating (Janis, 1986).

Another major type of constraint that policy makers are often keenly aware of when making vital policy decisions has to do with maintaining or enhancing their power, compensation, and status within the organization and eliciting social support from their personal network. Even a very wise policy decision that beautifully solves a major problem created by threats from a hostile nation could get a group of top-level policy makers in serious trouble if the decision fails to take account other powerful leaders' objections, no matter how mistaken those objections might be.

Various decision rules are typically used to take account of such constraints. I call them "affiliative" decision rules because they pertain to the policy maker's affiliation with others in his or her organization. Most often, these rules are used to preserve the policy maker's relationship with a primary group, such as a small planning committee that meets face to face to work on major policy recommendations. Sometimes affiliative rules are also used to deal with constraints arising from the policy maker's membership in secondary groups, such as a corporation, a political party, the State Department, or the U.S. government. Our understanding of the functioning of affiliative rules comes mainly from the work of social psychologists in the field of group dynamics. Later on I shall give several examples of over-reliance on such rules that I have encountered in my studies of presidential decision making during international crises.

One of my main assumptions concerning the various constraints I have been talking about is represented in Figure 2, which is based on Alexander George's (1980) insightful analysis of presidential decision making. George calls attention to the "ever present constraints" that often require the chief executive to consider "trade offs" in "the search for high quality decisions in foreign policy, as in domestic policy" (p. 1). He discusses many different kinds of constraints, almost all of which can be classified into one or another of the three main types in Figure 2.

Commenting on affiliative constraints, George points out that on some occasions the necessity arises "for the president to make a difficult trade- off decision whether to sacrifice some of the quality of a preferred policy for greater acceptability" (p. 2). Related to the need for acceptability are the needs for consensus within the administration and for policy legitimacy involving constitutional legal considerations.

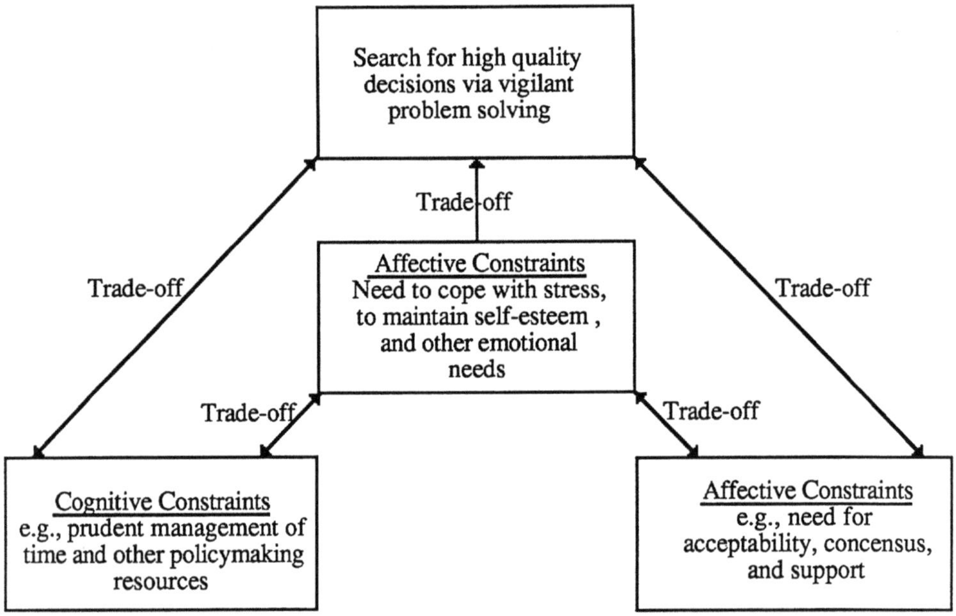

Figure 2. Constraints creating trade-off dilemmas in policy making.
Based on Figure 1 in George (1980, p. 2), with category added for affective constraints.

George asserts that the president also "must be sensitive to the constraints of time and the proper use of available policy making resources" (pp. 1-2). The president usually realizes that he cannot allow "the search for a higher quality decision on one policy question . . . to consume a disproportionate share of the manpower, and the analytical and intelligence resources that must be available to attend to other urgent policy questions" (p. 2). Other cognitive constraints become salient when the enormous complexity of the policy issues implicated in the decision and uncertainties about the expected consequences of alternative courses of action make a chief executive realize that he or she or anyone else who might work on the decision is handicapped by the limitations of the human mind.

George (1980) mentions a number of considerations pertaining to the president's emotional needs and feelings of personal well-being, which can also play an influential role in the policy-making process even though they are "extraneous to the values associated with even a very broad conception of the national interest" (p. 3). These affective constraints include essential needs that chief executives share with other people -- to counteract frustrations, to avoid damage to self-esteem, and "to cope with the anxiety, fear, or guilt that they experience from time to time" when dealing with decisional dilemmas (p. 4).

Whenever a policy decision involving vital interests of the organization is under consideration, a policy maker is likely to be keenly aware of at least a few of the powerful constraints that could be obstacles to a reflective problem-solving approach, such as the need to avoid overusing the time of top-level personnel within the organization. Any of the constraints shown in Figure 2 can become so salient when a policy decision is being made that the policy makers adopt a simple decision rule to deal with it, which gives it top priority at the cost of failing to carry out an adequate information search and failing to carry out the other essential steps of reflective problem solving.

Adopting a seat-of-the-pants approach involving reliance on simple decision rules usually is not a matter of deliberate conscious choice. Most often, the cognitive, affective, and affiliative rules are used by policy makers without deliberate intent and are not fully verbalized. That is to say, they are likely to operate at a preconscious level.

One of the main theoretical assumptions made repeatedly throughout the foregoing discussion can be summarized as follows: Within the repertoire of most (if not all) policy makers are various different approaches for making decisions in response to crises or other challenges, ranging from the high quality decision making procedures that characterize reflective problem solving to the low quality procedures that characterize crude seat-of-the-pants approaches that rely upon one or two "quick and dirty" decision rules.

Coping with Psychological Stress

Among the major constraints that interfere most drastically with reflective problem solving are those arising from psychological stress. The conflict theory model presented in *Decision Making* (Janis & Mann, 1977) describes the way different motivational states arising from low, moderate, and high levels of psychological stress adversely affect the quality of information search, the processing of whatever information is received, and contingency planning. The main postulates of the conflict model are summarized here because they provide part of the answer to a key question posed by the preceding discussion: What determines whether policy makers will use reflective problem solving rather than a seat-of-the-pants approach in response to events or communications that convey an important threat or opportunity requiring a policy decision?

One of our postulates is that stress engendered by decisional conflict frequently is a major determinant of failure to meet the criteria for high quality decision making. One source of stress arises from the decision makers awareness of their own limited knowledge and problem-solving capabilities. This is well illustrated by what President Warren Harding told a friend when he was struggling with a major domestic problem:

> John, I can't make a damn thing out of this tax problem. I listen to one side and they seem right, and then, God, I talk to the other side and they seem just as right, and there I am where I started... I know somewhere there is an economist who knows the truth, but I don't know where to find him and haven't the sense to know him and trust him when I did find him. God, what a job! (George, 1974, p. 187, quoted from Fenno, 1959, p. 36).

Other sources of stress include fear of suffering from the known losses that will be entailed by whichever alternative is chosen, worry about unknown things that could go wrong when vital consequences are at stake, concern about making a fool of oneself in the eyes of others, and losing self-esteem if the decision works out badly. Vital decisions involve conflicting values, and often decision makers realize that any choice they make will require trade offs that sacrifice ideals or other goals. As a result, the decision makers' anticipatory anxiety, shame, or guilt is increased, which adds to the level of stress.

From our review of the extensive research on behavioral effects of psychological stress, we discerned a number of different coping patterns that people use to deal with the stress of making major decisions (Janis & Mann, 1977, Chapter 3). Certain of those patterns interfere drastically with the essential steps of problem solving. Our analysis postulates five basic patterns of coping with the stress generated by any realistic challenge (threat or opportunity) requiring a person to make a highly consequential choice. Each pattern is associated with a specific set of antecedent conditions and a characteristic level of stress.

When decision makers display one particular pattern, which we label "vigilance," they tend to go about the tasks of decision making in a careful manner, carrying out to the best of their ability the steps of problem solving shown in Figure 1. They search painstakingly for relevant information, assimilate information in a relatively unbiased manner, appraise alternatives carefully before making a choice, and do everything else required. As a result of being so vigilant, a decision maker who is at least moderately competent will display relatively few symptoms of defective decision making (none at all if he or she is highly competent). But if the very same decision maker adopts one of the other four coping patterns, he or she will fail to carry out the steps of problem solving adequately and will display many symptoms of defective decision making, often all seven of the symptoms listed in Figure 1. (Of course, if the decision maker is extremely incompetent or chronically suffers from a mental disorder, he or she may practically always display all seven symptoms, irrespective of whether his or her motivational state is vigilance or one of the non-vigilant patterns.)

Four of the coping patterns lead to a seat-of-the-pants approach. Although the four have an important feature in common -- namely, that the policy maker relies on simple decision rules without carrying out the steps of vigilant problem solving -- they differ markedly in antecedent conditions and in the content of the dominating decision rules:

Unconflicted Adherence: Occurs when there is a weak challenge (unimpressive warning about expected losses or opportunities that will be missed). It results in reliance on simple cognitive heuristics, such as those embodied in adages like "let sleeping dogs lie," or "don't mess with it unless you run into real trouble," which foster inertia.

Unconflicted Change: Occurs when there is a strong challenge but little prior knowledge of and no impressive warnings about potential losses from changing. It results in reliance on simple cognitive decision rules, such as doing whatever is recommended by the first expert you talk with.

Defensive Avoidance: Occurs when there are impressive warnings that create intense conflict (by evoking expectations of severe losses if the current policy remains unchanged but also if it is changed) and when there are few signs of available resources that foster hope of finding a satisfactory solution. It results in a defensive type of emotional state with complete reliance on an affective rule, which, if it could be verbalized, would go something like this -- "don't beat your head against the wall; avoid thinking about the distressing problem." Along with this general rule, there are supplementary decision rules concerning ways to avoid thinking about it: "Procrastinate if you can; otherwise pass the buck; or, if necessary, select the least objectionable alternative and bolster it by focusing on supporting arguments and ignoring opposing arguments."

Hypervigilance: Occurs when there are impressive warnings that create intense conflict (by evoking expectations of severe losses if the current policy remains unchanged but also if it is changed) and when there are signs of available resources fostering hope of finding a satisfactory solution, but information about an imminent deadline or other time pressures evoke expectations of insufficient time to search and deliberate. It results in a panic-like emotional state with loss of cognitive efficiency (manifested by reduction in immediate memory span and repetitive thinking) and reliance on a simplistic affective rule, which if verbalized, would go something like this - - "try anything to get the hell out of this dilemma as fast as you can."

When policy makers use a sheer seat-of-the-pants approach in any of the above four coping patterns, the simple decision rules they rely upon can in certain circumstances be quite good guides to appropriate action. For example, the inertia decision rule that dominates in unconflicted adherence can be highly functional in handling the large number of warning messages that cross the desk of a top-level executive every day -- provided that the warnings pertain to trivial, improbable, on nonexistent dangers. Even the affective rule that dominates when decision makers are in a panicky state of hypervigilance can occasionally be more adaptive than vigilant deliberation by enabling them to escape an imminent catastrophic blow before it strikes.

Nevertheless, as indicated in my discussion of an earlier key theoretical assumption, reliance on decision rules without engaging in information search, critical appraisal, or any of the other essential features of the reflective problem-solving approach increases the likelihood of avoidable errors, which generally will prove to be highly detrimental for major policy decisions that affect vital interests of the organization. Because they generally are expected to reduce the decision maker's chances of averting serious losses, the four nonvigilant coping patterns are regarded as defective ways of making major policy decisions.

How do the conditions that make for vigilance differ from the conditions that make for each of the four defective coping patterns? Preliminary answers to this question, which were presented in the above discussions of coping patterns, are summarized in Figure 3, which is based on the Janis and Mann (1977) conflict model of decision making. This model, derived from extensive research findings on psychological stress, specifies the psychological conditions that mediate the five coping patterns and the level of stress that accompanies them.

Figure 3 indicates that the coping patterns are determined by the presence or absence of three psychological conditions: (1) awareness of serious risks for whichever alternative is chosen (i.e., arousal of conflict), (2) hope of finding a satisfactory alternative, and (3) belief that there is adequate time in which to search and deliberate before a decision is required. Although there may be marked individual differences in preference for one or another of the coping patterns, all five are assumed to be in the repertoire of practically every person when he or she functions as a decision maker. In different circumstances, the same person will use different coping patterns depending on which of the three crucial conditions are present or absent. Only when all three conditions are present will the vigilance coping pattern be the dominant one.

The model represented in Figure 3 appears to be compatible with Herbert Simon's (1967) analysis of the role of motivation and emotion in controlling cognitive behavior. Drawing on concepts formulated by cognitive psychologists, Simon views man as "a basically serial information processor endowed with multiple needs [who usually] behaves adaptively and survives in an environment that presents unpredictable threats and opportunities" (p. 39). He postulates that the arousal of anxiety or any other unpleasant emotion interrupts ongoing activity to enable the person to respond adaptively to urgent needs. But he also emphasizes that when emotional arousal is extremely intense and persistent, it becomes disruptive and produces nonadaptive behavior. The Janis and Mann theory attempts to specify the contrasting conditions that determine whether the stress engendered by decisional conflict will facilitate or interfere with effective search for and processing of relevant information that could be essential for survival.

According to the conflict model, the level or intensity of stress generated by decisional conflict is one of the determinants of the coping pattern that will be adopted. Extremely low stress and extremely intense stress are likely to give rise to defective patterns, whereas intermediate levels are more likely to be associated with vigilant information processing. Whenever a decision maker's dominant coping pattern is unconflicted adherence or unconflicted change, the person is so unaroused by the risks to be expected that he or she resorts to "lazy" ways of making judgments because of lack of motivation to engage in careful information search, appraisal of alternatives, and implementation planning before deciding what action to take. It is these unconflicted coping patterns that are likely to be evoked whenever the issue is regarded as relatively unimportant or as easily managed by resorting to organizational routines, which make for "mindlessness."

In contrast, the vigilant coping pattern is linked with a moderate level of stress, which is optimal in that it motivates the person to think carefully about the decision and to search for relevant information. When the decision maker experiences a very high level of stress from decisional conflict, the dominant coping pattern tends to be either defensive avoidance or hypervigilance, which give rise to gross misperceptions and miscalculations that are avoidable errors.

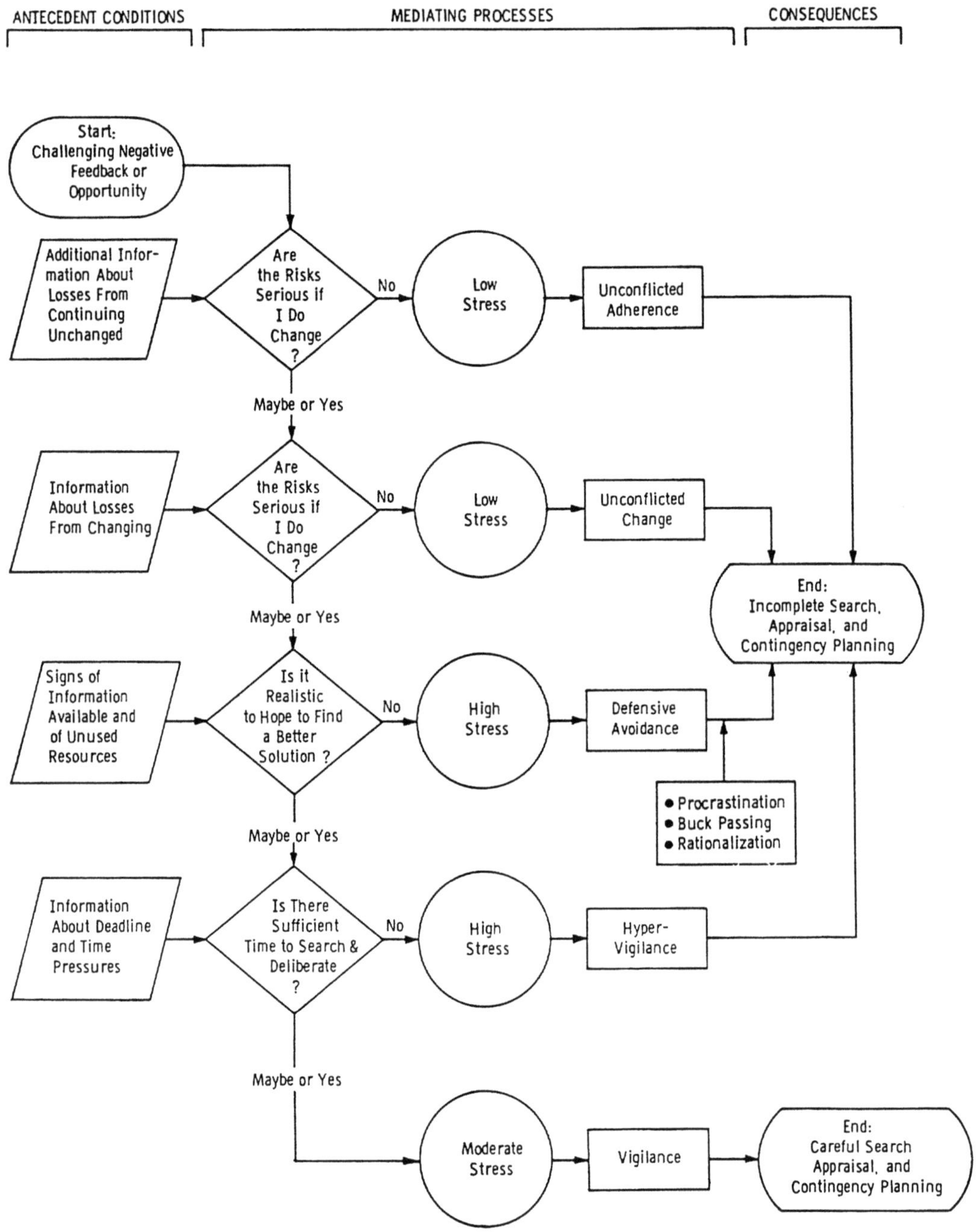

Fig. 3 – The conflict-theory model of decision making. (After Janis & Mann, 1977)

Importance of Feedback from Implementers and Decision Monitors

One final theoretical assumption pertains to the crucial role of informational feedback from lower to upper levels of the organization. Social scientists who study organizations have become increasingly aware of inadequacies of the so-called "classical" hierarchical model of power in policy making. According to that model, policy formulaters at the top choose and instruct policy implementors who, as subordinates, proceed to carry out the directives obediently in a "nonpolitical" way. In a detailed critique of the "classical" model, Nakamura and Smallwood (1980) cite a large number of research investigations indicating that the policy process frequently is not "unidirectional" as this model claims and that implementors are not "faceless automatons" but "flesh and blood" participants who can play a crucial role in shaping the policy process (p. 14). These authors summarize several impressive studies indicating that the process of implementation often does not unfold in a sequential fashion such that policy formulation precedes policy implementation.

The circular rather than linear view of the sequence, as these authors describe it, calls attention to feedback loops between policy formation and implementation. Feedback comes not only from implementers but also from other evaluators, some of whom are likely to be interested parties who are self- appointed monitors. As Lindblom (1980, p. 4) says, "one group's solution becomes another group's problem."

Social scientists who pursue the full implications of the feedback concept find themselves rediscovering the virtues of representative democracy. Of particular interest here are the implications for effective leadership that can be discerned when we consider the change from a linear to a circular conception of policy formation in conjunction with the conflict model.

One obvious implication is that innovative leadership with regard to introducing policy changes is not limited to top-level management. It can sometimes come from the bottom of the organization, from persons who have not been assigned any authority at all to formulate policy. The Janis and Mann conflict model is directly applicable to lower level administrators, provided that their decision making behavior can be observed sufficiently to enable their coping patterns to be diagnosed. For any new policy directive that comes down from the top, implementers at all levels of the organization have the choice as to whether to carry it out as they think is intended, to ignore it, or to modify it in one way or another. The conflict model is applicable to instances where middle level or lower level personnel function as innovative leaders, either by implementing an official policy in a way that changes it or by introducing new practices to fill in policy gaps, which enable them to cope with recurrent problems that had been ignored by the official leaders at the top. The conflict model indicates the conditions under which people at any level in the hierarchy who engage in innovative decision making will use a vigilant problem-solving approach or one of the seat-of-the-pants approaches, and the consequences of these alternative coping patterns, including whether or not they will make constructive use of informational feedback.

EVIDENCE BEARING ON KEY THEORETICAL ASSUMPTIONS

For most of the foregoing theoretical assumptions, excellent reviews are available of the supporting evidence. For example, the findings on the various types of constraints that interfere with reflective problem-solving and that make for reliance upon simple decision rules are summarized in numerous books in several different social science disciplines -- including works by Abelson and Levi (1985), George (1980), Nisbett and Ross (1980), Kahneman, Slovik, and Tversky (1982), Pennings and associates (1985). Supporting evidence for the theoretical assumptions that enter into the conflict theory analysis represented in Figure 3 above is extensively reviewed in Janis and Mann (1977, Chapters 4-12).

In our review of pertinent empirical studies, we have called attention to case studies of organizational as well as individual decisions and to numerous findings from social psychological experiments and correlational field studies that are consistent with pro positions about the determinants and behavioral consequences of vigilant versus non-vigilant coping patterns. From all of these various studies, we concluded that our theoretical analysis is plausible. We have also described a few of our own social psychological experiments that were designed to test prescriptive hypotheses derived from the conflict model (Janis & Mann, 1977, Chapters 13 and 14). These include studies of the effectiveness of a balance sheet procedure, stress inoculation, and a number of other interventions that counteract the beliefs and perceptions specified in Figure 3 as being responsible for defective coping patterns.

Of particular relevance to the present discussion are research monographs by political scientists who have investigated the relationship between level of stress and defective coping patterns in decisional performance of government leaders responsible for foreign policy during major international crises. Holsti and George (1975) conclude that a high level of crisis-induced stress among top-level policy makers produces cognitive rigidity and reduces their time perspective, which lead to over concentration on immediate goals, restricted search for alternatives, and perfunctory evaluation of alternatives and their consequences.

An analysis by Michael Brecher (1980) of Israel's decision making during the 1967 and 1973 crises lends support to Holsti and George's (1975) main conclusions. Brecher analyzed 57 decisions made by the Israeli cabinet and Kitchen cabinet during the 1967 and 1973 crises. Psychological stress was assessed in terms of perceptions of threat, time pressure, and perceived probability of war. In line with the postulates of the Janis and Mann conflict model, Brecher found a curvilinear (inverted U) relationship between stress and group performance with regard to the consideration of policy alternatives. None of the five decisions taken under conditions of low stress showed careful evaluation of alternatives. As stress rose to a moderate level, evaluation of alternatives became more careful. Beyond that, as stress became more intense, there was a decline in care taken in evaluation of alternatives and the overall quality of the deliberations decreased.

There is no need to repeat here the voluminous discussions and summaries of pertinent evidence to be found in the available literature. There is one key assumption, however, that needs to be examined in some detail because it is highly controversial. I am referring to the assumption concerning the relationship between symptoms of defective decision making (listed in Figure 1) and undesirable outcomes. If, as some social scientists maintain, that assumption is empirically unwarranted, it would follow that little or nothing could be gained from basic research on the causes and consequences of the symptoms of defective decision making n or from applied research on procedures designed to reduce or eliminate those symptoms.

Relationship Between Symptoms of Defective Decision Making and Undesirable Outcomes

Does it really make any difference whether the policy makers show few or many symptoms of defective decision making? Some social scientists think that it does not. It seems quite fashionable these days, especially among leading theorists in management studies and political science, to take a very jaundiced view of the prospects of improving policy making in government, business corporations, and other large organizations. For example, William Starbuck (1983, 1985), Professor of Management at New York University, argues in favor of two very pessimistic generalizations about policy-making on the basis of his own observations and his surmises from other research in the fields of organizational behavior and political science, drawing especially upon the critique of analytic problem solving by Charles E. Lindblom (1980).

One of Starbuck's generalizations is that policy makers seldom engage in reflective problem-solving, even though the executives may pay lip service to the value of this approach. Another of Starbuck's generalizations is that the reflective problem-solving approach would not be effective even if it were often used by executives for major policy decisions. He gives several reasons why the quality of decision-making processes would be unrelated to outcomes. For example, organizational problems are much too complicated to solve and the use of vigilant problem-solving gives policy makers the illusion that their solution is a good one, which leads them to develop strong rationalizations that make for more inflexibility in response to policy failures than when they take action without thinking carefully about the consequences of alternatives.

Starbuck cites a number of empirical research studies to support his two generalizations, but it seems to me that the evidence from those studies is weak and inconsistent. From my own review of the available evidence, I think his first generalization is fairly correct -- that is, policy makers seldom use a problem solving approach. But despite the cogency of Starbuck's arguments, I do not think that his second generalization is warranted. My opposing view concerning the outcomes to be expected from reflective problem solving, incidentally, is in line with the position taken by a number of theorists and investigators of policy making processes - for example, (Alexander George 1980), (Paul Lawrence 1985), and (Richard Ned Lebow 1981). The specific hypothesis that I have proposed as an alternative to Starbuck's generalization can be formulated as follows: For consequential decisions that implicate vital interests of the nation or organization, deliberate use of a problem solving approach, with judicious information search and analysis that takes due account of the limited organizational resources, will generally result in fewer miscalculations and therefore better outcomes than any other approach. A more refined, research oriented version of this hypothesis can be formulated in terms of the symptoms of defective decision making: The fewer the steps of reflective problem solving that are carried out and the more incompletely those steps are carried out -- as manifested by symptoms of defective decision making -- the higher the probability of undesirable outcomes from the standpoint of the nation's or organization's goals and values.

It is an important theoretical as well as practical question whether Starbuck's generalization about the lack of any relationship between process and outcome or my opposing hypothesis is closer to the truth most of the time for most policy making.

In looking over the research literature on policy making, I find some indirect evidence that I can cite to support my position, but it is far from conclusive. In order to obtain more direct evidence, Greg Herek collaborated with me on a systematic study of U.S. policy making in international crises, (Herek, Huth, and Janis, in preparation). The main purpose of our study was to determine the extent to which unfavorable outcomes in international crises affecting the United States are related to poor quality decision making by the nation's leaders. In order to investigate the relationship between outcome and quality of decision making processes, we assessed the U.S. governments's decision in each of 19 international crises by making detailed ratings of the presence or absence of each of the seven symptoms of defective decision making that I listed earlier.

One of the main methodological problems of comparative case studies of this type has to do with the selection of cases. If investigators were to pick the cases themselves, they could easily rig the selection -- consciously or unconsciously -- in such a way that the comparative study would yield the results that are expected. In order to avoid any such bias, we deliberately arranged to have the sample of cases selected for investigation independently by three leading social scientists who have studied international crises. None of the three was aware of the purpose of this study at the time he made the ratings. They were asked to select the most important crises in each of five administrations since the end of World War II (Truman, Eisenower, Kennedy, Johnson, and Nixon). One of the crises they selected had to be eliminated from the final sample of twenty crises because insufficient material was available in published sources to carry out the process ratings. Our sample, then, consists of 19 major crises.

In addition to having our sample of crises selected by social science scholars who were unaware of the hypothesis we were investigating, we also arranged for outside experts to select the best scholarly books and articles available to be used for analyzing the decision making processes of U.S. policy makers, and also to rate the outcomes of the crisis decisions made by the U.S. policy makers.

Greg Herek and I developed a set of systematic content analysis procedures to identify the presence or absence of each of the seven symptoms of defective decision making. With the help of Paul Huth, a graduate student in political science at Yale, we used these procedures to carry out a content analysis of the accounts that the three outside experts had selected as the best available sources concerning the way the U.S. policy makers arrived at their crisis decisions.

We found that there was considerable variability in the quality of decision making by United States government leaders in their management of major international crises during the period of almost three decades following the end of World War II. In 42 percent of the crises the policy makers obtained ratings indicating a relatively high quality decision making process (displaying only one or no symptoms of defective decision making). In 21 percent, they obtained medium scores (displaying two to three symptoms). In 37 percent, they obtained ratings indicating relatively low quality decision making (displaying four to seven symptoms). We turn next to the main hypothesis investigated in the Herek, Huth and Janis correlational study, namely that the symptoms of defective decision making, which reflect failures to carry out the essential steps of reflective problem solving, are predictive of unfavorable outcomes.

Outcome ratings were obtained from two political scientists who had done extensive research on international crises. They were kept blind with regard to the process ratings and also with regard to the hypothesis being investigated. Taking account of the possibility that outcome ratings might be influenced by the ideological views of the raters, we deliberately chose two experts who come from opposite ends of the conservative-liberal continuum in their personal views about the cold war (Professor Paul Seabury of the University of California and Professor Richard Ned Lebow of Cornell University). We found a fairly high degree of agreement in their outcome ratings, indicating a satisfactory degree of inter-analyst reliability. We averaged their ratings for our correlational analysis.

Each rater made two separate ratings of outcome. One rating involved the effectiveness of crisis management with regard to U.S. objectives, interests, and influence. The second rating involved the effect on international conflict. Each rater was asked to judge whether there was an immediate increase or decrease in tension, instability, hostility, and/or likelihood of war between the United States and the Soviet Union or China.

The results show a substantial relationship between quality of decision making as manifested by number of symptoms of defective decision making and unfavorable outcome (based on the average ratings of the two experts). The correlation between number of symptoms and failing to satisfy U.S. objectives and interests was .64; the correlation between number of symptoms and increase in international conflict was .57.

Because we took the precaution of obtaining independent blind ratings of the symptoms of defective decision making and blind ratings of outcomes so as to avoid contaminated judgments and other potential sources of artifactual results, it seems unlikely that the significant correlations we found are spurious. We also checked carefully on plausible third factor explanations that might conceivably account for the observed correlations -- such as severity of the crisis (as rated by independent experts). When these potential sources of confounding information were eliminated in partial correlational analyses, we still obtained significant relationships between quality of decision making and outcomes.

Our findings bear out the surmises of those social scientists who have concluded that poor quality procedures used in arriving at a decision give rise to avoidable errors that increase the likelihood of obtaining an unsatisfactory outcome. Stein and Tanter (1980), for example, assert an equivalent proposition in terms of favorable outcomes in their comments about case studies of policy decisions made during international crises. "Other things being equal," they assert, "'good' procedures are more likely to produce 'good' outcomes" (p. 8). Their account of 'good' procedures includes the key components of reflective problem solving: The decision makers identify their options, estimate the likely consequence of the options, consider "trade offs" as they examine expected costs and benefits prior to making their selection of "that option which promises them the greatest gain or the smallest loss" (p. 8). Stein and Tanter add, however that an imperfect positive correlation is to be expected between quality of procedures and outcome: "Those who recommend systematic procedures recognize, of course, that they cannot assure good outcomes in each case" (p. 8).

Implications of the Research Findings

Although the evidence currently available is rather meager, there appears to be sufficient empirical basis for regarding as plausible the major assumptions stated earlier. We have seen that the Herek, Huth and Janis study found that a small but substantial percentage of the policy decisions investigated were of high quality, as manifested by the absence of symptoms indicating failure to carry out the various steps of vigilant problem solving. It seems plausible, therefore, to assume that many, if not all, policy makers are capable of using a vigilant problem solving approach, even though for the majority of policy decisions they may use it only partially or not at all. Another assumption that appears to be plausible in light of the existing empirical evidence is that failures to carry out the steps of reflective problem solving, as manifested by symptoms of defective decision making, are predictive of unfavorable outcomes.

These two assumptions are embodied in Table 1, which lists four main procedural approaches to policy making decisions that are assumed to be used from time to time by most policy makers in national governments, big business corporations, and other large organizations. The table indicates in comparative terms the expected consequences of four types of approach. At one extreme is a sheer seat-of-the-pants approach. This approach entails the least amount of effort. But because it relies uncritically on decision rules, it involves the use of such poor quality procedures that all seven symptoms of defective decision making are likely to be manifested. The probability of avoidable errors resulting from this approach is very high.

Table 1.
CONSEQUENCES OF THE REFLECTIVE PROBLEM SOLVING APPROACH TO DECISIONMAKING AS COMPARED WITH OTHER APPROACHES

Type of approach in policy maker's repertoire	Amount of time and effort required	Quality of decisionmaking: No. of defective symptoms	Probability of avoidable errors
1. Reflective problem solving	Very large	0 - 1	Low
2. Quasi problem solving	Moderate to large	2 - 3	Moderate
3. By and large seat-of-the-pants	Small to moderate	4 - 6	Fairly high
4. Sheer seat-of-the-pants	Very small	7	Very high

At the opposite extreme is the reflective problem solving approach, which involves carrying out fairly well the essential steps shown in Figure 1. This approach has the disadvantage of being more costly in time and effort required on the part of the policy makers and their organizations than any of the other approaches. But it has the advantage of leading to higher quality decisions that are less likely to fail as a result of avoidable errors -- such as miscalculating the relative gains to be expected from the chosen course of action as a result of attending selectively to supportive information, ignoring clear-cut warnings that the chosen policy will be much more costly than the policy makers anticipated, or failing to obtain information available from implementers in advance indicating that there are such insurmountable obstacles to any genuine attempts at implementation that the new policy could not be expected to succeed.

The generalization that fewer *avoidable* errors are to be expected from a reflective problem-solving approach than from any of the other approaches does not overlook the well known reasons why the best laid plans of men and mice can go awry. Some *unavoidable* errors are bound to occur as a result of "unresolvable ignorance" as well as unforeseeable accidents and other chance occurrences commonly referred to as "bad luck." Obviously, unavoidable errors will occur no matter how the policy decision is arrived at. Such errors are always to be expected, and their probability of occurrence will not be affected in any way by the type of decision making approach used. The absolute level of unavoidable errors depe nds upon the level of ignorance of the culture concerning the consequences of alternative courses of action and other factors, some of which are not as yet well understood. But, as general knowledge about consequences increases, some hitherto unavoidable errors will become avoidable errors.

Because of chronic limitations of the human mind in dealing with the complexities of organizational policy issues, some avoidable errors in decision making are to be expected, no matter how carefully the policy makers carry out all the steps of reflective problem solving and no matter how creatively and intelligently they make use of decision rules in combination with critical thinking. But reflective problem solving, which requires the fullest use of available resources, is more likely than any of the other approaches to keep avoidable errors down to a minimum.

Obviously, a vigilant problem solving approach is not an all-or-none affair. Sometimes the decision maker carries out quite completely some of the steps shown in Figure 1, but deals with the tasks required by other steps in a superficial or skimpy manner or not at all. In such instances, the decision maker's procedures are likely to be of intermediate quality, not as good as when all steps are taken, but not as poor as when no steps are taken at all. These intermediate approaches are likely to be manifested by a number of symptoms of defective decision making, although not as many as when none of the steps of vigilant problem solving are carried out.

The table represents the main theoretical assumption that appears to be warranted by the fragmentary evidence currently available. It asserts, in effect, that the fewer the symptoms of defective decision making procedures manifested, the lower the probability of avoidable errors. According to this assertion, we can expect to find for sizable samples of almost all types of genuine i.e., non-ceremonial and non-deceitful) policy decisions a substantial correlation between the number of symptoms of defective decision making procedures and the degree to which the policy decision fails to achieve its objectives as the result of avoidable errors.

A CURRENT RESEARCH PROJECT ON DECISION MAKING DURING INTERNATIONAL CRISES

Insofar as the two key theoretical assumptions appear to be warranted in light of the empirical evidence now at hand, the next step in our inquiry is to examine the major question posed earlier: Under what conditions are policy makers most likely to resort to a sheer seat-of-the-pants approach (or a mixed approach) rather than adopting a reflective problem-solving approach when confronted with the necessity for making a policy decision? By directing our inquiry toward answering this question, we should learn more about major sources of avoidable errors in policy making because it requires us to specify the circumstances that are conducive to engaging in a decision making process dominated by reliance on simple decision rules instead of carrying out the essential steps of problem solving. In the discussion that follows I shall describe my current research project, which attempts to arrive at a preliminary set of answers to this key question by focussing on situational factors that appear to inhibit or promote reflective problem solving during international crises.

Research Goals

At present I am working on comparative case studies involving about thirty major international crises faced by the United States government since the end of World War II. This research includes comparative analyses, using the approach elaborated by Alexander George and Richard Smoke (1974) in their applications of the "focused comparison" method. Much of what I am doing is in the discovery phase of research - formulating hypotheses in terms of testable relationships concerning circumstances that promote errors in policy making. Some of the hypotheses are derived from previously formulated theoretical models, such as those that specify the detrimental effects of psychological stress. I am also trying to discern new causal sequences concerning the detrimental effects of various constraints on policy making that are suggested by case studies and that cannot be fitted to known sequences. (More about that later.) My research also includes carrying out systematic correlational analyses, using fairly large samples in order to test hypotheses that appear to be promising.

As a strong proponent of the experimental method for testing hypotheses about causal relationships, I am keenly aware of the shortcomings of case studies and correlational analyses based on historical narratives and archival documents. But despite all the methodological difficulties, it seems possible to obtain some valuable evidence bearing on when, how, and why national leaders make serious errors (see Oskamp, 1984; Tetlock, 1983).

The main goal of my research project on international conflicts is to develop empirically based theory to elucidate the conditions under which misjudgments and miscalculations are made by national leaders in situations of crisis. Ultimately, this line of research may provide some useful guidelines for policy makers who want to avoid gross errors when making crucial decisions about momentous issues.

Methodological Considerations

A great deal needs to be said about the methodological problems that confront the investigator in attempting to carry out this type of research. At this point, I shall summarize briefly some of the main methodological considerations concerning the values and limitations of case studies that I take into account.

When we are trying to describe a typical source of error and to identify the conditions under which it is most likely to occur, we obviously cannot rely upon anecdotes that seem to conform to a pattern expected on the basis of one or another current hypothesis or theory. Instead of merely picking up salient bits and pieces of case material that suggest a memorable story, we need to examine the available records intensively and systematically to find out what story, if any, is really warranted by the evidence.

In the early discovery phase of research, some genuine advances can be made by carrying out systematic case studies. When studying sources of error in policy making, we should certainly look for whatever explanatory determinants we can find. But whenever we think we have spotted one of them, we should treat it as an untested hypothesis that needs to be checked out as thoroughly as possible. The available sources for intensive case studies include official documents, unofficial observational reports, minutes of meetings, memoirs, and testimony by the executives and aides who participated in the policy-making deliberations. These are the main source materials that I am using in my current research.

In later stages of the research, the hypotheses derived from intensive case studies can be examined by diverse methodological approaches, including content analyses of archival records, questionnaire surveys of participants, laboratory simulations, and field experiments on the effects of changing the antecedent conditions that are expected to affect the quality of decision making procedures (see Janis, 1982b; Tetlock, 1983). Since each method has its unique strengths and weaknesses, converging findings from studies using different methodological approaches provide the most dependable evidence for verifying hypotheses about psychological determinants of defective policy making.

Social scientists have good reasons to be reluctant to take seriously the conclusions derived from case studies. Numerous examples can be cited from the recent social science literature to illustrate gross methodological deficiencies of case studies. Four of the most common deficiencies are (1) lack of standards for evaluating evidence, (2) selective attention to seemingly confirmatory items and neglect of contradictory evidence, (3) failure to examine rival hypotheses, and (4) exploitation of the "certainty of hindsight" tendency described by Fischhoff and Beyth (1975).

But case studies do not need to be that bad. Distinctions should be made among case studies in terms of the degree to which efforts are made to avoid obvious sources of bias and over interpretations of fragmentary evidence. At one extreme is the purely casual impressionistic type of case study in which anecdotes and details are selected by the story teller. At the other extreme is the highly systematic type of case study in which the behavioral scientist conscientiously answers a series of pertinent questions by carefully examining all the available evidence bearing on the questions and takes account of the judgments about the case materials by independent observers. I have attempted to provide an example of the latter type of case study in my analysis of the Watergate coverup (Janis, 1982a).

One of the main features of the systematic type of case study is that the behavioral scientist commits himself or herself to be non-selective in examining and appraising the available records and is answerable to the community of scholars for overlooking or distorting any important piece of evidence. This expectation of critical scrutiny acts back upon the behavioral scientist as he or she carries out and reports the case study.

It can be expected to have the effect of increasing conscientiousness in dealing with evidence, just as it does for experimental scientists and professional historians when they know that fellow experts will examine their methods, will be watching for errors of commission or omission in the evidence cited, and will challenge dubious inferences from the evidence.

I am strongly emphasizing the need to use the best possible methods of case study analysis because it seems to me that relying on such methods can advance the study of errors in decision making, which at present is still in a stage of infancy. Intensive case studies can provide valuable clues to the antecedent conditions -- including leadership practices and standard operating procedures -- that reduce the chances of inadequate utilization of available information and gross miscalculations when executives in a large organization arrive at a major policy decision. As knowledge of the antecedents increases, prescriptive hypotheses for improving the quality of policy making can be formulated and tested using other methods of research, including large-scale correlational studies and controlled experiments.

THE "MODUS OPERANDI" APPROACH

Michael Scriven (1976) has described a methodological approach that I am attempting to apply in my current research because I believe it will prove to be fruitful for investigating policy fiascoes resulting from defective decisions. Scriven has analyzed in detail the logic of the coroner who investigates the probable cause of death of each individual case brought to the morgue, using non-experimental evidence to draw fairly dependable inferences about the causal sequences. He describes the coroner's *modus operandi* approach, which is also used by detectives, historians, and structural engineering troubleshooters, and also by some of the aeronautics specialists who investigate the probable causes of death and injury in air crashes. The coroner and others who use this approach are able to determine probable causes of specific undesirable events with some degree of success, even though the data available to them do not come from controlled experiments or even quasi-experiments. Essentially the same modus operandi approach used by the coroner, it seems to me, can be applied in research designed to determine the probable causes of disastrous decisions by top-level executives that can lead to severe losses or the death of an organization.

Basically, the approach involves formulating all the known causal sequences that might account for the observed outcome -- whether it be a death, a crime, or a policy fiasco -- and then attempting to find out which of them appears to be implicated. This requires careful scrutiny of every shred of evidence that might provide telltale signs indicating the step by step ways in which each of the alternative causes is most likely to operate. Working backwards from the observed effect, the investigator watches for a pattern of indicators of mediating processes that connect antecedent conditions with the observed outcome.

Rules for Drawing Inferences About Probable Causes

Scriven points out that historians and others who use this *modus operandi* approach seldom formulate explicitly the entire list of causal sequences they have in the back of their minds, that they often they work with lists that are quite incomplete. But he argues that even with non-exhaustive lists that include speculative causal chains that have not yet been shown to have the postulated effect, valid inferences about the probable causes can still be made. From his reconstruction of the logic of the modus operandi approach, he offers four rules for drawing a causal inference after the investigator has checked carefully for the presence of indicators of each of the known possible or potential *modus operandi* sequences:

1. If there are indicators for only one *modus operandi* sequence, that is the most probable cause. The more complete the evidence for all the steps in the sequence, the higher the certainty.

2. If there are indicators for more than one causal sequence, the one that is most complete is the most probable cause.

3. If two complete causal sequences are detected, the two are to be regarded as co-causes.

4. If no indicators of any of the known causal sequences are found, none of the sequences in the original (obviously incomplete) list can be regarded as the probable cause.

The fourth rule can undoubtedly be expected to hold sway for initial investigations of puzzling outcomes when we do not know enough as yet to construct rich lists of causal sequences to work with. But it seems to me that the frustration of the case study investigator when he encounters the type of negative results specified in the fourth rule can nevertheless be a goad to creative discovery. The investigator can re-examine all the case material once again, looking for a new causal sequence that is not yet on the list of known ones.

Implications of Scriven's Reconstruction of the *Modus Operandi* Approach

Scriven points out several implications for the social sciences that follow from his reconstruction of the *modus operandi* approach. One is that new investigations can benefit from making explicit the implicit knowledge of causal lists possessed by research workers and practical specialists who have accumulated a great deal of tacit knowledge from a rich backlog of experience. Another is that research workers should be alert to opportunities to observe "telltale signs" that expose distinctive characteristics of the causal sequences they are investigating; whenever they use experimental designs, they should incorporate appropriate "tracers" or "signature arrangements." Another important implication is Scriven's conclusion that "the time has come to change our orientation in the development of social science away from the goal of abstract, quantitative, predictive theories toward scientific, qualitative, explanatory checklists and trouble shooting charts." Like the coroner, the detective, and the engineering troubleshooter, the social scientist in new fields of research has the opportunity, according to Scriven, to draw plausible and useful conclusions from arrays of relatively weak qualitative bits of data by using the *modus operandi* approach in a sophisticated way.

A number of recent methodological developments in the social sciences, summarized by Cook, Leviton, and Shadish (1985) tend to support the premises of Scriven's *modus operandi* approach along with other arguments by those who advocate more extensive use of qualitative methods of causal analysis:

> ...This advocacy has been legitimized by Campbell's (1975) apparent revocation of the hard line against one shot case studies taken in (Campbell and Stanley 1966), for Campbell now argues that case studies can under some conditions provide strong causal inferences. His argument is based on an analogy with the degrees of freedom concept in statistics. He contends that case studies usually generate multiple sources of cause-relevant evidence that, when linked to background knowledge from experience, prior research, or grounded theory, can permit all the identified threats to internal validity to be ruled out.
>
> ...An epistemological framework much like Campbell's buttresses Scriven's (1976) thinking about the relevance of *modus operandi* methods in evaluation.
>
> ...Guba and Lincoln (1981) use investigative journalists rather than detectives as a stimulus to thinking about how causal inferences are made. They invoke the journalists' use of networks of contacts with known credibility, their persistent questioning of sources, the requirement that all information be validated across more than one source, and the further requirement that it pass the critical scrutiny of senior (and experienced) editorial staff.
>
> ...In the monitoring context, hardly anyone would dispute that qualitative methods are useful for generating hypotheses about causal agents, for outlining possible explanatory processes, and for uncovering more obvious project effects.
>
> ...While the arguments in favor of case study methods complementing quantitative methods are compelling, the arguments for qualitative methods substituting for quantitative ones are more questionable (see Reichardt and Cook, 1979, 1981). This is not to deny, though, that qualitative methods will normally facilitate inferences about causal agents by uncovering some possible unanticipated effects and by making some interpretations of the cause seem implausible. Indeed, in a small minority of cases only a single cause will remain in contention and strong causal inference will be warranted (Cook, Leviton & Shadish, 1985, pp. 736-737).

Potential Application for Failure Analysis

Should the approach recommended by Scriven be taken seriously in the fields of research bearing on failures of information systems and other sources of error in policy making? I believe that it should be, particularly when working on problems that cannot be investigated by means of controlled experiments.

With regard to the central problem with which I have been concerned -- sources of error in crisis management -- we already know a fair amount about a few of the causal sequences, and these should be formulated much more explicitly than has been done so far. One well-known sequence involves the unproductive bickering and lack of attention to pertinent information that occurs when a decision making group is composed of hostile factions within an organization who are engaged in a power struggle. A second involves the collective monologues that fail to make use of the cognitive resources of the group when it is composed of members with non-overlapping expertise who do not use the same kinds of vocabulary and rhetoric and therefore cannot communicate adequately with one another. Other causal sequences involve over-dependence of the members of the group on the leader. The most familiar one occurs when a powerful, autocratic leader induces conformity to his or her idiosyncratic position, stifling all dissent and skepticism, which results in failure to take account of warnings about dire consequences of the chosen course of action because anyone who does not completely go along with the leader's foregone conclusions expects to be fired or subjected to other punitive recriminations.

The few sequences that are well known, like the ones just mentioned, may occur only rarely, especially since they are frequently described as avoidable pitfalls. They are often featured in textbooks on effective management and in cautionary tales that are recounted among executives as part of the local culture of many large organizations.

In line with Scriven's recommendations, I believe that social scientists should make a concerted effort to specify as fully as possible the causal sequences that are the most frequent sources of error in policy making. We need to elaborate, it seems to me, the faint glimmerings we have obtained so far concerning such sources of error as reliance on simple rules-of-thumb and stress coping patterns that interfere with effective problem solving. Intensive case studies are needed initially, incorporating the *modus operandi* approach in order to specify step-by-step sequences that link antecedent conditions (such as structural features of the organization, design and composition of the decision ma(king-group, and norms of the organization affecting leadership practices) with the quality of decision making procedures used in arriving at strategic decisions. A series of such case studies coul help us to formulate descriptive hypotheses concerning the conditions under which each source of error is most likely and least likely to occur. Insofar as subsequent research findings are confirmatory, they could be used to formulate prescriptive hypotheses, which specify when, where and how to reduce the likelihood of each of the various types of error. In the later phases of the research designed to test both the descriptive and prescriptive hypotheses, as I have already indicated, the full range of behavioral research methods could be brought to bear, including comparative case studies using blind ratings of the variables.

CAUSAL SEQUENCES OBSERVED IN STUDIES OF FIASCOES

In order to illustrate the type of findings that emerge when the *modus operandi* approach is used, I shall describe briefly two causal sequences that I observed in my research on fiascoes. These two sequences are less well known as sources of error in group policy decisions than the ones I have just mentioned (involving internecine warfare, nonoverlapping expertise, and conformity out of fear of recriminations induced by an autocratic leader).

The "Groupthink" Trap

In my prior research on fiascoes resulting from foreign policy decisions made by top-level governmental advisory groups (Janis, 1972, 1982a), I call attention to a "groupthink" tendency among moderately or highly cohesive groups. When this tendency is dominant, the members use their collective cognitive resources to develop rationalizations in line with shared illusions about the invulnerability of their organization or nation and display other symptoms of concurrence seeking (referred to as "the groupthink syndrome").

A number of historic fiascoes appear to have been products of defective policy planning on the part of misguided government leaders who obtained social support from their ingroups of advisors. My analysis of case studies of historic fiascoes suggests that the following groups of policy advisors were dominated by "groupthink": (1) Neville Chamberlain's inner circle, whose members supported the policy of appeasement of Hitler during 1937 and 1938, despite repeated warnings and events indicating that it would have adverse consequences; (2) Admiral Kimmel's ingroup of naval commanders whose members failed to respond to warnings in the fall of 1941 that Pearl Harbor was in danger of being attacked by Japanese planes; (3) President H. S. Truman's advisory group, whose members supported the decision to escalate the Korean war in 1949 despite firm warnings by the Chinese Communist government that

United States entry into North Korea would be met with armed resistance from the Chinese; (4) President J. F. Kennedy's advisory group, whose members supported the decision to launch the Bay of Pigs invasion of Cuba in May 1961 despite the availability of information indicating that it would be an unsuccessful venture and would damage the United States' relations with other countries; (5) President L. B. Johnson's "Tuesday luncheon group," whose members supported the decision to escalate the war in Vietnam during the mid- 1960's despite intelligence reports and other information indicating that his course of action would not defeat the Viet Cong or the North Vietnamese and would entail unfavorable political consequences within the United States.

In all these "groupthink" dominated groups, there were strong pressures toward uniformity, which inclined the members to avoid raising controversial issues, questioning weak arguments, or calling a halt to softheaded thinking. Other social psychologists (Green & Conolley, 1974; Raven, 1974; Raven and Rubin, 1977) have noted similar symptoms of groupthink in the way Nixon and his inner circle handled the Watergate coverup during the early 1970's. Drawing on their work and my own detialed analysses of the unedited Nixon apes, I have carried out an intensive case study of the Watergate coverup and have used it to elaborate on the theory of groupthink (Janis, 1982a).

In assessing whether any one of the above seven symptoms is absent, it is necessary to examine the available evidence in the decision-making process carefully in order to detect instances whee the apparent absence of the symptom is continual rather than a genuine manifestation of vigilant problem solving. Most high-leel policy makers know about the criteria for sound decision making, and sometimes deliverately contrive activities to make it look as though they had carried out the essential tasks of information search, appraisal, and planning.

Taking account of prior research findings on group dynamics, I have formulated a set of hypotheses concerning the conditions that foster concurrence seeking on the basis of inferences from my case studies of groupthink and from comparative case studies of well worked out decisions made by similar groups whose members made realistic appraisals of the consequences. One of these counterpoint case studies is of the main decision made by the Kennedy administration during the Cuban missile crisis in October 1962. Another deals with the hardheaded way that planning committees in the Truman administration evolved the Marshall Plan in 1948.

These two case studies indicate that policy making groups do not always suffer the adverse consequences of group processes, that the quality of the group's decision making activities depend upon current conditions that influence the group atmosphere, including leadership practices.

My main hypotheses concerning the causes and consequences of concurrence-seeking are summarized in Figure 4. This chart is a condensed representation of the causal sequence that links defective decision making with a set of antecedent conditions (involving a cohesive group of decision makers in an organization characterized by certain types of structural faults when the members are exposed to certain types of provocative situations). It can be used in subsequent case studies of policy fiascos to carry out the type of *modus operandi* analysis described by Scriven. In order to determine whether the groupthink hypothesis can help to account for any given policy fiasco, the investigators must examine all the available evidence to determine whether the entire pattern represented in Figure 4 is present. It does not suffice merely to see if a few of the eight telltale symptoms of groupthink can be detected. Rather, it is necessary to find out if practically all the symptoms were manifested and to see if the antecedent conditions and the expected immediate uences Dthe symptoms of Sefective decision Yaking) Pre also present.

In line with Scriven's (1976) recommendations, my current research on international crises

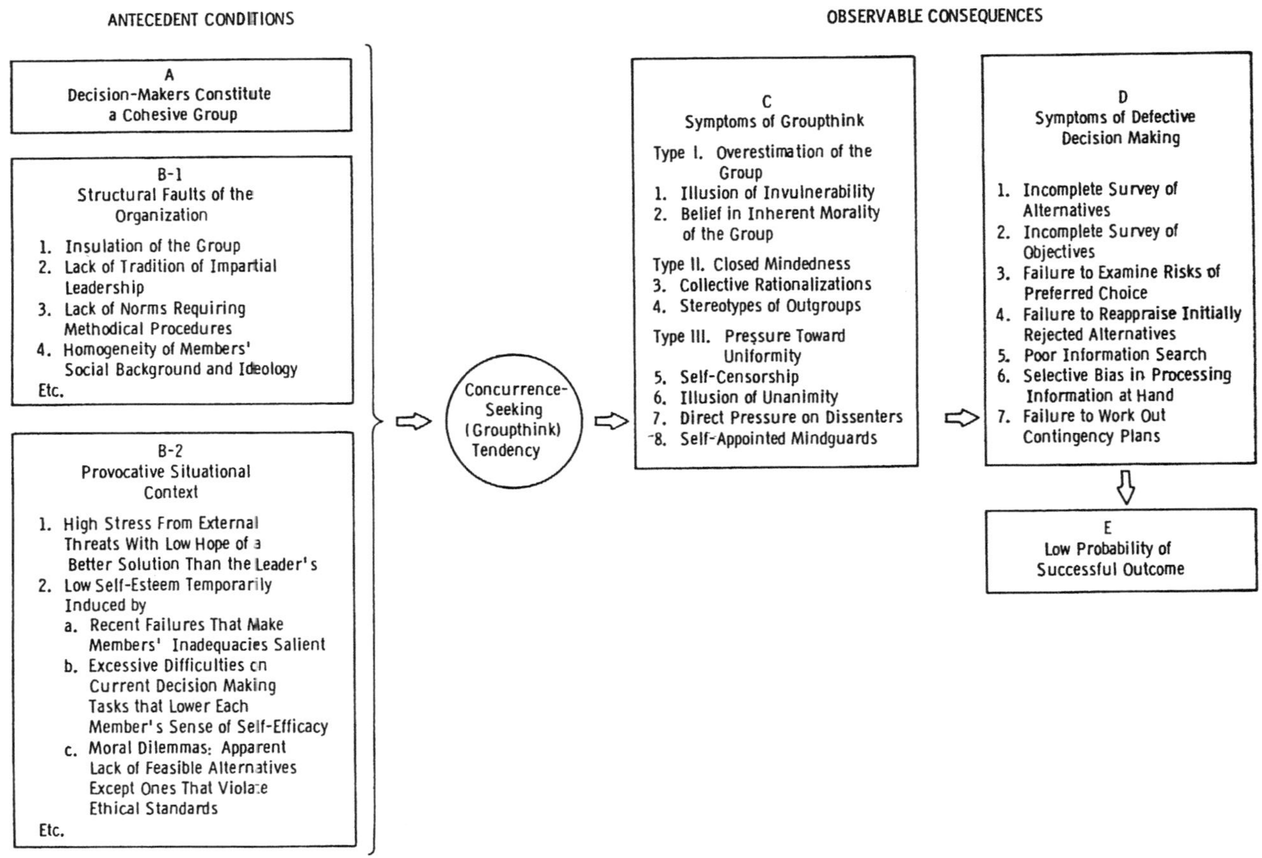

Fig. 4 – Theoretical analysis of groupthink based on comparisons of high-quality and low-quality decisions by policy-making groups.
(From Groupthink: Psychological Studies of Policy Decisions and Fiascoes by I.L. Janis. Reproduced with permission of Houghton Mifflin.)

attempts to specify as fully as possible each of the causal sequences that are major sources of error. In this research I am trying to focus on whatever faint clues can be discerned concerning such sources of error as bureaucratic politics, misperceptions arising from cognitive consistency strivings, misuse of analogies, over-simplified decision rules, faulty heuristics used to estimate probable outcomes, and stereotypes of opponents or other out-groups, as well as groupthink and the various defective coping patterns that interfere with effective problem solving.

For purposes of investigating sources of error in policy making, I have begun to work out analytic flow charts (like the one in Figure 4) showing the antecedent conditions, the telltale signs of the mediating processes, and the specific effects on quality of decision making of each of the known sources of error. Once the sequence charts are available for a variety of cognitive, affective, and affiliative sources of error, it should be possible for policy analysts as well as social scientists to use them whenever they are analyzing any major policy failure. Such charts might help investigators and practitioners to extract more valid "lessons of history" than can be inferred on the basis of the impressionistic analyses now being used. In some instances, as I mentioned earlier, more than one causal sequence may be found to play a contributory role as sources of error. In other instances, we can expect to find that the list of causal sequences is incomplete because none of them fit the case at hand, which can lead to the discovery of causal sequences that have not yet been explicated.

The "Can Do" Rule and the "Unsqueaky Wheel" Trap

My second example is a causal sequence involving serious defects in implementation planning, which I had not encountered in any of my prior research or in the social science literature until I carried out a case study of the ill-fated decision by President Carter and his advisors in 1980 to use military force to rescue the American hostages in Iran. The outcome has been called a fiasco comparable to the Bay of Pigs.

Government spokesmen claimed that it was all a matter of unforeseeable accidents and bad luck that helicopter failures forced the military commander to abort the entire mission on the desert sands of Iran hundreds of miles from Teheran. Some informed critics, however, say that the plan was ill conceived.

In this case, a "can do" rule appears to have dominated the thinking and planning of the key personnel in the White House and in the military establishment. This affective rule appears to be used by decision makers who pride themselves upon being capable and courageous leaders. It is used to overcome deterrents in circumstances that evoke apprehensiveness about the considerable chances of failure of a desperate course of action that they feel obliged to carry out (Janis, 1986). The "can do!" rule goes like this: "Don't be chicken; concentrate all of your available decision making resources to find ways of overcoming the deterrents." When this rule is used in a seat- of-the-pants approach, the decision makers are likely to focus their attention almost entirely on the enormous obstacles and risks that make them apprehensive about carrying out the course of action that is deemed necessary. From the available accounts, I surmise that the most obvious risks of the Iranian rescue mission that evoked a "can do" stance to overcome apprehensiveness were in the later phases -- the enormous chances of the invading U.S. forces being detected and attacked when they approached Teheran, of the hostages being killed before their would-be rescuers could get to them, and of U.S. aircraft being shot down as they attempted to depart from Teheran.

One of my major sources of information was the memoirs of Zbigniew Brzezinski (1983), the President's National Security Advisor, who for many months had been urging President Carter to restore America's honor by launching a military attack rather than continuing to try to negotiate. Brzenzinski's claim that the rescue mission was thoroughly worked out is contradicted to some extent by his own detailed account of how it failed. From what he says, it is apparent that the major risks of the first crucial phase of the rescue plan -- the rendezvous and refueling of he U.S. aircraft in the Iranian desert -- had hardly been examined at all by

President Carter and his key advisors. For example, Brzezinski acknowledge that "my concern over the possible failure of the mission did not pertain to something like what finally happened -- its early abortion -- but rather to...its execution in Teheran" (p. 488).

Brzenzinski's memoirs make it clear that the top policymakers in the White House group had not envisaged the possibility that the mission might have to be aborted either because several of the helicopters might malfunction or that the mission might be compromised as a result of being observed by Iranians in passing vehicles on the nearby road. Apparently, certain of the crucial implementation plans were not examined carefully enough by Brzezinski and others in the president's advisory group. A high level military review group set up by the Joint Chiefs of Staff to investigate what went wrong concluded that the flaws could have been corrected if the small secret circle of planners had asked for their plan to be reviewed by qualified experts within the government (Shaw, 1980). For example, arrangements could have been made to carry out a weather reconnaissance flight in advance of the helicopter flights, and more helicopters could have been added to the mission. The same report also faults the planners for having failed to arrange to have the desert rendezvous "fully rehearsed" in advance and also for having neglected to set up an identifiable on-scene command post with adequate communication and control in the event of unanticipated difficulties (Middleton, 1981, p. 112).

If these additional implementation and contingency plans had been worked out, the chances of failures requiring that the mission be aborted, as well as the chances of a disaster, such as the accidental crash of U.S. aircraft in which eight men were killed, might have been considerably reduced. Evidently, as the old saying goes, it is the squeaky wheels that get the oil; any unsqueaky wheel is neglected. When grappling with a complicated multi-stage plan, p olicymakers who rely upon the can do" rule are likely to focus their implementation and contingency planning on the steps that pose the almost insurmountable risks. There appears to be a contrast effect whereby the least risky steps tend to be labeled as safe and easy to handle -- "that's not where our troubles will be."

In summary, my main hypothesis is this: Even if decision makers are quite vigilant in their approach to the problems of working out a feasible plan that takes account of the obvious risks, there is a strong tendency, if they adopt a "can do" orientation, to focus their information search, appraisals, and implementation planning entirely upon the steps that are known from the outset to be fraught with danger, the "squeaky wheels." Consequently, they tend to overlook the more subtle or hidden risks in any relatively nondangerous or routine step in the plan. This is what I refer to as "the unsqueaky wheel trap."

Perhaps the tendency to neglect the unsqueaky wheel could be counteracted if every policy making group adopts a new procedure that requires one or more of its members to take on a monitoring role toward the end of the group's deliberations to make sure that none of the important risks are being overlooked. In order to fulfill this role, they would have to look carefully into all the risks entailed by each step of the plan, making the assumption that every wheel is squeaky and could lead to failure if one complacently assumes that it will be trouble free. Perhaps in this way the "can do" rule could sometimes be counteracted and replaced by a more comprehensive problem- solving approach whereby the policy makers do not ignore the potential dangers from the apparently unsqueaky wheels, even though they originally concentrated on the squeaky ones. In some instances, the monitors might raise crucial questions that make the policy making group realize that their entire "can do" type of plan is too dangerous to be acceptable.

KEY QUESTIONS FOR BASIC AND APPLIED RESEARCH

Taking account of my earlier comments about using Scriven's *modus operandi* approach, I suggest that one of the directions for future research is that of seeking answers to the following general question: What are the steps in the causal sequence for each of the known sources of failure in policy making? Adequate answers require specifying the telltale signs for identifying the mediating processes that connect the observable antecedent conditions with the observable indicators of the policy makers, failure to work out adequate solutions to the problems that confront them. Our knowledge of effective decision making will be advanced, it seems to me, if we encourage (and carry out) this type of inquiry for each of the seven symptoms of defective decision making that I described at the beginning of this paper.

Of course, as I have already indicated, certain psychological sources of error increase the chances of displaying all seven of the symptoms. Examples of such causal sequences from my own research were presented in my discussion of patterns of coping with the stresses of decisional conflict (Figure 3) and of groupthink (Figure 4). Obviously, much more work needs to be done on these general sources of error as well as on the more specific sources pertaining to each of the seven criteria. We might then be able to develop somewhat more comprehensive middle-range theories that take account of the interacting antecedent conditions.

The development and testing of such theories should enable us to formulate more sophisticated -- and I hope more valid -- prescriptive recommendations for preventing error in policy making (which, in turn, will need to be tested to find out how valid they actually are). For example, it follows from the analysis of stress-coping patterns summarized in Figure 3 that in order to facilitate reflective problem solving among middle and lower level executives who have to make decisions about when, where, and how to implement a new major policy decision, the top-level policy makers need to accompany the announcement of each such decision with supplementary communications containing three main messages.

One such message should attempt to counteract complacent inertia by highlighting instances of continuing business as usual in the face of whatever new threats or opportunities constitute the challenge that induced the policy makers to work out the new policy decision. A second message should be designed to counteract procrastination, passing the buck, and other defensive reactions by conveying a sense of optimistic hope and self-confidence in the implementers. The message could contain hints about how to use the organization's resources as well as their own resources to work out good solutions to whatever implementation dilemmas they may encounter. Above all, it could give cogent arguments to refute any specific reasons that the implementers might have for feeling pessimistic about being able to live with the new policy. Eliciting participation of the implementers during the planning stages and taking account of information about potential implementation problems obtained from them (by pre- testing a new policy before putting it in operation) contributes to their readiness to accept such arguments (and also may eliminate some of the worst problems of implementation that otherwise would have a demoralizing effect).

The third message should try to prevent an undue sense of time pressure that can evoke panic-like feelings, impulsiveness, and unproductive vacillation. Often this can be done by emphasizing in a reassuring way what can be accomplished before deadlines are at hand. The message could also help to counteract hypervigilance by calling attention to the flexibility of those deadlines that are likely to be negotiable and extendable if information obtained from monitoring the decision reveals that many implementers feel there is not sufficient time to arrive at sound implementation decisions.

Ultimately it should be possible to arrive at sound prescriptions for counteracting all the various psychological and sociological sources of error mentioned in this paper -- including not only the defective patterns of coping with the stresses of decisional conflict, but also internecine warfare, bureaucratic politics, conformity out of fear of recrimination, group-think, and reliance on over-simplified decision rules to deal with cognitive, affective, and affiliative constraints that give rise to defective decision making.

From my studies of decision making during international crises, I believe that it is possible to learn a great deal about the conditions under which cognitive, affective, and affiliative constraints lead to over-reliance on simple decision rules that curtail reflective problem-solving. This type of research (using comparative case studies and correlational analysis), it seems to me, is starting to contribute answers to a fundamental theoretical question: Under what conditions are policy makers most likely to adopt a reflective problem-solving approach rather than resorting to a seat-of-the- pants approach when confronted with a crisis requiring a major policy decision? The answers that are emerging -- some of which I have alluded to in this paper -- are likely to have some value in suggesting ways of preventing able errors in policy making, especially during international crises.

ACKNOWLEDGMENTS

Much of this paper is from a forthcoming book that I am now preparing for publication. In the sections on my current research, the studies I describe were supported by Grant No. SES-821351 from the National Science Foundation and by a grant from the Carnegie Corporation of New York. The statements made the view expressed are solely the responsibility of the author.

FOOTNOTE

1. Janis and Mann (1977, pp. 10-14) extracted the following seven criteria of sound decision making from the extensive research literature on effective decision making by individuals and by groups of policy makers in business corporations, public welfare organization, and government. To the best of his or her ability, the decision maker performs the following activities:

a. Surveys a wide range of objectives to be fulfilled, taking account of the multiplicity of values implicated by the choice
b. Canvasses a wide range of alternative courses of action
c. Intensively searches for new information relevant to evaluating the alternatives
d. Correctly assimilates and takes account of new information or expert judgments to which he or she is exposed, even when the information or judgment does not support the course of action initially preferred
e. Reconsiders the positive and negative consequences of alternatives originally regarded as unacceptable, before making a final choice
f. Carefully examines the costs and risks of negative consequences, as well as the positive consequences, that could flow from the alternative that is preferred
g. Makes detailed provisions for implementing the chosen course of action and monitoring how well it is working out, along with elaborations of contingency plans that might be required if various known risks were to materialize.

In a series of case studies of highly consequential policy decisions made by the U.S. government during major crises since the end of World War II, which I have recently analyzed, I found that it is usually relatively easy to rate each decision in terms of the seven criteria. I compared my ratings with those made independently by my colleagues and found fairly high agreement, indicating an adequate degree of inter-analyst reliability (Herek, Janis, and Huth 1986; Janis, 1986). Our disagreements usually had to do with judgments as to whether the group of policy makers who made a given decision during a crisis should be given a moderate rather than a high score on intensiveness of information search or on one of the other criteria. But in most of these ambiguous cases, we found no disagreement in judging whether or not the rating should be low. In other words, even for the cases most difficult to evaluate with regard to how high the rating of quality of decision making should be, it was usually possible to judge reliably and with high certainty whether or not the score should be so low as to be regarded as a symptom of defective decision making.

Accordingly, I propose that ratings of quality of decision making should focus on the lower end of the scale for each criterion. Each rating, then, specifies whether or not the quality is so low that the policy makers can be described as manifesting the given symptom of defective decision making. In short, the seven symptoms of defective decision making represent the lower end of the continuum for the seven criteria of sound decision making.

In order to assess the frequency and intensity of the symptoms displayed by members of a policy planning group for any given decision, it is necessary to examine the available records of the group's formal meetings and informal discussions. The best records, of course, would be verbatim recordings (such as the White House tapes made by Presidents Kennedy, Johnson, and Nixon). In the absence of such recordings, it is necessary to rely on documents, such as minutes of the meetings and secondary source materials that describe the policy makers' discussions.

The following breakdown is intended to specify some of the main indicators that can be used in order to assess each of the seven symptoms of defective decision making, when an analyst has access to either verbatim recordings of the meetings or descriptions of the meetings by participants or by secondary sources.

A. Incomplete survey of alternatives:
   a. Amount of time devoted to formulating alternatives.
   b. Number of alternatives mentioned by any of the participants.
   c. Number of alternatives discussed, i.e., commented upon by one or more other participants after initially being mentioned.
   d. Number of requests (or demands) for making a more complete survey of alternatives (i.e., comments to the effect that "we should look into other possible courses of action").

B. Incomplete survey of objectives:
   a. Amount of time devoted to discussion of objectives.
   b. Number of objectives mentioned by any of the participants.
   c. Number of objectives discussed.
   d. Number of requests (or demands) for making a more complete survey of objectives (i.e., comments to the effect that "we should consider what else we want to achieve").

C. Poor information search.
   a. Number of requests for information or advice from others within the group (e.g., "Does anyone here have any special knowledge or experience to inform us about what is likely to happen when you try to carry out a policy like this?)
   b. Number of requests for information or advice from outside the group (e.g., "Could you find out what those experts recommend?")
   c. Number of references to deliberate information search activity (for example, "Yesterday I asked an expert [or read] about this problem we have been discussing").
   d. Number of information sources mentioned (for example, "I read in the New York Times that ..."; "my assistant informed me that ...").

D. Failure to take account of (or selective bias in processing) information or expert judgments already received.

This category pertains to instances of one or more members of the policy making group ignoring, discounting, or refuting information or a persuasive message that opposes the policy referred to by the group. Selective bias can best be detected by special probing interventions, which can sometimes be used in field experiments, to see how members of a group differentially respond to and assimilate information from communications that support the group's preferred course of action as compared with communications that do not. In the absence of probing interventions, this category must be assessed by indirect evidence of closed mindedness, which require careful (and unbiased) judgments by the analyst. The main indirect indicators that could be picked up in an analysis of records of meetings include the following:

   a. Number of requests to terminate exposure to opposing communications (for example,

"You don't need to read the rest of that letter to us; we know what those people have to say.")

b. Number of times opposing communications under discussion are refuted by counter-arguments.

c. Number of times opposing communications under discussion are rephrased in a distorted and biased way (e.g. making them sound ridiculous or too trivial to bother with).
d. Number of times opposing communications under discussion are said to be wrong, unimportant, irrelevant, or for some other reason not worthy of consideration.
e. Number of episodes reported by observers revealing selective bias (not covered by the above sub-categories).

E. Failure to re examine initially rejected alternatives in light of new information before making final choice.

The following indicators can be rated for the period after the various unchosen alternatives were either tentatively or definitively rejected. Some of the same bits of evidence may also be scored under one of the items listed for category 3.

a. Amount of time devoted to discussing positive consequences (or relative advantages) of formerly rejected alternatives.
b. Number mentioned of different positive consequences (or relative advantages) of formerly rejected alternatives.
c. Number discussed ...
d. Number of requests (or demands) for more discussion of possible advantages (or pros and cons) of formerly rejected alternatives.

F. Failure to examine and weigh expected consequences -- especially the known unfavorable consequences and the potential risks of the preferred policy.

The following items could be rated for the meetings held both before and after a consensus (or straw vote) indicates which one of the alternatives under discussion is preferred.

a. Amount of time devoted to discussion (i) pro and (ii) con consequences of each alternative.
b. Number of different (or partially different, i.e., not completely overlapping) consequences mentioned -- (i) pro and (ii) con.
c. Number of different consequences discussed -- (i) pro and (ii) con.
d. Number of requests (or demands) for more discussion of consequences (for example, "We should talk more about what could go wrong with that plan") -- (i) pro and (ii) con.

G. Failure to work out implementation, monitoring and contingency plans. Some well elaborated implementation plans are also monitoring andcontingency plans. In some instances, a given plan can be rated as pertaining to all three in rating the following indicators:

a. Amounts of time devoted to discussing (i) implementation (ii) monitoring and (iii) contingency plans.
b. Number mentioned of different (or partially different, i.e., not completely overlapping) problems involving (i) implementation (ii) monitoring or (iii) contingencies that need to be taken into account.
c. Number discussed ...
d. Number of requests (or demands) for more discussion of (i) implementation, (ii) monitoring, and (iii) contingencies to be aken into account.

REFERENCES

Abelson, R. P. and A. Levi. (1985). Decision making and decision theory. In G. Lindzey and E. Aronson. *The Handbook of Social Psychology, Vol. I* . New York: Random House, pp. 231 - 310.

Brecher, M. (1980). *Decisions in Crisis: Israel 1967 and 1973.* Berkeley: University of California Press.

Brzezinski, Z. (1983). *Power and Principle: Memoirs of the National Security Advisor, 1977 - 1981.* New York: Farrar, Straus, Giroux.

Campbell, D. T. (1975). "Degrees of freedom" and the case study. *Comparative Political Sudies,* 8, 178 - 193.

Campbell, D. T., and J. C. Stanley. (1966). *Experimental and Quasi-experimental Designs for Research.* Chicago: Rand McNally.

Cook, T. D., L. C. Leviton, and W. R. ShadishJr. (1985). Program evaluation. In Lindzey, G. and Aronson, E. *The Handbook of Social Psychology, Vol. I.* New York: Random House. pp 699 - 777.

Cooper, C. L. (1970). *The Lost Crusade.* New York: Dodd, Mead.

Fischhoff, B., and R. Beyth. (1975). I knew it would happen: Remembered probabilities of once future things. *Organization Behavior and Human Performance,* 13, 1 - 16.

George, A. (1974). Adaptation to stress in political decision making: The individual, small group, and organizational contexts. In G. V. Coelho, D. A. Hamburg, and J. E. Adams (Eds.), *Coping and Adaptation.* New York: Basic Books.

George, A. (1980). *Presidential Decision Making in Foreign Policy: The Efective Use of Information and Advice.* Boulder, Colorado: Westview.

George, A., and Smoke, R. (1974). *Deterrence in American Foreign Policy: Theory and Practice.* New York: Columbia University Press.

Green, D., and Conolley, E. (1974). "Groupthink" and Watergate. Paper presented at the Annual Meeting of the American Psychological Association.

Herek, G., P. Huth, P., and I. L. Janis. (In preparation). Decision making during international crises: Is quality of process related to outcome?

Holsti, 0. R. and George, A. L. (1975). The effects of stress on the performance of foreign policy makers. *Political Science Annual,* 6, 255-319.

Janis, I. L. (1982a). *Groupthink: Psychological Sudies of Policy Decisions and FIascoes.* (Revised and enlarged edition of *Victims of Groupthink,* 1972). Boston: Houghton Mifflin.

Janis, I. L. (1982b). Counteracting the adverse effects of concurrence seeking in policy planning groups: Theory and research perspectives. In H. Brandstatter, J. Davis, and G. Stocker Kreichgauer (Eds.) *Group Decision Making.* New York: Academic Press.

Janis, I. L. (1985). Sources of error in strategic decision making. In J. M. Pennings (Ed.). *Organizational Strategy and Change.* San Francisco: Jossey Bass. pp. 157 - 197.

Janis, I. L. (1986). Problems of international crisis management in the nuclear age. *Journal of Social Issues.*

Janis, I. L., and Mann, L. (1977). *Decision Making: A Psychological Analysis of Conflict, Choice and Commitment.* New York: Free Press.

Kahneman, D., Slovic, P., and Tversky, A. (Eds). (1982). *Judgment Under Uncertainty: Heuristics and Biases.* Cambridge: Cambridge University Press.

Lawrence, P. R. (1985). In defense of a rational approach to change. In J. M. Pennings (Ed). *Organizational Strategy and Change.* San Francisco: Jossey Bass. pp. 373 - 382.

Lebow, R. N. (1981). *Between Peace and War: The Nature of International Crisis.* Baltimore: Johns Hopkins University Press.

Lindblom, C. E. (1980). *The Policy Making Process (Second edition).* Englewood Cliffs, NJ: Prentice Hall.

Middleton, D. (1981). Going the military route. *New York Times Magazine,* May 17.

Nakamura, R. T. and Smallwood, F. (1980). *The Politics of Policy Implementation.* New York: St. Martin's.

Nisbett, R.E., and Ross, L. (1980). *Human Inference: Strategies and Shortcomings of Social Judgment.* Englewood Cliffs, NJ: Prentice Hall.

Oskamp, S. (1984). *Applied Social Psychology.* Englewood Cliffs, NJ: Prentice Hall.

Pennings, J. M. (Ed.). (1985). *Organizational Strategy and Change.* San Francisco: Jossey Bass.

Quinn, J. B. (1980). *Strategies for Change: Logical Incrementalism.* Homewood, IL: Richard D. Irwin.

Raven, B. H. (1974). The Nixon group. *Journal of Social Issues,* 30, 297 - 320.

Raven, B. H., and Rubin, J. Z. (1977). *Social Psychology: People in Groups.* New York: John Wiley.

Reichardt, C. S., and Cook, T. D. (1979). Beyond qualitative versus quantitative methods. In T. D. Cook and C. S. Reichardt (Eds). *Qualitative and Quantitative Methods in Evaluation.* Beverly Hills, CA: Sage.

Scriven, M. (1976). Maximizing the power of causal investigations: The modus operandi method. In G. V. Glass (Ed.). *Evaluation Studies.* Review Annual Volume I. Beverly Hills, CA: Sage.

Shaw, G. (1980). Secrecy, ill iuck cited in failure of Iran rescue raid. *Los Angeles Times,* August 24, pp. 1, 18.

Simon, H. A. (1967). Motivational and emotional controls of cognition. *Psychological Review,* 74, 29-39.

Simon, H. A. (1976). *Administrative Behavior: A Study of Decision-Making Processes in Administrative Organization.* (2nd ed., New York: MacMillan, 1957); 3rd ed., New York: Free Press.

Starbuck, W. H. (1983). Organizations as action generators. *American Sociological Review*, 48, 91 - 102.

Starbuck, W. H. (1985). Acting first and thinking later: Theory versus reality in strategic change. In J. M. Pennings (Ed.) Organizational Strategy and Change. San Francisco: Jossey Bass. pp. 336 - 372.

Stein, J. G., and Tanter, R. (1980). *Rational decision making: Israel's security choices, 1967*. Columbus, OH: Ohio State University Press.

Taylor, S. E. (1982). The availability bias in social perception and interaction. In D. Kahneman, P. Slovic, and A. Tversky (Eds.). *Judgment Under Uncertainty: Heuristics and Biases*. Cambridge: Cambridge University Press, pp. 190 - 200.

Tetlock, P. E. (1983). Psychological research on foreign policy: A methodological overview. In L. Wheeler (Ed.), *Review of Personality and Social Psychology. Vol. 4*. Beverly Hills, CA: Sage.

Tversky, A., and Kahneman, D. (1973). Availability: A heuristic for judging frequency and probability. *Cognitive Psychology*, 5, 207 - 232.

FALLIBLE HUMANS AND VULNERABLE SYSTEMS: LESSONS LEARNED FROM AVIATION

Earl L. Wiener

University of Miami
Coral Gables, Florida, USA

INTRODUCTION

Is this destruction [of Korean Airlines Flight 007] the consequence of man's absolute confidence in the robots he has built, and is this a warning on the dangers of the coming robotics era? (Machado 1984, p. 5)

In 1959, the commercial jet transport era in the United States began with the Boeing 707. Soon after, a rather serious incident occurred over Newfoundland. A 707, flying at 35,000 feet, experienced an autopilot disconnect and began a downward spiral. The captain was in the cabin, and the copilot and flight engineer did not detect the loss of control, even after the aircraft was descending rapidly, and by then were overcome by g forces and unable to make a recovery. Somehow the captain made his way back to his seat and recovered the plane at about 6000 feet above the Atlantic (Civil Aeronautics Board, 1959). With all of its electronic sophistication, the autopilot was capable of uncoupling, and perhaps more important, the crew was seemingly incapable of detecting and remedying the situation.

Ten years later, a similar occurrence led to less fortunate results west of Miami. While the crew was attempting to diagnosis a landing gear unsafe warning light, a Lockheed L-1011 experienced an autopilot disconnect, probably due to one of the crew members bumping the control column, and slowly descended into the Everglades. Although the descent from level flight at 2000 feet was gradual, the crew was totally unaware of what was occurring (NTSB - National Transportation Safety Board, 1973; Wiener, 1977).

And in 1985, a B-747, flying at 41,000 feet, bound for Los Angeles from China, suffered a partial loss of power in its Number 4 (right outboard) engine. The crew took no action, and with the autopilot lacking sufficient authority to correct for the right yaw, the aircraft rolled, almost inverted, to the right, and the nose fell through the horizon into an almost vertical dive. At 9500 feet the crew was finally able to save the aircraft. The crew believed that the unusual attitude shown on their instruments was due to instrument failure (NTSB, 1986).

What do these three incidents tell us? Mainly, that automatic devices can fail or be inadvertently disengaged; that when this occurs, the crew may be not be monitoring the instruments that would indicate the failure as well as the immediate consequences of the failure, and finally may not be able to revert to a manual mode in time to effect a recovery. Clearly, the humans are fallible and the machines are vulnerable. There are endless other examples of problems with automatic flight (see Wiener and Curry, 1980; Wiener, 1985a). For whatever reason, the flight crews did not have situational awareness about the status of the automatic devices or of the vehicle's status and flight path.

The author takes the position in this paper that the problems being experienced in complex, automatic systems are essentially due to the failure of information management and communication. This failure covers the entire spectrum: display devices and techniques, coding of information (symbolic or verbal) so as to reduce human error, and an area of escalating importance, information economy -- i.e., resisting the temptation to bombard the operator with unlimited information simply because the system possesses the capability to do so. In short, there has been great progress since World War II in the human engineering of

NATO ASI Series, Vol. F32
Information Systems: Failure Analysis
Edited by J. A. Wise and A. Debons

hardware, of devices. Now it would appear that further attention is needed in the "soft" side of systems. The approach should focus on two lines of defense: 1) preventing (or making less probable) human cognitive slips in the first place; and 2) making the systems less vulnerable to such slips when they do occur.

Finally, while this paper will not directly address the techniques of post-accident investigation under consideration at this workshop, it should alert the investigators to the importance of information management and the vulnerability of the systems to even the slightest lapse in communication and information transfer.

Most of the examples and principles come from the author's studies of cockpit automation, but the problems of information management in complex systems are ubiquitous. The reader should find it easy to strike out "cockpit" and fill in the blank with his own area of application or interest. For examples of human factors problems in the introduction of automation into other systems, See Adler (1984) on banking, Pew (1984) on nuclear power plants, and Hoffman (1971) on hospitals.

BASIC ISSUES IN MODERN SYSTEMS

The Role of the Human

The central issue that we confront in designing systems is age-old: the allocation of functions to human or machine. This hardly needs to be said, but what is essential is to realize that the rapid growth of computational capability forces the issue to be considered anew. Traditional engineers have long dreamed of removing the one system component that they mistrust, replacing him with inanimate devices that they do trust, or at least understand. Automation (broadly defined) has been seen as the answer to human error, work overload, and lack of dependability.

The problem with this reasoning is that often, when humans are replaced by devices, the system vulnerability increases. It is fairly well understood now that systems cannot operate for long in an automatic mode; that with the slightest unanticipated perturbation, human rather than device intervention is essential. Systems designers have turned to the human because he is not rigidly programmed ("hard-wired"), but can fall back on his remarkable ability to analyze, seek novel solutions, and extrapolate beyond his "program." The examples from space flight and aviation are endless. Pew (1984) reports that NASA has stated that no space flight to date could have been completed successfully without crew intervention at some point.

One celebrated case from aviation involves an L-1011 that totally lost control of its vertical stabilizer in flight, and hence the ability of the pilot to pitch the aircraft as needed for landing. The pilots recalled that the two wing engines had a tendency to pitch the aircraft one way, the center engine the other, but remarkably, they were able to use this probably unintended property of the plane to control pitch angle and land safely at San Francisco. Ironically, it is this very property, his flexibility and adaptability, that makes the human susceptible to error.

As systems became more complex, the "doer versus monitor" distinction became design doctrine, the assumption being that devices should provide moment-to-moment control of processes, and the humans would be relegated to keeping watch. The machines would do, and the humans would watch them. Wiener and Curry (1980) have discussed this dichotomy with respect to cockpit automation, as shown in Figure 1. The two axes are manual and monitoring tasks, running from full manual to full automatic. Presumably any system, depending on how it is configured at any given time, could be described as a point in this two-dimensional space.

A vast literature on vigilance, as well as examples too numerous to cite, have shown that the human is not a very effective monitor, and research in aviation has confirmed that when the pilot is operating in an automatic control mode, he is less likely to detect system faults than when operating manually (Kessel and Wickens, 1978) and is more likely to commit large blunders (Bergeron and Hinton, 1983). The unfortunate examples of failures of human vigilance following autopilot disconnect just cited clearly confirm these laboratory experiments. For these reasons, Wiener (1985a) recommends that the traditional doer versus monitor doctrine be re-examined. The faulty vigilance of the human emphasizes the need for effective warning and alerting systems (the machine monitors the operator), which will be discussed later. This will take us to the doorstep of machine intelligence, with self-diagnosing machines and smart warning and alerting systems.

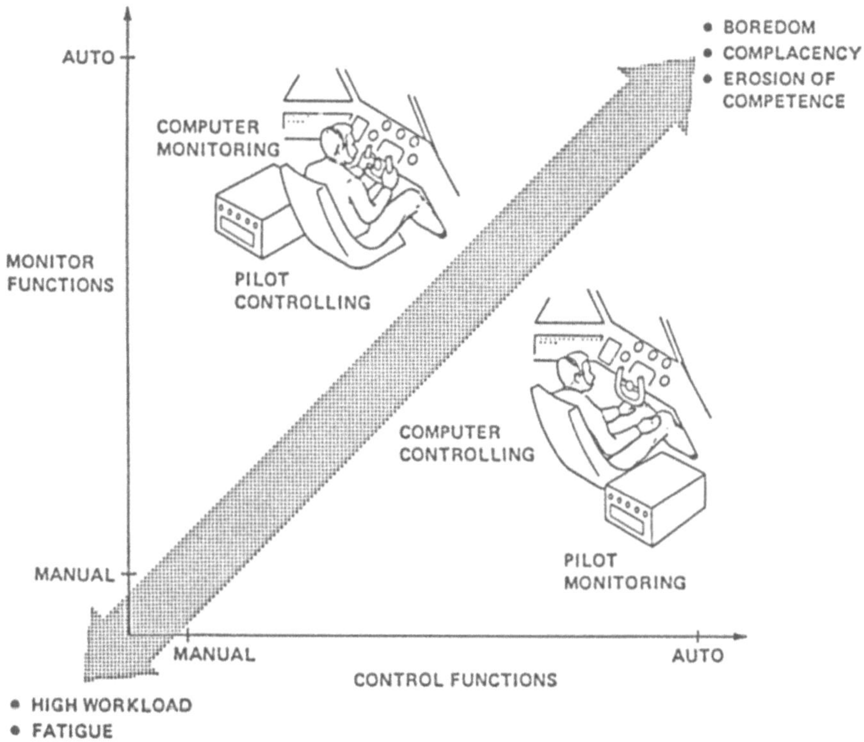

Figure 1. Two dimensions of automation: Control and monitoring.

Piloting an aircraft (any many other tasks in modern systems) involves both. This figure depicts the possibility of different levels of automation in these subtasks. (From Weiner and Curry, 1980, p. 1004)

We can complete the discussion of the role of humans with a brief examination of a developing application, the National Air Space Plan (NASP), the blueprint for air traffic control (ATC) for the end of the century. Describing it somewhat superficially for the sake of brevity, the NASP envisions a data link between ground-based computers that develop flight plans and on-board flight guidance computers. Under this plan, the ATC computers will determine the flight plan route and altitudes, and transmit this via data link to the aircraft. In a semi-automatic mode, the crew would take the flight plan (probably displayed either on a CRT or a hard-copy printer) and implement it by programming their flight guidance computer (FGC), just as they presently do after receiving the clearance by voice link. Obviously, the logical next step would be to eliminate the human link, with computer-to-computer communication, the ground-based computer directly programming the on-board FGC.

This forces us to ask a basic question: in such a system, what will be the role of the controller and the role of the pilot? The controller's job will probably be designed in obedience with the doer versus monitor doctrine. He will be relegated to the position of monitor, and presumably will "review" machine-generated clearances before they are transmitted to the aircraft. Removed from active participation in the loop as a generator and communicator of clearances, can the controller be depended on to passively "watch the world go by" -- will he have the vigilance, and the cognitive skill, to detect erroneous clearances as they are displayed at his station for his review? We must ask the same question of the pilot, especially in the fully automatic role. At this point, he may indeed feel that he is, as pilots interviewed by Wiener (1985c) described it, along for the ride.

In a less elegant but no less important application of urban rapid transit, Soto (1986) paints a dark picture of an automatic speed control and braking system on the Miami Metrorail: Driver Pam Walden, who describes the automatic train operation (ATO) as 'pretty good but not that great,' insisted that there's still room for the human touch. About every 10 stops, she said, ATO fails to receive the signal to stop, forcing her to hit a red button that slams on the brakes. 'Usually you end up overshooting the station, so you have to backtrack to let people out,' Walden said, noting that without her the train would keep going to the next station."

<u>The Role of the Machine</u>

With the development of microprocessor technology came a rapid expansion in machine capability. Wiener and Curry (1980) have pointed out that the question today is no longer what automation can do but what it should do. The human versus machine allocation decision is being reexamined in the light of expanding capability of the machine, as well as artificial intelligence on the horizon. Even the doer versus monitor doctrine is in question, not so much because of deficiencies in the human's ability to maintain vigilance, but more from the belief of the designers that alarms can be programmed to alert the operator to the need for action, or possibly to allow the machine to diagnose and remedy its own problem.

Currently function allocation decisions are being expanded beyond just "who does what" to what role the human has even when functions are "automatic." At one extreme it may be very little. For example, jet transports have yaw dampers that control rudder movement at high speed to prevent Dutch roll. The yaw damper is not absolutely necessary - in fact, an airliner can be dispatched with an inoperative yaw damper, but the control of yaw then reverts to the manual mode and increases pilot workload. Normally, the yaw damper does its job with absolutely no pilot intervention. There is nothing for the pilot to do but turn it on. A warning light will indicate its failure. This simple example of automation that reduces workload requires no decisions on the part of the crew, and has virtually eliminated the Dutch roll that plagued the early days of the swept-wing jet. It is unattended automation at its best.

However, the yaw damper has limited authority. There are no alternative modes, no interaction with other systems, no programming, and essentially no opportunity for human failure. It either damps yaw or it does not. At the other extreme, there are devices that take action on their own, not always inviting crew intervention or even informing the crew, and this seems to be the path down which the artificial intelligence community is traveling. For a refreshing exception, see Woods (1986).

A critical example is stall recovery. If a plane is approaching a stall, there is usually ample warning. There are ample instrument indications, tactile indications (the plane buffets, though this can be misinterpreted; NTSB, 1980), and at some point alerting systems sense and annunciate the impending stall. These alerts can be visual, auditory, and tactile (stick shakers). Finally, if the crew still has not remedied the situation, the machine takes action. For example, the MD-80 can automatically extend leading edge slats to break a stall.

In the extreme case of machines operating on their own, some aircraft are equipped with stick pushers that attempt to break the stall by actually forcing the control column forward to reduce the angle of attack. Note that it does not warn the pilot of what it is going to do, does not advise him to push the stick, and does not ask for his approval -- it makes the decision and implements it. At that moment, it has removed the primary control of the pitch from the hands of the crew.

Undoubtedly, stick pushers have at some times saved aircraft when the crew, for whatever reason, has failed to respond to cues of an impending stall. But as we know all too well, Type I (commissive) errors can also occur in systems, and recently there have been a number of crashes in which aircraft struck the ground with severe nose-down attitudes, suggesting inadvertent stick-pusher actuation (Aviation Week and Space Technology (AWST), 1984). Understandably, many pilots are unenthusiastic about any form of automation by which a device decides on its own to take over primary control of the aircraft, possibly leading literally to a battle between man and machine (Melvin, 1983, p. 320). Pew (1984) raises a similar question about automated power plants and what the role of automation vis-a-vis the operator should be.

If Melvin and other pilots are alarmed about mere stick-pushers and other relatively simple devices that seem to wrest control from the pilot and take action on their own, they would be even more chagrined to learn what the artificial intelligence enthusiasts have in mind for the future, in fact, the not-too-distant future for military aircraft (AWST, 1986). In such systems, there may be very little understanding on the part of the human crew about what the black box is doing, and after-the- fact-traceback of "explanation" modes may provide little comfort. If the pilot of today's modern aircraft feels that he is "along or the ride," one can hardly imagine how he will feel about crewing a plane governed by heuristics he neither sees nor understands.

Alternatives

The the stage is thus set for difficult philosophical and practical decisions. Given the immense computing capability of present-day machines, the promise of machine intelligence and the mounting evidence that the machines, however intelligent, require human supervision and decision-making, fallible though the human may be, the systems designer faces anew the traditional human-machine dilemma, and he faces it with relatively few principles to guide him. Wiener and Curry (1980) made an attempt to derive automation guidelines, but this was only a beginning. Wiener (1985b) lamented the lack of a philosophy of automation and the fact that most systems are still being developed one box at a time. He called for a "template against which to hold future designs"; more easily stated than done.

In the most advanced aircraft, such as the B-757/767 and the A-310, equipment capability has outstripped operational doctrine. Some of the airlines have taken an extreme position, insisting that the full capability of the equipment be exploited. This is called, informally, the "we bought it, you use it" doctrine. Pilots tend to resist this doctrine, basing their argument on the traditional view that equipment is subordinate to pilot discretion (and legal responsibility), and they wish to use equipment as they see fit.

The Wiener-Curry guidelines support pilot's discretion. They argue that the equipment is on board for one reason, to help the pilot do his job, and that within the bounds of safety and regulatory conformity, the pilot should fly in a manner he sees fit, using cockpit resources as he deems necessary. Such freedom requires that the crew be backed up by an enhanced warning and alerting system, a multivariate "electronic cocoon" around the plane. As long as the plane stays within the cocoon, let the pilot select his equipment and style of flight as he sees fit. If he punctures the cocoon, the system will warn him (possibly gently at first, with increasing stridency if he fails to respond).

Wiener (1985b) takes it a step further, recommending the implementation of forecasting models that will predict the penetration of the cocoon long before any parameter reaches a critical point, warning of possible violation of cocoon space so that the crew could evaluate the situation and make a decision before the system reaches criticality. Wiener borrows a term from modern business management, referring to the discretion-cocoon-prediction-warning sequence as "flight management by exception."

An excellent example, which stirs rather strong emotions on the part of pilots, is an on-board collision avoidance system, currently under development and soon to be tested on regular airline flights (Proctor, 1986). The present system is called Traffic Advisory Collision Avoidance System (TCAS). A TCAS- equipped aircraft receives transponder returns from other aircraft in the vicinity (in some cases altitude encoded, in some cases not), and by a filter logic determines which ones are collision threats and displays the position of those on a CRT. The first line of defense is to alert the crew to the presence of a potential threat and aid in visually acquiring the target. If the threat is not resolved and a hazard collision is predicted within a certain period (tau seconds), for altitude-encoded targets TCAS recommends a vertical escape maneuver (e.g., "climb"), non-maneuver ("do not climb"), or restriction ("do not climb faster than 500 fpm").

The pilot quite reasonably asks what he is supposed to do: has TCAS given him an order or a suggestion? and what of the divided authority between ground controller and on-board device? The operational doctrine has yet to be developed, and in the forthcoming test on airline flights, TCAS is supposed to be turned on only when flying in Visual Meteorological Conditions (clear weather). "Enhanced TCAS," currently under development, will also allow lateral escape maneuvers.

Pilots are, with good reason, apprehensive when they perceive that the next step is linking the TCAS logic to the flight guidance system, and directing the plane (without pilot intervention) away from the collision hazard. As in stall recovery, the machine could be allowed to take action on its own, based on information and logic which the pilot may distrust. If there were just one plane, it might be bad enough -- but in a terminal area, with potentially multiple conflicts, the planes may be going through a series of maneuvers not always comprehended by the crew or controller, and possibly two (or more) planes could enter into a "sidewalk shuffle," like so many pedestrians trying to avoid each other with self-aggravating maneuvers.

ISSUES OF INFORMATION MANAGEMENT

How Much Information Is Enough? Too Much?

In the few decades since the beginning of the systematic study of human-machine systems, one can sense a dramatic reversal in the problem of information quantity. In the past, the problem has been a paucity of information, or of timely information, being displayed to the operator. Operators in a variety of systems complained about needing this and that displayed to them, and accident reports frequently ended with statements that but for the lack of timely information, the accident would not have occurred.

Today we have come full circle, and the prevailing display problem is being called "information glut." Of course, it has been the computer and its associated peripheral display devices that have enabled the provision of such a wide variety and huge quantity of information to the operator. With vast storage capacity, fast computing algorithms, and rapid development of display techniques and devices, including high resolution color graphics, the temptation has been to deluge the operator. To an alarming extent, the programmer and computer graphics designer have set the tone for many display systems, and their credo all too often has been "the information is available, so let's throw it up on the display -- it can't hurt." Displays became a jungle of symbols, moving pointers, text, colors, and three-dimensional graphics, and in the auditory mode, tones, musical chimes, and more recently the human voice.

The early development of the head-up displays for pilots was plagued by over-display. The TCAS display was developed without the aid of human factors, and it shows it. Some aircraft were loaded up with human voice displays, usually employing a female voice for its putative attention-getting properties, to the point where pilots of one aircraft dubbed the system the "mother-in-law." In fairness to all, some of the new information displays are quite good. The color graphics in the B-757/767 are strikingly effective, particularly the color map display, one of the modes of the horizontal situation indicator, which can be reconfigured by the pilot into six different modes, including a map format with five range settings. One pilot interviewed by the author remarked, "You can take all this other stuff (digital flight guidance systems) away if you want, but don't touch my map display, and especially the green arc." The green arc is displayed along the course line during vertical maneuvers, predicting at what point the plane will reach the target altitude (Figure 2).

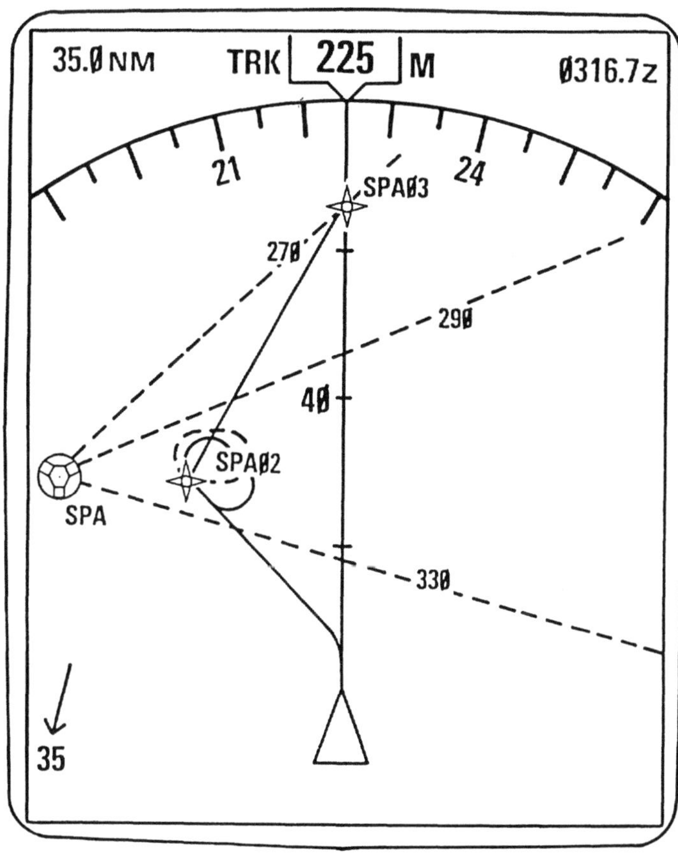

Figure 2. Map display option on the horizontal situation indicator on a Boeing 767/757.

The triangle in the center is the aircraft. SPA03 and SPA02 are "man-made" waypoints constructed from the navigation station SPA, which is stored in the data base. The full impact of this display cannot be appreciated without the color coding. (Copyright Sperry).

The success of the display systems in the 757/767 aircraft may be due in part to the heavy involvement of human factors engineers at Boeing and Sperry. What is required today is for human factors scientists to wrest control of information display decisions from the computer enthusiasts, and return to a fundamental question: what information is needed, and how should it best be displayed? Again, one is struck by the ubiquity of this problem -- pilots, business managers, government officials, health care professionals, and battlefield commanders alike are complaining about information deluge.

Coding of Information and Code-Induced Errors

Much of the problem can be solved not only by removing the unnecessary and often distracting portions of the displays, but by packaging the necessary information in non-traditional forms. For example, it is fairly clear that endless columns and rows of numbers may overwhelm the decision-maker, especially if he must make rapid judgments on real-time information. Assuming that ample data are stored and that computer graphics can give us whatever we want, the human factors specialist must discover ways to present multivariate information in graphic rather than numeric form (Wang, 1978). Figures 3, 4, and 5 illustrate some examples. I am not suggesting that the cockpit of a plane or the control room of a nuclear plant be an array of Chernoff faces, but only that the reader consider the potential of non-traditional multivariate displays.

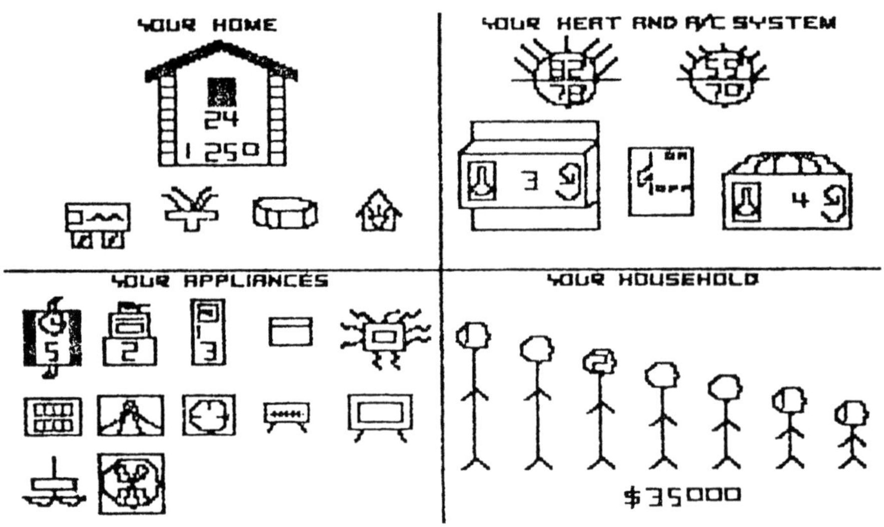

Figure 3. Graphic screen design to display multivariate information on household electrical equipment, occupants, and other residential information for electrical power company decision-makers. (Courtesy of H. Gitlow and J. Stewart, University of Miami)

So far the discussion has centered on visual displays, and it should be extended to include a rather elusive area, verbal (including oral) communication in human-machine systems. Here again, the flexibility of the human in understanding and interpreting what he sees and reads is both an asset and a liability. My point is that every standardized message, written or spoken, should be subjected to human factors scrutiny (see Wiener, 1980).

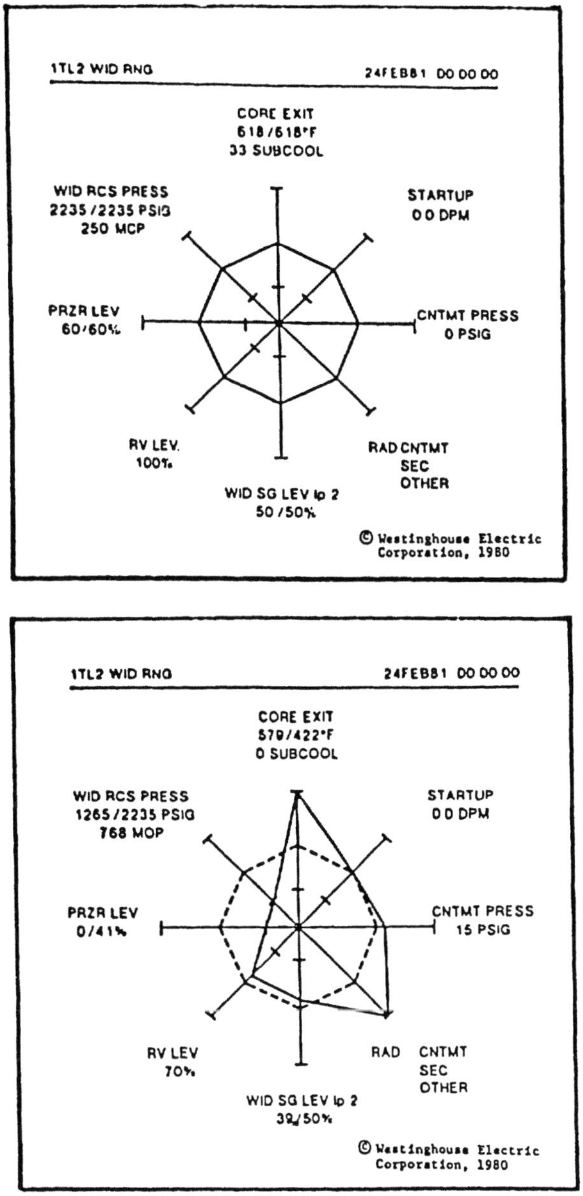

Figure 4. A multivariate display, capable of showing the magnitude of eight variables simultaneously.

The length of the spokes are scaled so that when each parameter is at its set point, a right octagon will appear (top figure). As a parameter changes, the length of a spoke changes, and a highly conspicuous irregular octagon is produced (bottom figure). Limits can also be shown as the length of the spokes. The display was designed for use in nuclear power plant control rooms. (Copyright Westinghouse).

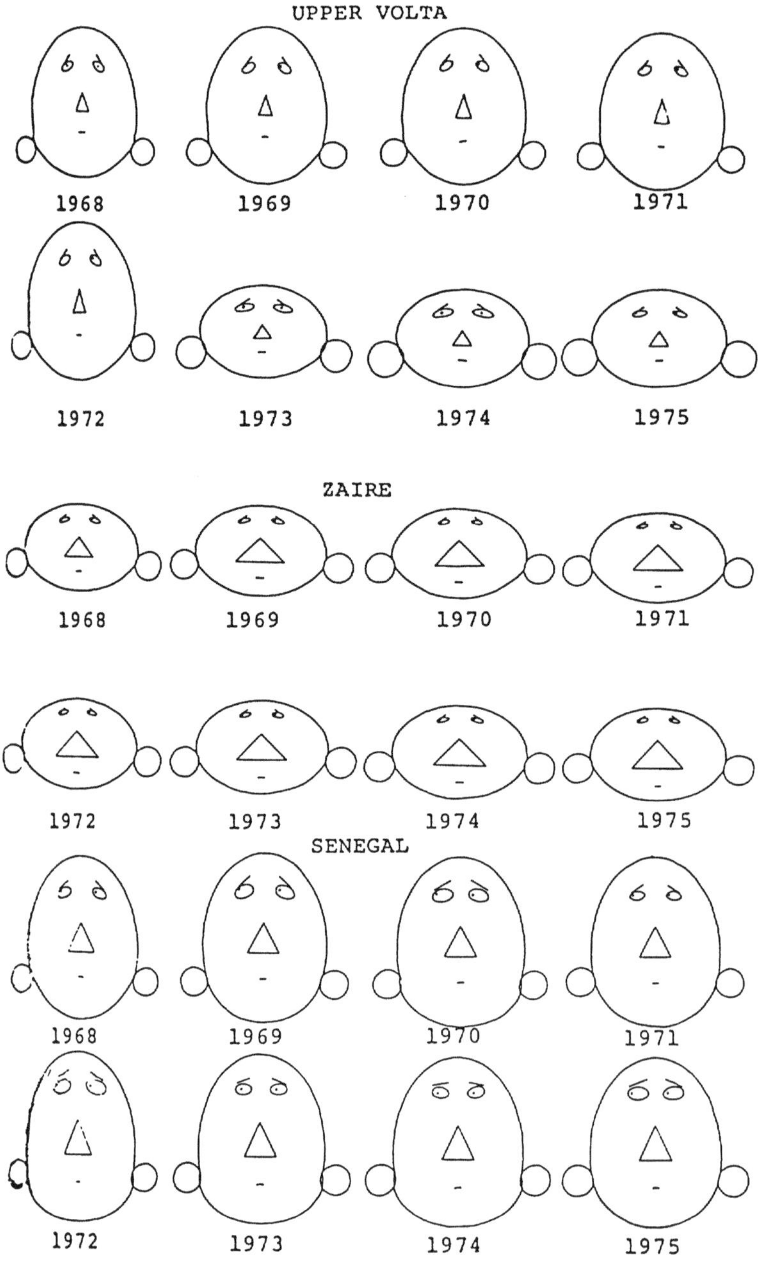

Figure 5. Chernoff faces.

A means of displaying multivariate information in a single graphic display. In this application, the relations between the U.S. and certain Third World nations, along various dimensions, are depicted in the faces. (From Wang and Lake, 1978).

Names on Maps

On aeronautical charts, intersections of airways or arbitrary points along airways were used for navigation and were assigned names (see Figure 6). In days gone by, the names were colorful and easily recognizable since they usually referred to geographic names in the local area. The author recalls with nostalgia some obscure points in Texas where he learned to fly, such as PAINT ROCK and FARMERS BRANCH. Later, in order to accommodate computers no doubt, the FAA shortened all the intersection names to five-letters. BARRICUDA intersection northeast of Miami became BARON. Sounds innocent enough; but Murphy's Law is not to be ignored. The stored data in the 757/767 made it possible for crews to establish waypoints (points on the surface of the earth for navigation purposes) by retrieving stored coordinates from the navigational database by typing the intersection name on the keyboard. Type in BARON and the coordinates will be retrieved and the waypoint established and placed on the map display.

For the first time, navigation required that the pilot know how to spell, and some of the five-letter names had been shortened to absurd spellings. For example, the "bridge" intersection near the San Mateo Bridge east of the San Francisco airport is spelled BRIJJ. Pilots now complain that when ATC clears them to an unfamiliar intersection, they must either search their charts for the spelling or ask the controller how it is spelled, which is not only humbling but time-consuming. How do you spell "Laker" (phonetic) intersection near Seattle? LACRE, of course.

Even worse, some of the names can lead to confusion when pronounced. An incident was reported to the Aviation Safety Reporting System (ASRS) in which a pilot was cleared to TOUTU. (I would guess that it was once Teutonic.) After a period of silence, the pilot called back to the controller and asked, "Am I cleared to 2-2 thousand feet, or twenty-two miles distance measured from a navigation station (DME)?"

This confusion may be amusing, but it could have resulted in a midair collision, and probably the investigators would never have known why. As a result of this report, the intersection was renamed.

This illustrates two points: first, the potential for seemingly trivial choices in phraseology and nomenclature to create hazards in real-time systems; and second, the great value of a voluntary incident reporting system such as NASA operates in the aviation world. Analysis of human error reports is also discussed by Rasmussen (1980). It might be worthwhile for the FAA to review all of the thousands of intersection names to remove such confusion. One name they might wish to examine, for the same reason as TOUTU, is LAVAN.

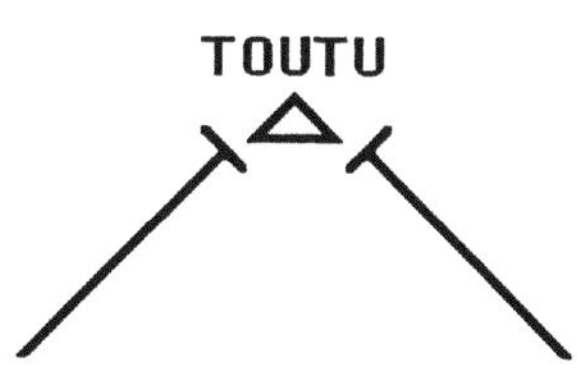

Figure 6. TOUTU airway intersection.

Confusion in air traffic control instructions resulted from the possible interpretation of the name as numerals.

Oral Commands: Solution by Redesign

The management of information also includes the scrutiny of oral messages. The importance of this was discovered during World War II, when the Allies found that certain critical words could be confused, especially when transmitted over noisy radio channels. For example, "repeat," a common radio- telephone word, could easily be mistaken for "retreat," so an invented term, "say again," replaced it. In the military, law enforcement, or commercial aviation, where standardization and authority prevail, these kinds of substitutions are fairly easy. It is less easy to standardize and to substitute engineered for "natural" terminology in the work-a-day world.

During the 1950's, the U.S. Air Force lost a heavy transport (C-124) when the aircraft commander, realizing that he was low and slow on final approach, gave the command "Takeoff power" to the flight engineer. The engineer did exactly what he was told and took off the power, and the plane crashed short of the runway. The problem was, of course, that the flight engineer could not distinguish between two reasonable interpretations of the oral message, "takeoff power" or "take off power," which would lead to opposite actions on his part. As a result of this accident, the USAF human engineered the solution -- the term "takeoff power" was eliminated and replaced by "maximum power."

Non-standard Terminology: Cognitive Failures

In 1977, two B-747 jumbo jets collided on the runway at Tenerife, Canary Islands (Spanish Ministry of Transport and Communications, 1978; Roitsch, Babcock, and Edmunds, undated). The KLM jet started its takeoff roll while the PAA jet was still on the runway. It had been following the KLM on taxi out, and was to turn onto a taxiway to clear the runway. As the KLM started its fatal takeoff roll, the captain radioed, "We are at takeoff." The tower, and possibly the PAA crew, interpreted this to mean at takeoff position: "at takeoff" is not standard aviation terminology.

As a result of this accident and its communication failures, some authorities recommended placing highway-like stop and go lights at the takeoff points of runways, to give crews a final signal when they were cleared for takeoff. Such a plan has not been implemented, and before it is, systems designers would be well advised to consider how easy it would be to make a switching error that could launch an unintended takeoff. The suggestion has been revived as a result of two near collisions in recent months at U.S. airports.

As another example of engineered information communication, following the Tenerife accident, the term "cleared" was dictated by U.S. ATC to be used only in two commands, "cleared to takeoff" and "cleared to land." No longer would a plane be "cleared to taxi" or "cleared to cross the active", etc. (For discussion of the runway incursion problem, (see Tarrel, 1985; and NTSB, 1986b).

Finally, imprecise language on the part of both the flight crew and the air traffic controllers figured in the crash of an Inex Adria MD-80 in 1981 (Lenorovitz, 1984).

Digit Inversion.

Inversion of digits is a common source of data error, known to accountants and accident investigators alike. It is an extremely common error, whether with the pen and pencil or keyboard, and one whose detection appears to be perceptually evasive. It is very likely an explanation of the navigational error (assuming that it was an error, conspiracy theoretic explanations aside) that led to the destruction of KAL 007 (International Civil Aeronautics Organization, 1983; Wiener, 1985a). The navigation plate in Figure 7 is illustrative.

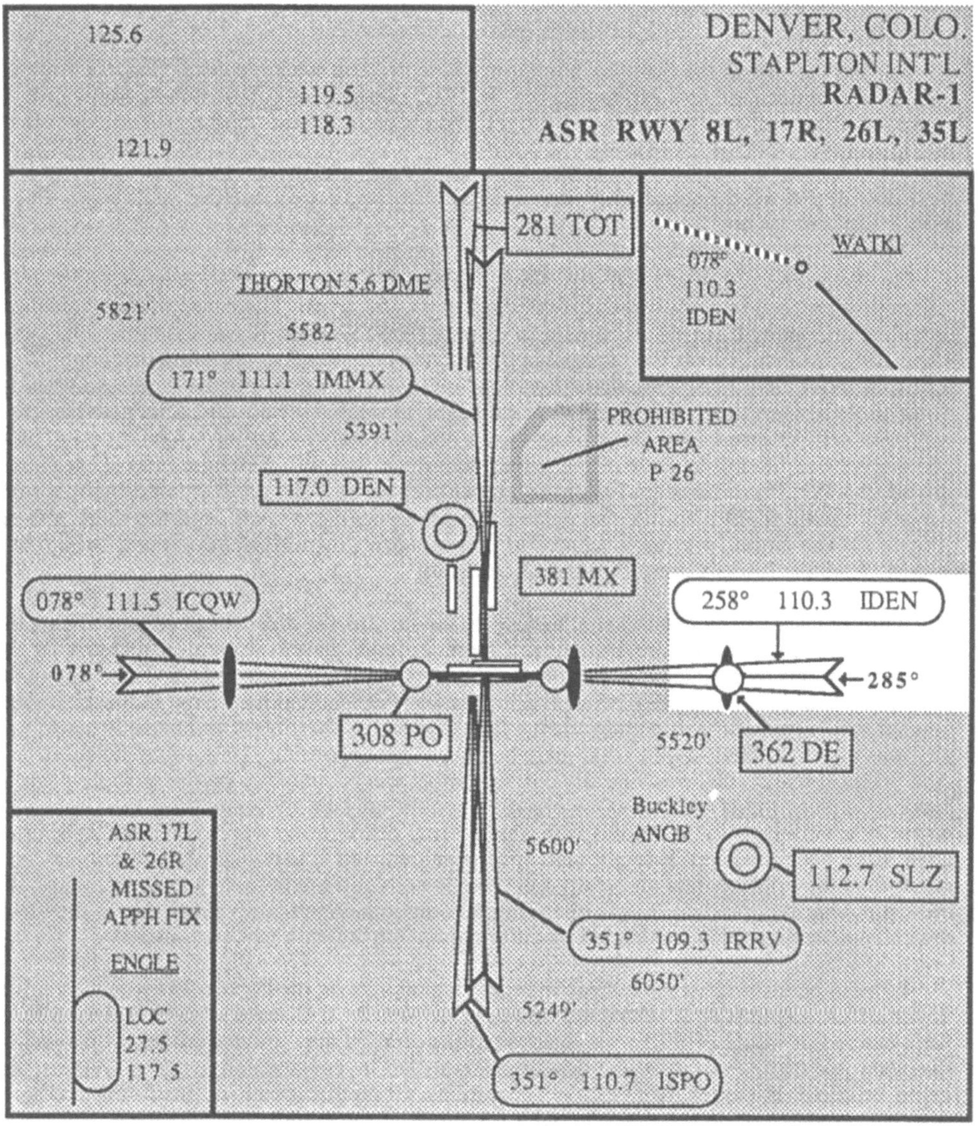

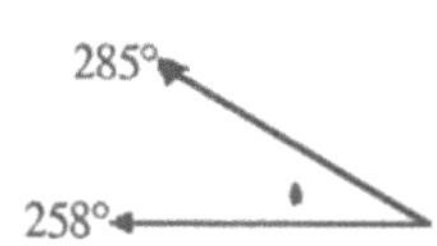

Figure 7. An illustration of an actual approach plate with a serious digit inversion error. *Notice that the inbound heading to Runway 26L is given as 285°. The correct heading should be 258°. The arrows below the chart show the relative difference between these two headings. (Redrawn from original plate, copyright Jeppeson-Sanderson, Inc.)*

Missing Words.

Needless to say, in the complex human-machine systems we are considering, the slightest lapse in communication can lead to disaster. In 1972, two airlines collided on the runway at Chicago O'Hare because a single word ("right") had been omitted from the ground controller's taxi instructions (NTSB, 1973).

INFORMATION MISMANAGEMENT:
THE CASE OF TE 901

On November 28, 1979, an Air New Zealand DC-10, on a sight-seeing flight over Antarctica, crashed into Mt. Erebus, with the loss of 257 lives. The cause of the accident is complex, and many of the facts remain in dispute (New Zealand Royal Commission, 1981; Vette, 1983; Mahon, 1984). On November 9, the crew had attended a briefing that contained a taped narrative and navigational information, including the coordinates of the waypoints. Some time between this briefing and the fatal flight, through a series of blunders, an incorrect waypoint was inserted into the ground-based computer at the airline's navigation center, 166.48 east (longitude), in place of the correct longitude, 164.48. At the latitude of McMurdo Sound, the slight error was equivalent to 26 nautical miles. The crew entered the incorrect waypoint on the flight plan into the plane's inertial navigation system, and the stage was set (Figure 8). The aircraft was programmed to fly not safely down McMurdo Sound but directly into Mt. Erebus. It was a pre-programmed collision.

In many installations of automatic navigation equipment, the most recent example being the B-767/757, the databases are transferred from ground- based computer systems to the on-board systems by maintenance personnel via a portable digital tape device. This was not the case in the fatal flight of TE 901. The flight plan was furnished in hard copy to the crew, who toggled in the waypoint coordinates via the INS keyboard. This author erroneously stated in a previous paper (Wiener, 1985a) that there had been "no opportunity for human intervention." As a practical matter, this was virtually correct: pilots accept computer-produced flight plans as accurate, unless there is an obvious error. It would only be by comparing the flight plan waypoint coordinates to their notes from the briefing nearly three weeks earlier that the error could have been detected. It might also have been detected if they had plotted the new data, rather than the briefing data on a map, but there is no requirement to do so. In summary, the crew did what thousands of pilots operating automatic navigators do every day -- they took their computer flight plan and loaded the information into the on-board computer.

It might be instructive to mention that the map display on the B- 767/757 provides a final line of defense against erroneous data, whatever the source. If an error of sufficient magnitude were entered, it would probably be apparent to the crew as they previewed the course line on the electronic map. For those who design, operate, and investigate non-aviation systems, this could be taken as model for a variety of systems. Computer graphics, including predictor displays, could provide the operator with a picture of the future "direction" of the system, based on the data entered. Many of the accidents of the type under discussion at this conference might have been and might be prevented if the operators had been provided with displays more communicative than rows and columns of numbers or a forest of gauges. Such a display might have saved KAL 007.

This would be consistent with the recommendation of Rouse and Morris (1985) and others that systems be made "fault tolerant." The first step in achieving fault tolerance should be to make information errors more conspicuous to human intelligent systems. As a next step, designers might consider intelligent machine recognition of the error.

INFORMATION MANAGEMENT IN COMPLEX SYSTEMS:
ARE THERE MEANS FOR IMPROVEMENT?

Thus far the author has outlined the problems of the management of information in

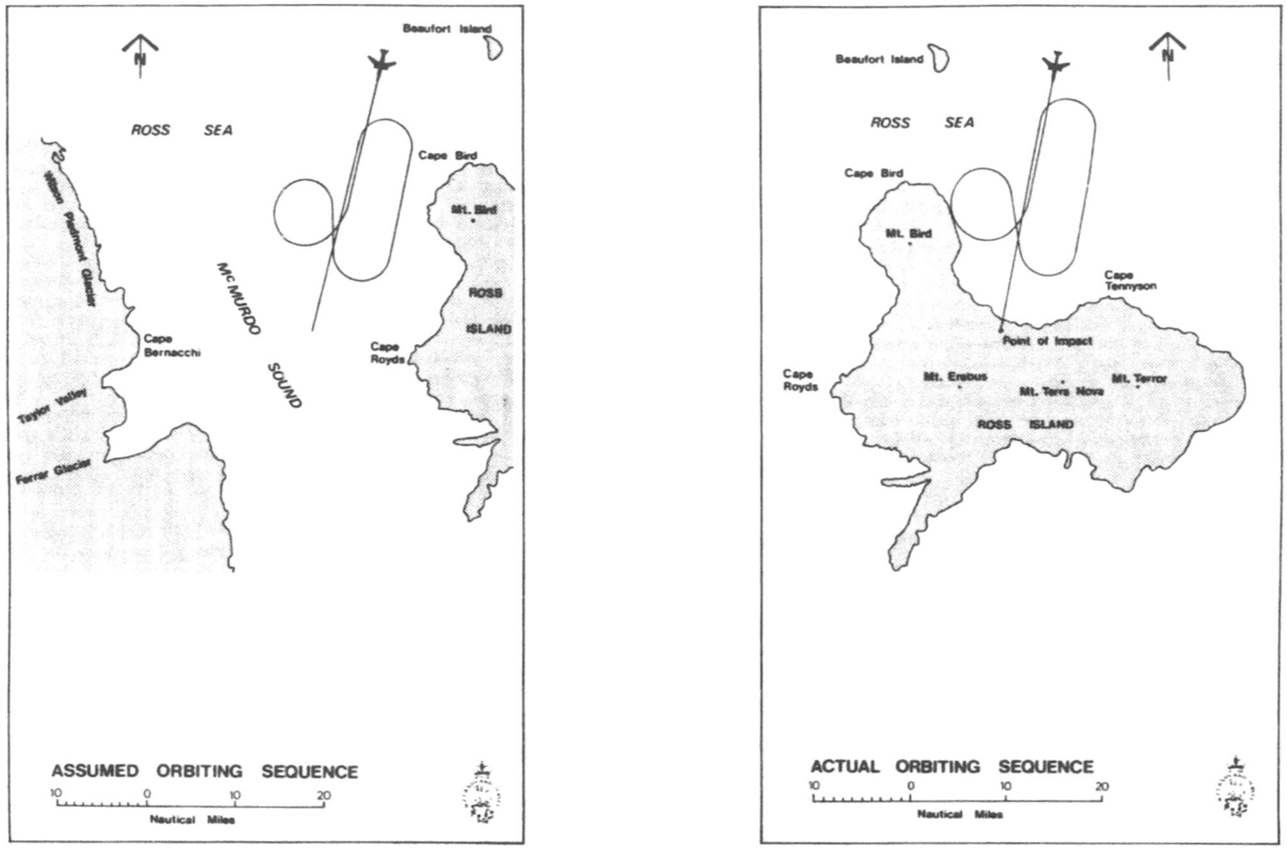

Figure 8. Planned and actual flight paths.

The correct course for TE 901 (left) would have taken the aircraft safely over McMurdoSound. The erroneous waypoint generated the fatal path to Mt. Erebus. (From Mahon, 1984)

complex systems, from the spoken word to the loading of data into storage devices, and has shown by a number of examples from aviation, the vulnerability of these elaborate systems to even the slightest lapses. Now let us turn to the question of the means available to make the systems less vulnerable.

Basic Human Factors

There is no substitute for time-honored human factors design based on time-honored methods. The author is distressed to hear some of these methods disparaged as (mere) "knobs and dials." There has been enormous progress since the early days of human factors engineering, much of it being nothing more elegant than knobs and dials. The influence of basic human factors engineering can be seen in the development of modern aviation (Miller, in press; Sexton, in press) and in military systems, and it is slowly spreading to other areas. Some remain virtually untouched.

The reader is invited to take a look at the work station of an anesthesiologist if he would like to see systems of the pre-ergonomic era and in an application where a minor "knobs and dials" error can lead to instant injury or death. In one attempt to apply basic task analysis, investigators found that the most frequent sub-task was data logging, mainly using a pencil and clipboard (Kennedy, Feingold, Wiener, and Hosak, 1976). The anesthesiologist pays the highest premium rate for medical malpractice insurance in the U.S. A look at his tools, methods, and work stations reveals why.

Error-Resilient Coding of Information

These examples point toward the need to examine written, spoken, and device-loaded communication procedures and codes. This is an area of information management barely touched outside of the military and aviation fields, and one that promises a high benefit/cost ratio. Incident collection tends to be haphazard. Examples from NASA's ASRS (e.g., TOUTU intersection) demonstrate the value of a systematic means of collecting not only accident but incident data from the operational world, as a means of implementing error resilience.

Error-Tolerant Systems

With rapid developments in machine intelligence has come considerable interest in designing systems to be error-tolerant. This activity is born partly of frustration with the difficulties of preventing the errors in the first place, as a first line of defense, through traditional means, such as training, selection, motivation, and basic human factors design. For those errors that slip through, partly because not every error is predictable and partly because a zero error rate is unattainable, error tolerance offers a "safety net." The methodologies would include everything from making the errors more conspicuous, as discussed in connection with aircraft displays, to warning and alerting systems (including the electronic cocoon, system management by exception, and predictive systems mentioned previously), to decision support systems, and ultimately machine recognition (and perhaps automatic remedy) of errors. This will no doubt be an arena of considerable activity in the years to come, but we have warned before (Wiener and Curry, 1980) about the tendency for backup systems to become primary systems.

CONCLUSION

In modern human-machine systems, the management of information takes on increased importance, with the dependence on the processing and communication of huge amounts of stored data. It is incumbent on systems designers, managers, operators, and accident investigators to recognize the vulnerability of these systems to information failures, especially those induced at the device-human interface, and to recognize as well that much of the vulnerability can be ameliorated by redesign.

The experience from commercial aviation shows that it is unwise to dream of automating human fallibility out of a system. Automation essentially relocates and changes the nature and consequences of human error, rather than removing it, and, on balance, the human operator provides an irreplaceable check on the system. The search should be directed toward the management of human caprice, not the elimination of its source.

ACKNOWLEDGMENTS

This research is supported by a grant, No. NCC2-377 from the NASA-Ames Research Center. The author gratefully acknowledges the cooperation of Eastern Airline and Northwest Orient Airline, and the Safety Committees of the Air Line Pilots Association at those companies. The author was assisted by Shiela Bennett and Sharon Sainz. The views expressed here are those of the author and not of any organization or institution.

REFERENCES

Adler, P. (1984). New technologies, new skills. Working paper, Harvard Business School, No. 9-784-076. Boston: Harvard University.

Aviation Week and Space Technology. (1984, July 23). NTSB urges review of Metro stall device, p. 31.

Aviation Week and Space Technology. (1986, February 17). DARPA's pilot's associate program provides development challenges, pp. 45-52.

Bergeron, H. P., and D. A. Hinton. (1983). Aircraft automation: The problem of the pilot interface. Paper presented at *Ohio State University Symposium on Aviation Psychology*, Columbus.

Civil Aeronautics Board. (1959). *Pan American Airways B-707 over Gander, Newfoundland.* (Report No. 1-0001). Washington, DC: Author.

Hoffman, P. B. (1971). Meeting resistance to hospital automation. *Hospital Progress*, April, 45-47, 60.

Hurst, R., and Hurst, L. (1982). *Pilot Error: The Human Factors.* London: Granada Press.

International Civil Aviation Organization. (1983). *Destruction of Korean Airlines Boeing 747 over the Sea of Japan, 31 August 1983.* Montreal: Author.

Kennedy, P. J., A. Feingold, R. Hosek, and E. L. Wiener. (1976). Analysis of tasks and human factors in anesthesia for coronary-artery bypass. *Anesthesia and Analgesia, 55*, 374-377.

Kessel, C. and C. D. Wickens. (1978). *The Internal Model: A Study of the Relative Contribution of Proprioception and Visual Information to Failure Detection in Dynamic Systems.* 27 NASA CP-2060, Ames Research Center, Moffett Field, CA.

Lenorovitz, J. M. (1984). Inex DC-9-80 crash involved imprecise terminology. *Aviation Week and Space Technology*, February 27, p. 34.

Machado, F. (1984). The destruction of KAL 007 (KE 007): How did it happen? *International Journal of Air Safety*, 2, 5-14

Mahon, P. (1984). *Verdict on Erebus.* Auckland: Collins.

Melvin, W. W. (1983). A philosophy of automation. In *Behavioral Objectives in Aviation Automated Systems Symposium*, pp. 319-325. Warrendale, PA: Society of Automotive Engineers.

Miller, C. O. (In press). System safety. In E. L. Wiener and D. C. Nagel (Eds.), *Human Factors in Aviation.* To be published by Academic Press.

National Transportation Safety Board. (1973a). *Eastern Airlines L-1011, Miami, Florida, December 29, 1972* (Report No. NTSB- AAR-73-14). Washington, DC: Author.

National Transportation Safety Board (1973b). *North Central Airlines McDonnell Douglas DC-9-31 and Delta Airlines Convair CV-880, O'Hare International Airport, Chicago, December 20, 1972* (Report No. NTSB-AAR-73-15). Washington, DC: Author.

National Transportation Safety Board. (1980). *Aeromexico DC-10-30, XA-DUH, Over Luxembourg, Europe, November 11, 1979* (Report No. NTSB-AAR-80-10). Washington, DC: Author.

National Transportation Safety Board. (1986a). *China Airlines Boeing 747-SP, N4522V, 300 Nautical Miles Northwest of San Francisco, California, February 19, 1985* (Report No. NTSB- AAR-86/03). Washington, DC: Author.

National Transportation Safety Board. (1986b). *Runway Incursions at Controlled Airports in the United States.* (Special investigation report No. NTSB/SIR-86/01). Washington, DC: Author.

New Zealand Royal Commission. (1981). *Report of the Royal Commission to Inquire into the Crash on Mt. Erebus, Antarctica, of a DC-10 Aircraft Operated by Air New Zealand, Ltd.* New Zealand: Government Printer.

Pew, R. Automation in supervisory control systems. (1984). Paper presented at the seminar of The Federation of Behavioral, Psychological, and Cognitive Sciences, Washington, May.

Proctor, P. (1986). T/CAS entering operational service on Piedmont 727-200s. *Aviation Week and Space Technology*, June 23, pp. 43-45.

Rasmussen, J. (1980). What can be learned form human error reports? In K. D. Duncan, M. M. Gruneberg, and D. Wallis (Eds.), *Changes in working life.* London: Wiley and Sons.

Roitsch, P. A., G. L. Babcock, and W. W. Edmunds. (undated). *Human Factors Report on the Tenerife Accident.* Washington: Air Line Pilots Association.

Rouse, W. B., and N. M. Morris. (1985). Conceptual design of a human error tolerant interface for complex engineering systems. In Proceedings of the *2nd IFAC/IFIP/IFORS/IEA Conference on Analysis, Design, and Evaluation of Man-Machine Systems.* Varese, Italy. 29

Sexton, G. A. (In press). Cockpit/crew systems integration. In E. L. Wiener and D. C. Nagel (Eds.), *Human Factors in Aviation*. To be published by Academic Press.

Soto, L. F. (1986). Metrorail now running on cruise control. *Miami Herald*, January 4, p. 2B.

Spanish Ministry of Transport and Communications. (1978). Report of the collision between PAA B-747 and KLM B-747 at Tenerife, March 27, 1977. Translation published in *Aviation Week and Space Technology*, November 20 and 27, 1978.

Tarrel, R. J. (1985) *Non-airborne conflicts: The causes and effects of runway transgressions.* NASA Contractor Report No. 177372, Ames Research Center, Moffett Field, CA.

Vette, G. (1983). *Impact Erebus.* Auckland: Hodder and Stoughton.

Wang, P. C. C. (Ed.) (1978). *Graphical Representation of Multivariate Data.* New York: Academic Press.

Wang, P. C. C., and Lake, G. E. Application of graphical multivariate techniques in policy sciences. In P. C. C. Wang (Ed.) (1978). *Graphical Representation of Multivariate Data.* New York: Academic Press.

Wiener, E. L. (1977). Controlled flight into terrain accidents: System-induced errors. *Human Factors,* 19, 171-81. Also published in R. Hurst and L. Hurst. (1982). *Pilot Error: The Human Factors.* London: Granada Press.

Wiener, E. L. (1980). Midair collisions: The accidents, the systems, and the realpolitik. *Human Factors,* 22, 521-533. Also published in R. Hurst and L. Hurst. (1982). *Pilot Error: The Human Factors.* London: Granada Press.

Wiener, E. L. (1985a). Beyond the sterile cockpit. *Human Factors,* 27, 75-90.

Wiener, E. L. (1985b). Cockpit automation: In need of a philosophy. In *Proceedings of the Fourth Aerospace Behavioral - Technology Conference,* Long Beach, CA. Warrendale, PA: Society of Automotive Engineers.

Wiener, E. L. (1985c). *Human Factors in Cockpit Automation: A Study of Flight Crew Transition.* (NASA Contractor Report CR-177333). NASA-Ames Research Center, Moffett Field, CA.

Wiener, E. L., and R. E. Curry. (1980). Flight-deck automation: Promises and problems. *Ergonomics,* 23, 995-1011. Also published in R. Hurst and L. Hurst. (1982). *Pilot Error: The Human Factors.* London: Granada Press.

Wiener, E. L., and Nagel, D. C. (Eds.) *Human Factors in Aviation.* Orlando: Academic Press. To be published 1987.

Woods, D. D. (1986). Cognitive technologies: The design of joint human-machine cognitive systems. *The AI Magazine,* 6, 86-92.

HUMAN RELIABILITY IN INFORMATION SYSTEMS

Balbir S. Dhillon

University of Ottawa
Ottawa, Ontario, Canada

INTRODUCTION

Modern technology has created a tendency to produce systems of greater capital cost, sophistication, complexity, and capacity. The consequences of unreliable behavior of information systems have become increasingly severe and have led to the desire for improved reliability. Some examples of complex systems failure are the Three Mile Island nuclear accident, the loss of the space shuttle Challenger, the chemical spills at Bophal, India, and the Chernobyl nuclear disaster. In all these cases, the information systems have experienced a major blow.

Nowadays, complex system reliability analysis is no longer restricted to the hardware but also takes into consideration other aspects, such as reliability of the human element and the software. Although during the years of World War II human factors began to be regarded as a discipline, not until the late 1950's was it clearly stated that realistic system reliability analysis must include the human aspect.

Many studies have indicated that human errors account for a large proportion of total failures. Findings of some of these different studies (Suchlon 1986) were as follows:

i) About 50-70% of the failures in electronic equipment were human initiated (Christensen et al., 1981).
ii) In aircraft and missile systems, the human initiated failures accounted for 60-70% and 20-53% (Christensen) of the total failures, respectively.
iii) About 10-15% of the total failures were directly due to humans (Hagen 1976).
iv) About 20-50% of the total failures resulted from human error (Shapero et al 1960).
v) About 23-45% of the total failures occurred because of humans (Levan 1960).
vi) About 25% of the maintenance events described in 213 problem reports from the field were due to human errors (Robinson et al 1970).

At least two major disasters -- the Three Mile Island nuclear accident and the DC-10 accident (Christensen 1981) at Chicago's O'Hare Airport in which 272 persons lost their lives -- were partially or fully due to human error in one way or the other. Nowadays, efforts are directed toward replacing the human functions with machines and monitoring the human with computers. The main objective of this effort is to reduce human errors. However, since even highly automated systems do not totally remove human involvement, it is not wrong to state that without proper consideration to human reliability during the information system design phase, the reliability analysis will be incomplete.

Human reliability and human error are defined as follows:

i) Human reliability: This is the probability of humans accomplishing a job or task successfully at any required stage in system operation within a specified minimum time limit (if the time requirement is specified).

NATO ASI Series, Vol. F32
Information Systems: Failure Analysis
Edited by J. A. Wise and A. Debons

ii) Human error: This is the failure to carry out a specified task or the performance of a forbidden action that could lead to disruption of scheduled operations or result in damage to property and equipment.

HUMAN ERROR

In information systems, human errors may occur in various ways and for various reasons. The ways in which a human error can occur are the following:

i) An incorrect decision in response to a difficulty.
ii) Failure to realize a hazardous situation.
iii) Carrying out a task that should not have been accomplished.
iv) Failure to carry out a necessary function.
v) Poor timing and poor response to a contingency.

Some of the reasons for error occurrence could be inadequate lighting in the work area, inadequate training or skill of the concerned manpower, poor system design, high temperature in the work area, crowded work space, poor motivation, poorly written maintenance and operating procedures, poor management, task complexity, poor verbal communication, and poor work layout.

Generally, human errors may be classified into six distinct categories: operating errors, design errors, inspection errors, maintenance errors, assembly errors, and installation errors. Each of these groups is described in detail in Suchlon (1986).

HUMAN ERROR PREVENTION METHODS

Two approaches can be useful in the prevention of human error in information systems. These methods are described below as Method I and Method II (Suchlon 1986).

Method I

This method is known as man-machine systems analysis. Credit for the original version of it is due to Robert B. Miller. He developed a technique for "man-machine task analysis" in the early 1950s (Miller 1953; Swain 1973). The man-machine systems analysis method (MMSAM) is used to lower unwanted effects caused by human error to some acceptable level in a system. The method consists of 10 steps:

i. Outline the functions and goals of the system.
ii) Outline the situational characteristics.
iii) Outline the characteristics of manpower associated with the system.
iv) Outline the tasks and jobs carried out by the system manpower.
v) Perform analysis of tasks and jobs to surface potential error-likely conditions and other associated difficulties.
vi) Obtain an estimate for the occurrence of each potential error.
vii) Obtain an estimate for the likelihood that each potential error will remain undetected and uncorrected.
viii) Obtain an estimate for the consequences of each undetected potential error.
ix) Make recommendations for changes to the system.
x) Re-evaluate each system change by repeating most of the above steps.

Steps 1 and 4-10 are considered to be self-explanatory. Their description is therefore omitted. Step 2 is concerned with those performance-shaping factors (i.e., situational characteristics) under which humans will have to carry out various tasks and jobs. Typical examples of such performance-shaping factors are illumination, union actions, quality of the air, general cleanliness, and so on. Step 3 is concerned with the identification and estimation of the characteristics of the manpower associated with the system. Examples of such characteristics are training, experience, motivation, and skills.

Method II

Method II, known as the "causal factors concept," can be used to improve human reliability in information systems (Suchlon 1986; Bailey 1973). This method was originally proposed for computer-based business information systems. It offers new flexibility to system designers as compared with alternatives of the past. These alternatives were either to eliminate or ignore the human component, or to develop a system that tolerated a level of human reliability. Some of the assumptions associated with Method II are as follows:

1) In a situation where one or more causal factors act to degrade performance, a human error is generated.
2) Human errors can be eliminated by acting on factors that cause them to occur.
3) It is possible to identify and control the factors that cause human error.

The major causal factors that act or interact to degrade the performance of humans in computer-based business information systems are system design, source data, documentation, training, man-machine interface, personal factors, and environmental factors. Therefore, one must pay close attention to these factors in order to improve reliability of information systems. Each of these factors is described in detail in Suchlon (1986).

HUMAN PERFORMANCE RELIABILITY EVALUATION

This section is concerned with human performance reliability modeling in a continuous time domain. Using the Markov technique, the human performance reliability function for time-continuous human tasks can be developed. Examples of time-continuous tasks performed by humans are scope monitoring and aircraft maneuvering.

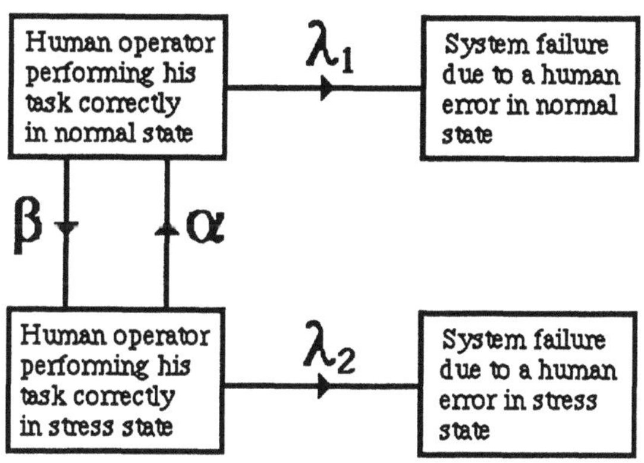

Figure 1. State-space diagram for the human working in fluctuating environments.

To predict the reliability of a human performing time-continuous task in information systems, a Markov model is proposed (Dhillon 1982). The state-space diagram of the model is shown in Figure 1. In this model, the human operator performing the time-continuous task fluctuates between normal work and stress states. The task is associated with a system. The system fails because of human error during either the normal work state or the stress state (i.e., when the operator is performing under normal conditions or under stress). The following assumptions are associated with this model:

1) Errors are statistically independent.
2) Human error rates are constant.
3) The human operator is performing a time-continuous task.
4) The rate of changing human operator condition from the normal state to the stress state and vice versa is constant.
5) Numerals in Fig. 1 denote corresponding states.

The following symbols are associated with the Figure 1 model:

- λ_1 is the constant human error rate from 0.

- λ_2 is the constant human error rate from state 2.

- α is the transition rate from the normal state to the stress state.

- β is the transition rate from the stress state to the normal state.

At $\beta = \alpha = 0$, this model represents a situation in which the human is performing the time-continuous task in normal state only. Equations for human reliability and mean time to human error are developed in Suchlon (1986) and Dhillon (1982).

HUMAN RELIABILITY ANALYSIS METHODS

This section presents three methods of performing human reliability analysis. These methods will be useful for analysis of information systems with respect to the human element.

Method I

Method I is known as the "failure modes and effects analysis (FMEA)" approach. It was first used in the development of flight control systems in the early 1950s (Countinho 1964; Dhillon 1983b and 1985). The FMEA method is concerned with determining design reliability by considering potential failures and their effect on the system under study. More specifically, it is concerned with listing each failure mode of an item on paper and its effects on the listed subsystems. This "bottom-up" approach can be utilized at any level, that is, from complete systems to components.

In broader terms, the following steps are involved in performing failure modes and effects analysis:

1. Define the boundaries of the system in question and its requirements.
2. Establish ground rules for performing FMEA.
3. List the system's subsystems and parts.
4. List the identification, failure modes, and description of the part under consideration.
5. For each failure mode, estimate the failure rate.
6. List the effects of each failure mode on subsystem, system, and so on.
7. For each failure mode, enter remarks.
8. Investigate the serious failure modes and take desirable steps.

The FMEA approach offers several advantages:

1. It is a visibility tool; that is, FMEA tabulations are visible.
2. It is a systematic procedure.
3. It can easily be understood.
4. It identifies weaknesses in the system design.
5. It becomes useful in design comparison.

The method is described in detail in Dhillon (1983b).

Method II

Method II is known as the fault tree technique. In Suchlon (1986), it is applied to human reliability analysis. This method, which makes use of various symbols, is commonly used in industry to evaluate the reliability of a system. It was developed in the early 1960s to evaluate reliability of the Minuteman Launch Control System. Since then, it has gained favor over other techniques, especially for analyzing complex systems.

Fault tree analysis begins by identifying the top event, known as the "undesirable event" of the system. This undesirable event is caused by events generated and connected by logic gates, such as AND, OR, etc. The following basic steps are involved in performing fault tree analysis (Dhillon 1985):

1. Establish system definition.
2. Construct the fault tree.
3. Evaluate the fault tree qualitatively.
4. Collect basic data, such as component failure rates, repair rates, and failure occurrence probability.
5. Evaluate the fault tree quantitatively.
6. Recommend corrective measures.

Some of the advantages of this technique (Dhillon 1981) are as follows:

1. It provides insight into the system behavior.
2. It requires the reliability analyst to understand the system thoroughly and to deal specifically with one particular failure at a time.
3. It provides a visibility tool to designers, users, and management to justify design changes and trade-off studies.
4. It provides options for quantitative or qualitative reliability analysis.
5. It can handle complex systems more easily than other techniques.

The method is explained in Dhillon (1981 and 1983a).

Method III

Method III takes into consideration the occurrence of critical and non-critical human errors in a redundant system. It can be used in reliability evaluation of redundant information systems with critical and non-critical human errors. It is a block diagram method that can be used to compute the reliability of a parallel system. In this system, it is assumed that all the units are active and at least one unit must function normally for system success. The system fails when all of its units fail.

Each unit of the system may fail because of a hardware failure or a noncritical human error. Human errors are classified as either critical or noncritical. A critical human error causes all of the system units to fail simultaneously, whereas only one unit fails when there is a noncritical human error.

The method assumes that the system units fail independently. The block diagram of the proposed method, shown in Figure 2, is composed of n units. Each unit's failure probabilities are separated into probabilities of hardware failures and human errors. In addition, failure of all system units is due to the occurrence of a critical human error. A hypothetical unit representing critical human errors is also shown in the diagram. This method is described in detail in Suchlon (1986).

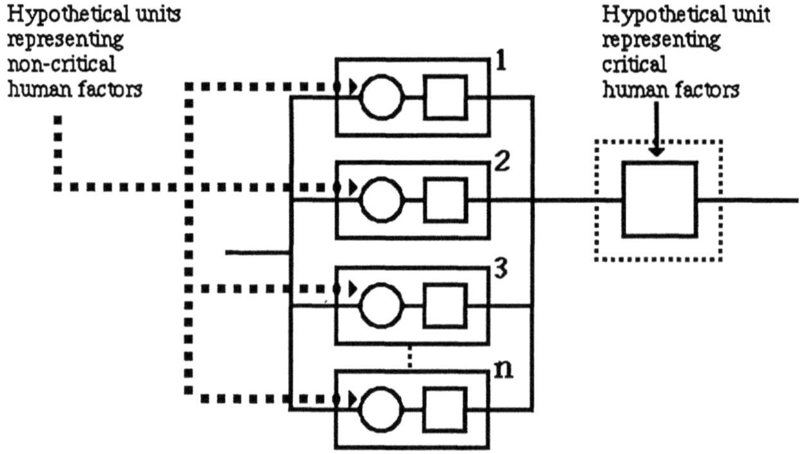

Figure 2. Block diagram of the proposed method.

HUMAN RELIABILITY DATA

Human reliability data play a vital role in reliability studies. In fact, human reliability data are the backbone of any human reliability study. Moreover, the accuracy of these studies will be directly affected by the quality of the input data. In other words, poor data will engender misleading results. Therefore, as much attention as possible must be paid when collecting and analyzing reliability data. In addition, one has to be sure of what, when, why and where to collect human-reliability-related data because of the high price tag associated with the data collection.

At present, it may not be unreasonable to state that there is a lack of data in human reliability work in comparison with the availability of techniques and methods for predicting human reliability. Furthermore, the lack of human data is probably the most critical impeding factor to the development of a human performance reliability index.

Various methods are used to collect human-related data. Common standards, such as reliability validity, automation, economy, objectivity, and quantifiability, are to be applied to data-collection methods. In Suchlon (1986), several aspects of the human reliability data are discussed. In particular, various sources of human reliability data are listed. These sources could be quite useful in obtaining data for failure analysis of information systems.

CONCLUSION

Some of the concepts described in this paper are well known in reliability engineering and are being used to perform analysis of engineering systems. In the author's opinion, these approaches could also be quite useful in performing information system failure analysis. Finally, it is contended that this paper will be useful to professionals who study information system failures.

REFERENCES

Bailey, R.W., S.T. Demers, and A.I. Lebowitz. (1973). Human reliability in computer-based business information systems. *IEEE Transactions on Reliability*, 22, pp. 140-147.

Christensen, J.M., J.M. Howard, and B.S. Stevens. (1981). Field Experience in Maintenance. In J. Rasmussen and W.B. Rouse. *Human Detection and Diagnosis of System Failures*. New York: Plenum Press. pp. 111-133.

Countinho, J.S. (1964). Failure-effect analysis. *Transactions of New York Academy of Sciences*, 26 (Series II), pp. 564-584.

Dhillon, B.S. (1982). Stochastic models for predicting human reliability. *Microelectronics and Reliability*, 22, pp. 491-496.

Dhillon, B.S. (1983a). *Reliability: Engineering in Systems Design and Operation*. New York: Van Nostrand Reinhold.

Dhillon, B.S. (1983b). *Systems Reliability, Maintainability and Management*. New York: Petrocelli Books.

Dhillon, B.S. (1985). *Quality Control, Reliability and Engineering Design*. New York: Marcel Dekker.

Dhillon, B.S. and C. Sin. (1981). *Engineering Reliability: New Techniques and Applications*. New York: John Wiley.

Suchlon, B.S. (1986). *Human Reliability: With Human Factors*. New York: Pergamon Press.

Hagen, E.W. (ed.). (1976). Human reliability analysis. *Nuclear Safety*, 17, pp. 315-326.

LeVan, W. L. (1960). *Analysis of the Human Error Problem in the Field*. Bell Aerosystems Company, Report No. 7-60-932004.

Miller, R.B. (1953). *A Method for Man-Machine Task Analysis*. Wright Air Development Center Technical Report No. 53-137, Wright-Patterson Air Force Base, Ohio.

Robinson, J.E., W.E. Deutsch, and J.G. Rogers. (1970). The field maintenance interface between human engineering and maintainability engineering. *Human Factors*, 12, pp. 253-259.

Shapero, A., J. I. Cooper, M. Rappaport, K.H. Shaeffer, and C.J. Bates. (1960). *Human Engineering Testing and Malfunction Data Collection in Weapon System Programs*. Wright Air Development Division Technical Report, 60-36, Wright-Patterson Air Force Base.

Swain, A.D. (1973). *Design Techniques for Improving Human Performance in Production*. London: Industrial & Commercial Techniques Ltd.

ERROR AUDITING IN AIR TRAFFIC CONTROL

Jacob Empson

University of Hull
North Humberside, U. K.

Recently developed methods of classifying errors in psychological terms have been applied to ergonomics(e.g., Reason, 1977). Langan-Fox and Empson (1985) applied the technique of error auditing in the setting of air traffic control. This paper summarizes their findings and assesses the implications for improving safety in air traffic control, in the contexts of training, maintenance of standards, and improvement of the working environment. Other potential uses for the technique will also be assessed.

The study of errors, or "actions not as planned," has made it clear that common slips of the tongue and hand are not only very frequent but classifiable in psychologically meaningful ways. Reason (1979), for instance, found that about 95% of all the 433 errors collected by 35 subjects over a two-week period were classifiable using five major categories of error and 18 subcategories. Norman (1981) developed a similar number of categories, although his system has not been field- tested quite as fully. Both these approaches have shown how easy it can be to collect errors and to categorize them in terms of psychological processes rather than their consequences.

Central to this new approach to the analysis of error is the notion that errors are not merely accidental. A totally unforseeable Act of God that disrupts otherwise well-constructed and effectively executed plans is of little psychological interest. Errors on the other hand, are seen as the consequence of poor planning or of the inadequate execution of reasonable plans. Three models of error in terms of information processing (Reason, 1977, 1979; McRuer et al, 1980; Norman 1981) have been formulated, and Table 1 shows the classification system given by Reason (1979). Without discussing the relative theoretical merits of these models in detail, it can be noted that this system evolved as the outcome of a series of trials, including a two-week diary study of 35 volunteers as well as the post hoc analysis of accidents and near misses.

Table 1. Classification of Motor Program Error

A. Discrimination Failures
1. Perceptual Confusion
2. Functional Confusion
3. Spatial Confusion
4. Temporal Confusion

B. Program Assembly Failures
1. Spoonerisms
2. Confusion between current programs
3. Confusion between stored and ongoing programs

C. Test Failures
1. Stop-rule overshoots
2. Stop-rule undershoots
3. Branching Error
4. Multiple side-tracking

D. Sub-routine Failure
1. Insertions
2. Omissions
3. Misordering

E. Storage Failure
1. Forget previous action
2. Forget discrete items of the plan
3. Forget substance of the plan
4. Refer to earlier plan

Adapted from Reason (1979).

NATO ASI Series, Vol. F32
Information Systems: Failure Analysis
Edited by J. A. Wise and A. Debons

Commercial flying remains the safest mode of travel, and even in military aviation, a good degree of safety is maintained, considering the inherent instability of some military aircraft and the peril of such manoeuvers as low flying and mid-air refueling, which are commonplace in and intrinsic to military flying. The levels of safety achieved have been a result of attention to procedures, human factors ,and training as well as purely engineering considerations. Highly important in this respect is the complex system of air traffic control that governs both military and civil flying. Danaher (1980) has pointed out the success of A.T.C. in preventing accidents, although he also reports that... "more than 90% of all the system errors that do occur stem from human mistakes in attention, judgment, and communications by controllers and their supervisors."

All A.T.C. controllers in the U.K. are highly trained and periodically examined in their professional knowledge and practical controlling skills. As well as local examining officers, a team of Examiners tours the units to validate all controllers annually. If a particular controller remains unexamined for 3 years, his or her license automatically becomes invalid, unless special dispensations are sought and granted.

A selection of the errors made by military controllers that were noted by Examiners in 1981/1982 are shown in Table 2. It can be seen that errors were not uncommon in all three groups of units visited. In the context of the analysis of error mentioned above, it is also evident that the categories of error used sometimes refer to the consequences of a number of possible errors. For instance, the erosion of Separation Standards could be the consequence of errors of commission, in directly instructing aircraft to fly too close, or of omission, in failing to instruct aircraft to alter course to avoid one another.

Table 2. Some Criticisms of R.A.F. A.T.C. Units 1981/82

	Percent of Units Criticized		
	Command 1	Command 2	Command 3
Item	1982 ('81)	1982 ('81)	1982 ('81)
Separation Standards	46 (32)	35 (28)	20 (40)
Avoiding Action	46 (32)	35 (28)	20 (60)
Procedure Adherence	18 (20)	29 (36)	60 (20)
Co-ordination	14 (28)	29 (14)	20 (60)
R/T Phraseology	85 (76)	57 (64)	60 (60)
Landline Identification	73 (72)	14 (28)	20 (40)
No. of Units assessed	26 (25)	14 (14)	5 (5)

Source: Ministry of Defence (1982)

The Joint Airmiss Working Group (J.A.W.G.) is a committee of the Civil Aviation Authority in the U.K. which has been collating and investigating reports of near-misses since 1972. About 40 incidents per year have directly implicated air traffic controllers. While the incidents are carefully evaluated and causes attributed, this has been done in a relatively unsystematic way. The examples below illustrate imprecise attributions of causation:

(a) "There appears to have been a management problem" (C.A.A. Document 265, p.9).
(b) "Failure of the ... sector controller to provide the prescribed separation". (C.A.A. Document 261, p. 18).
(c) "Sector Controller ... should have ensured that he was aware of the interaction of all traffic before he assumed responsibility for the sector". (C.A.A. Document 247, p. 20).

An example of a more definite attribution of error:

> In spite of the ... controller's having recognized the confliction and planned appropriate action, his foresight was negated by then using the wrong callsign to pass the information to the pilot of the F111 - and that was the cause of this incident. (C.A.A. Document 265, p. 15).

Both the A.T.C. examiners (the A.T.C.E.E.B.) and the J.A.W.G. log and report errors, either as they occur or post-hoc, but neither of them has a standard method of doing so except in terms of their consequences. It is clear that this lack of precision is unsatisfactory, both in terms of feedback for controllers and of the understanding of the causation of dangerous near-misses. The critical importance of understanding errors, however trivial, in A.T.C. suggests that here is a potentially useful application of Reason and Norman's classifying techniques.

THE FIELD STUDY

Air Traffic Control Tasks

This study involved eight air traffic controllers (the total complement at this station) at a Royal Air Force fast-jet airbase in Lincolnshire. All eight subjects were licensed to perform two distinct roles:

(1) Radar Director - responsible for directing aircraft to the airfield either for visual or instrument approaches. The director gives heading and altitude instructions to aircraft and keeps them separated from other traffic and sequenced for approaches to the airfield.

(2) Radar Approach - responsible for all aircraft within a 30 mile radius which request A.T.C. services, and the initial identification of aircraft inbound to the airfield. The approach controller also provides a climb-out service for departing aircraft. (For readers unfamiliar with this terminology a potential source of confusion will probably need clarification - the Director is responsible for aircraft approaching the airfield, while Approach is responsible for aircraft within the approach area.)

Both Director and Approach controllers use identical equipment, at times using the equipment of the other during maintenance periods. Their tasks are ostensibly very similar, requiring that aircraft under control are separated from other traffic; that co-ordination with other control units is undertaken and adhered to; that information is passed and received; and that magnetic heading height instructions are given. In order to achieve these ends, both task controllers must take account of the speed and vectors (direction of travel relative to 'true' north) of their aircraft with respect to other radar contacts using the same criteria of separation, and both must pre-plan and predict on a continual basis. Both receive aircraft from free-call (an unannounced radio call from an uncontrolled pilot to a controller) and handover from others requiring the same range of identification procedures.

The controlling task is virtually identical. The difference lies in the intent. For the Director, aircraft have a single destination - the home base. The Approach controller's aircraft have a multitude of destinations, not normally reached while under approach control. Inputs, outputs and processing of information are therefore markedly similar, as are the skills required to execute them. Thus, their error performance would be expected to be equivalent under similar conditions of workload. Approach controllers typically handle twice as much traffic as Directors. One might expect on this basis that there would be more opportunity for error in the Approach task.

The Self-Report Study

Fifteen controllers, comprising eight fully licensed and seven trainee controllers, participated in a two-week self-report scheme, similar to the one reported by Reason (1979). Unlike Reason's subjects, however, they did not notice many of their own slips - admitting to a total of 13 instances of error being. To anticipate the findings of the main study, which involved direct observation of performance - ten times as many errors were recorded in half as many subjects over one-seventh of the time. Self-report levels of errors in performance must obviously be regarded with some scepticism in the future.

Direct Observation Method

All 8 fully qualified controllers at this station were observed over two periods of 2 hours and 30 minutes, once as Approach Controller, once as Director Controller. Order was counterbalanced, as was time of day (involving the morning shift from 09.30 to 12.00, and the afternoon shift from 14.30 to 15.00). Prior to each session, the subject was briefed as to the specific intention of the study.

Subjects were required to indicate when they felt they had erred. A note was made immediately and the nature of the error was clarified as soon as traffic conditions allowed. In addition ,the observer noted any errors and clarified them with the progenitor as soon as practicable. This dialogue with the controllers made it possible to take their intention into account in making a classification. Many errors were readily apparent, e.g., incorrect use of equipment, incorrect phraseology, and confusions queried by other controllers and pilots. Errors that went unnoticed by the controller (but not the observer) were verified by a third party, either by another controller or by reference to tape-recorded radio transmission exchanges. At the end of each session, each error was examined in a debriefing session with the subject that lasted up to 30 minutes, and the observer completed his observation notes prior to the next session.

It should be made clear that the observer (C. L-F) was well known to all the controllers at this station as a recently retired colleague, and thus not only had their confidence but was an expert in their somewhat arcane world.

Results of the Observation Study

A total of 131 errors were observed, all but eight of which were categorizable using Reason's system - a similar percentage to that obtained in his 1979 study (94%). Table 3 enumerates the errors for each category. These are generally very few. However, the five major categories can be merged into three psychologically distinct categories of error type to make statistical analysis more manageable.

Grand Categories of Error Type:

(1) Category A, the discrimination error, is perceptually based.
(2) Categories B and D can be combined as program based.
(3) Categories C and E relate specifically to memory functions.

By reducing the number of categories and ignoring the subcategory distinctions (and the unclassified residue), some indication can be given of the relative frequencies of errors related to three broad but distinct cognitive functions.

The total number of errors made by Directors was more than the Total made by Approach controllers despite the fact, that Directors generally have fewer aircraft on frequency (under their control) than Approach controllers. A sensible comparison should take this into account.

Table 3. Number of Errors in Reason's Categories

Category	Director	Approach	Total
Discrimination Failures (A)	19	13	32
Program Assembly Failures (B)	21	19	40
Test Failures (C)	10	6	16
Subroutine Failures (D)	11	8	19
Storage Failures (E)	10	6	16
Non-categorizable	4	4	8
Total	75	56	131

To enable reasonable comparisons to be made, three levels of workload were distinguished:

(1) 0 SU - no aircraft on frequency
(2) 1 SU - a single speaking unit (SU) on frequency
(3) 2 SU - 2 or more speaking units on frequency

Using error rates per hour as a measure for each level of workload, it is feasible to analyze the relationships between error rate, workload, and task variables. Table 4 shows these error rates under three levels of workloads.

Table 4. Error Rates per Hour under Three Levels of Workload

Workload (Speaking Units)	0			1			2+		
Error Type (Reason)	A	BD	CE	A	BD	CE	A	BD	CE
Director	.79	.54	.48	1.18	4.32	1.21	1.35	3.39	2.41
Approach	.40	1.24	.77	.87	1.11	.11	.61	2.00	.84

Analysis of variance taking task, broad error type (A, BD and CE) and workload as three repeated measures factors showed the following:

(1) Error rate was greater for Director than Approach control tasks ($F = 21.24$, $df = 1,7$, $p<.01$).
(2) Error rate increased significantly with workload ($F = 5.36$, $df = 2,14$, $p<.05$).
(3) An interaction between workload and task ($F = 7.53$, $df=2,14$, $p<.01$) reflects the increase in error, with workload being predominantly in the Director task. Furthermore, the increase in errors made by Directors is predominantly between conditions 0 and 1. A post hoc analysis here showed that the slight increase in error rate with workload in Approach did not even approach statistical significance ($F = .38$, $df = 2,23$).
(4) The BD (program) errors were more numerous than CE (memory) errors Reply received from user and A (discrimination) errors ($F = 5.74$, $df = 2,14$, $p<.05$). (e)

DISCUSSION OF THE RESULTS AND THEIR IMPLICATIONS

Error Rates and Workload Assessment

In the more demanding tasks, the number of errors produced per unit time increased dramatically with workload. Despite Directors having aircraft on frequency only 36% of the time, compared with 79% for Approach controllers, they made more errors and made 68% of them in the time that aircraft were on frequency, while Approach controllers seemed to produce errors at about the same rate whether aircraft were on frequency or not.

With hindsight, it is possible to see how the Director's job was more difficult, in terms of pacing and time stress, than the Approach controller's job. The Director's job was typically to receive aircraft from free-call at a distance of 12 miles from the P.A.R. handover position, and direct them to an appropriate place for handover to P.A.R. for landing at Binbrook at a rate allowing them to be dealt with one at a time. Since these fast jets returning from reconnaissance duties were normally very short of fuel, no prolonged delays were feasible.

Approach controllers were similarly asked to receive aircraft from free-call, but since these were not heading for the home airfield, the controller cold refuse to receive them until he or she was prepared to. When taken on, they were guided to positions determined by the pilot's intentions, but without the high priority to fuel economy and operational criteria typical of the Director's task. Handover to another control agency or to enroute under his own navigation was similarly a relatively leisurely business, with no equivalent pressure on Approach controller to hand over in a precise position at a precise heading and height, each pre-determined and planned.

The error count alone gave a good indication of workload levels, understandable in terms of task demands. Regardless of any fine grain analysis of the types of error produced, it seems that this method alone could be used in the development of the organization of work in other demanding environments. An obvious application would be in the study of pilot and navigator workloads, and it should be feasible to collect error data from performance in simulators.

Types of Errors

Program-based (BD) errors predominated, with 45% (59) of all errors. Of these 59 errors, 50, were of type B - spoonerisms both verbal and behavioral, and the confusions of stored, current and ongoing programs. There were 24% each of CE (memory-based) and A (perception- based) errors. There was no evidence of any interaction between workload or task type and the proportions of errors of different types, despite the wide variations in error rates with workload (with Directors' error rate increasing by 3.6 times with aircraft on frequency, compared with no aircraft on frequency). Of the 131 errors, 23 could be considered as potentially serious in their consequences. However, their production did not mark them out in any way from the remaining 108, and there was no significant bias towards any one error type.

The analysis of error type can provide useful information in two contexts - the identification of poor engineering ergonomics, and of procedures in need of improvement either by additional training or redesign. Again, this technique would appear to be readily transferable to other working environments, ranging from the office with new microprocessor equipment or to the high-tech environment of avionics.

A-Type Errors

In general, type A errors (perceptual-based) tended to pinpoint particular flaws in equipment design. For instance, misidentification of landline keys accounted for 10 errors - one third of all type A errors. These keys are common to both controller tasks although the usage of individual lines is divided between them - i.e., all lines call to both controllers, but the Approach controller is responsible for answering some, the Director the others. As a line calls, both controllers are obliged to identify the relevant line and decide if they should answer it. Control towers with more up-to-date equipment have filtering devices so that a controller can pre-select which lines will call on his own panel and which will not.

Two serious A-type errors involved the direction-finding equipment, both due to the controller misinterpreting the source of the stored data. That is, the device that indicated the source of a radio transmission was not co-located with the controller's transmitter so that the controller's transmissions could be treated as a source just as easily as the aircraft's.

The radar control knobs and the wind clocks were also the source of A-type errors, which, although not as potentially serious as some of those mentioned above, also highlighted some rather obvious (with hindsight) ergonomic failings.

Type A errors increased with workload as the others did, showing how operators cope with awkward, incompatible working arrangements when under little pressure but tended to fail when under high pressure of work.

BD-Type Errors

These errors were program assembly failures. They included spoonerisms such as saying "The minima procedure for this approach is ..." instead of "The procedure minima for this approach ..." which resulted in time lost through repetitions. Another type is the misordering of programs, where the correct actions are carried out but in the wrong order.

No particular failings in equipment design or procedure seem indicated by these sorts of error, although like the others they increased with increasing workload. Should they be regarded as inevitable? One of the B controllers in this study made no errors of any sort while under observation, indicating that perfect performance is indeed possible. Type BD errors can therefore be taken to indicate the need for more training, more careful selection, or possibly reduction of workload either by employing more operators or delegating some of the work to a machine.

CE-Type Errors

CE errors, or test failures, were typically associated with the use of the memory aids. The Director used a plaque that listed cues for information to be given and requested. The errors made using this plaque were all trivial, yet indicative of "poor housekeeping." For instance, information calls were not ticked off or filled in, or information on the back of the plaque was forgotten. The Approach director used no plaque, but wrote abbreviated information on a formica worktop below the radar screen.

The additional memory aid (the plaque with cues) provided for Director controllers indicated an implicit acknowledgment that their job was more demanding than the Approach's job. This mnemonic aid did seem to help, although 15 errors were associated with its use. Four of them involved Directors forgetting about information on the back of the plaque. Approach controllers used the formica worktop surface below the radar screen, adopting a standard method and shorthand scheme for this purpose.

Of the Approach errors, 10%, involved failing to inform the local controller that previously notified aircraft had now cleared the Military Air Traffic Zone. That is, they had correctly notified the local controller that an aircraft was about to transmit over the airfield so that he could warn aircraft under his control (typically, just about to take off), but failed to say when the aircraft had passed. This is an indication of poor housekeeping - well established as a symptom of slack controlling, predisposing system failure induced by human error (Danaher, 1980). While poor housekeeping should always be promptly brought to the attention of the progenitor, controllers might be helped by additional mnemonic devices, and a possible recommendation would be that all controllers should be provided with plaques like those provided for Directors but with all the mnemonic cues on the front.

POTENTIAL CONTRIBUTIONS OF ERROR AUDITING

At the beginning of this paper, it was pointed out that the routine assessment of air traffic controllers involves extensive monitoring of errors, normally in terms of their consequences.

While it would be absurd to claim that the consequences of errors are unimportant, it is also true that all slips may potentially contribute to danger buildup. In addition, some consideration of the types of error being committed could provide useful information about the causes underlying unsatisfactory performance. A.T.C. examiners might find this sort of analysis very revealing.

The post-hoc analysis of air-misses by bodies such as the Joint Airmiss Working Group could also benefit from distinguishing more carefully between errors and their consequences, although in their case the evidence is frequently no longer available to determine precisely what the error was.

More generally, the techniques of error auditing may find a useful role in a variety of working environments. The simple error count is a sensitive index of high workload - useful in that the increase in minor errors occurs long before a frank breakdown in performance. The analysis of perceptually based errors can identify poor ergonomic design, and the analysis of test failure, or memory- based, errors can pinpoint failings in procedures or training. Error auditing may well find applications in the nuclear industry, in chemical process, and in the modern office as well as the aviation industry.

REFERENCES

Danaher, J.W. (1980). Human error in A.T.C. systems operations. *Human Factors*, 22, 535-545.

Langan-Fox, C. P. and J. A. C. Empson. (1985). 'Actions not as planned' in military air-traffic control. *Ergonomics*, 28, 1509-1521.

McRuer, D. T., W.F. Clement, and R. W. Allen. (1980). *A Theory of Human Error.* NASA Ames Research Center, Moffat Field. NASA contract NASA-10400.

Ministry of Defence. (1982). *Analysis of Criticisms.* A.T.C.E.E.B./190/AIR (file) Royal Air Force, Shawbury, Shropshire, U.K. School of Air Traffic Control.

Ministry of Defence. (1983). Major flying accidents 1981/1982. *Air Clues*, 37, 100-101.

Norman, D. A. (1981). Categorization of action slips. *Psychological Review*, 88, 1-15.

Reason, J. T. (1977). Skill and error in everyday life. In Howe, M. (ed). *Adult Learning.* London: Academic Press.

Reason, J. T. (1979). Actions not as planned: the price of automation. In G. Underwood and R. Stevens (eds). *Aspects of Consciousness (Volume 1).* London: Academic Press.

FAILURE ANALYSIS OF INFORMATION SYSTEMS: REFLECTIONS ON THE USE OF EXPERT SYSTEMS IN INFORMATION SYSTEMS

Erik Hollnagel

Computer Resources International
Copenhagen, Denmark

INTRODUCTION

One of the specific products of modern information technology is the expert system. Expert systems have been developed to support humans in handling the large amount of information on which more and more tasks depend for their success. The expert systems do that by applying techniques of artificial intelligence, mainly various types of reasoning, to cases where human intelligence is insufficient - either because it is too slow or error-prone or because it is a scarce resource (cf. Wiener, 1986). Expert systems thereby more and more often become part of large information systems - in addition to being Information Systems themselves. It is therefore necessary to consider the implications of the increased use of expert systems for the general approach to failure analysis of information systems. There are two completely different aspects of this:

- The expert system as an information system that itself may fail, i. e., as a target.
- The expert system as an aid for the prevention and diagnosis of failures in information systems, i.e., as a tool.

The latter aspect has been alluded to by Wagenaar (1986) and has also been treated generally in a number of cases (e.g., Hollnagel, Mancini and Woods, 1986). I will therefore not consider that here - apart from noting the interesting possibilities for recursion that arise when an expert system becomes the tool as well as the target. I will instead consider the principles and problems of failure analysis of expert systems as information systems.

DEFINING FAILURES IN EXPERT SYSTEMS

System failures are normally detected only from how the results or products appear, either by the absence of the expected consequences or by the presence of unexpected and unacceptable consequences. This, of course, requires that the system user is able to determine whether one or the other is the case. In technological systems that deal with matter and energy (for instance, transport systems or process industries), it is normally easy to determine that a failure has occurred -- although it may be difficult to find the cause of it. In systems that deal exclusively with information, such as expert systems, it may be difficult even to determine that a failure has occurred - either because the expected product is less well defined than for a technological system or because the consequences do not show themselves as clearly. A decision support system may, for instance, produce a recommendation, but the user has no quick and simple way of determining whether the system has functioned correctly or has failed.

The difficulty in defining failures in expert systems is very much related to the fact that they, like information systems, are cognitive rather than technological. Hence, failures must be defined not so much from the product as from the process that led to it (cf. Hollnagel 1986). Finding the failure in a process is always difficult, as experiences from technological accident analysis show. Finding a failure in an expert system is further complicated by the fact that the

NATO ASI Series, Vol. F32
Information Systems: Failure Analysis
Edited by J. A. Wise and A. Debons

process is based on (simulated) reasoning that neither develops in an orderly, linear fashion nor leaves any clear trace behind.

CURATIVE FAILURE ANALYSIS

Normally, a failure analysis is started only when a failure has been found, i.e., as part of a cure. The analysis is therefore basically a type of troubleshooting or diagnosis with the purpose of finding the true cause of the failure as a step towards preventing its recurrence. There are many ways of doing this, but all of them require that some sort of prior analysis of the system has been made. A good example is the so-called Modus Operandi approach, proposed by Scriven (in Janis, 1986), which entails the following steps:

- Formulate all known causal sequences that may account for the observed outcome. This may be done in advance during a phase of system analysis.
- Backtrack from the observed outcome, and watch for a pattern of indicators that connect antecedent conditions with the observed outcome.
- If there only are indicators for one Modus Operandi sequence, accept this as the most probable cause. The degree of certainty depends on the completeness of the evidence (indicators).
- If there are indicators for more thán one Modus Operandi sequence, accept the one with the most complete evidence as the most probable cause.
- If the indicators support two Modus Operandi sequences equally much (or both completely), accept them as co-causes.
- If there are no indicators for any of the known (formulated) causal sequences, then none of the Modus Operandi sequences can be regarded as the probable cause. Instead, the list must be considered incomplete and consequently extended/revised.

While this procedure seems reasonable and certainly is more relaxed than the engineering analysis in terms of fault-trees or cause- consequence-trees, it nevertheless still suf\fers from the same type of limitations in the sense that the outcome of the failure analysis depends on these three factors:

- The quality of the prior analysis (the Modus Operandi sequences). If this is ambiguous or incomplete in the description of the individual cases, the quality of the result will suffer.
- The exhaustiveness of the prior analysis. Even though it must be accepted that a prior analysis never can be complete unless the system is extremely simple (cf., the last step in the above mentioned procedure), it is crucial that the analysis covers the most frequent and the most important cases.
- The description of the cases. The failure analysis is basically a form of pattern matching or recognition and therefore requires that the descriptions of the cases and the actual failure are compatible.

For expert systems there is the further complication that the system's line of reasoning is impossible to predict because in a complex way it depends on the extent and quality of the available information. In addition to these limitations, the failure analysis is restricted to situations where a failure has been recognised. An extension to other situations would require that the analysis of Modus Operandi sequences was carried out for all possible normal sequences as well - clearly an impossible task.

PREVENTIVE FAILURE ANALYSIS

While it may be economical and defensible from a practical point of view to analyse only situations where failures have occurred, it is clearly insufficient for expert systems - and probably for most other types of pure information systems. With expert systems, the all-important issue is the quality of reasoning, and that must be ensured before the system is actually used. One can distinguish three categories of situations:

- Expert systems that have failed and where the failure has been detected.
- Expert systems that have failed but where the failure has not been detected. Here one might make a further distinction between whether a failure is suspected or not.
- Expert systems that have not failed.

The second category, which contains the hidden failures, is certainly as important as the first - and possibly just as frequent. But in order to deal with these cases, the failure analysis must change from being curative to being preventive. This can obviously not be done just by enlarging the scope of the analysis to include normal and abnormal cases alike. An alternative is to consider the characteristics of a correctly functioning expert system. This can be done, for instance, by using a set of standard criteria, such as the following proposed by Assemat & Bonnet (1986):

- Correctness of the final outcome.
- Accuracy of the final outcome.
- Sensitivity, i.e., the minimum variation in input needed to change the outcome of the system.
- Robustness, i.e., the ability of the expert system to absorb and compensate for non-standard input (noise, disturbances, incompleteness, contradictions, etc.).
- Quality or correctness of the reasoning - or, generally, of the information processing of the system.
- Additional criteria are response time and correctness and accuracy of intermediate results. These are particularly relevant with regard to expert systems that can explain their own performance.

These criteria can be used to characterise the process whether a failure has occurred or not. They are global in the sense of being independent of a prior formulation of criteria and cases. The first pair, correctness and accuracy, must be developed in further detail before they can be used and will, naturally, depend on the context. The next pair, sensitivity and robustness, are more general and can serve as operational criteria for the functional testing of an expert system. The last, quality of reasoning, is rather unique and requires further explanation.

THE QUALITY OF REASONING

The reasoning employed by an expert system is obviously crucial for its proper functioning. One would therefore expect that some precautions were taken to ensure the correctness of the reasoning, but surprisingly enough, this is almost never the case. There are several apparent reasons for the lack of such precautions:

- The reasoning mechanisms of expert systems are hidden in the tools.
- Even if the reasoning mechanisms were made explicit, humans would have problems in grasping them.

Expert systems are normally designed and implemented, using very advanced hardware and software tools, by a technique referred to as "rapid prototyping." It is rapid because the detailed specification of the logic, i.e., the principles of reasoning, are provided by special software systems - either full-blown AI development systems or programming environments such as KEE, KnowledgeCraft or ART, or the smaller so-called expert system shells. In either case, the system developer normally lacks several essentials:

- Motivation to look into the details of the reasoning.
- Access to the type of reasoning that goes on.
- Control over the details of the reasoning.
- A reliable way of checking that the reasoning actually is correct.

The situation is not much better with one of the declarative programming languages, such as PROLOG or LISP. In contrast to the procedural languages that have hitherto dominated, the declarative languages by their very nature do not reveal to the application programmer how they work. Their built-in control structures and resolution techniques are the source of their power but also an obstacle to understanding how they work (e.g., Coombs & Stell, 1984).

It may, of course, be assumed that the reasoning mechanisms that are part of the software tools are correct. If the reasoning was restricted to elementary logic, such as Boolean algebra, this assumption might be warranted. After all, the mechanisation of this type of logic is a trivial matter and the very foundation of digital computers. The experience from other software systems, particularly the statistical packages, nevertheless indicates that errors may occur and remain undetected for long periods of time even in fundamental functions. But moving on to other and more sophisticated types of logic and reasoning, or even just using more complex combinations of the elementary operations, almost certainly will introduce failures - simply because the programs are written by fallible humans.

When it comes to validating a reasoning mechanism in a program, there are no practical formal methods. The standard approach is to validate by using a number of pre-defined test cases, but that has obvious limitations. The alternative is to have a qualified person scrutinize the reasoning mechanism. This may be done for each rule in isolation, but not when they interact. As many psychological experiments have demonstrated, people are very bad at reasoning in a consistent way if the problem is even moderately complex. That goes for reasoning about syllogisms and certainly also for understanding how a reasoning mechanism works. We are thus trapped in an impasse: expert systems are introduced to compensate for the shortcomings of human reasoning, but these very shortcomings make it impossible to determine whether the reasoning mechanisms of the expert systems work correctly.

Let me summarise the preceding arguments in the following way. Expert systems are becoming more and more important as parts of information systems. Failures may occur in expert systems, but it is insufficient only to analyse the cases where failures have been detected because that will (1) miss the non-detected failures and (2) give only an incomplete understanding of the expert system. The reasoning mechanism is crucial both for understanding the normal functioning and for the failure analysis. In fact, an adequate understanding of the normal functioning is a precondition for the failure analysis - exactly as for any other information system. We must therefore consider how the understanding of reasoning mechanisms can be improved.

ARCHITECTURE, COMPLEXITY, AND TRANSPARENCY

A general problem in failure analysis is to attribute the cause of the failure to the right part of the system, which in the case of expert systems means the right step in the reasoning mechanism. In most cases, failure analysis seems to follow the principle of satisficing - i.e., the analysis is terminated when the first reasonable cause has been found. This is, of course, highly insufficient because there may be (1) multiple causes or (2) another cause, one or more steps further removed, which better explains the failure.

Defining the correct stop-rule for the analysis requires that the functioning of the system is properly understood and described. The system, in other words, must be transparent rather than opaque, i.e., the user must be able to look inside it (and, speaking metaphorically, see the reasoning mechanism) and understand what is going on. This means that the system must be of a very limited complexity. The evidence from experimental cognitive psychology and information science seems to indicate that the comprehension should not require reasoning of more than a few steps or cover more than a few alternatives for each step, particularly if they include conjunctions (AND gates) or negations.

A system can be transparent to the user only if its architecture has a low level of complexity. This does not imply that the system must be very simple and able to do only very simple things, but it means that the architecture must be based on considerations of the user's cognitive characteristics rather than on technological considerations. The architecture of information systems in general and of expert systems in particular is dictated by the technological imperative. The systems are designed to be as efficient as possible using existing technology. Their functions are thus determined by what the technology can offer and dominated by a tendency to achieve technological goals, e.g., faster CPUs, larger memories, high resolution graphics, etc. A recent example is the growing use of parallel processors to design machines that not even system programmers can use or debug efficiently.

Rather than being dictated by what the technology can offer, information systems should use the flexibility of the technology to keep the architecture functionally simple. Or put differently, efficiency should be defined not from technological considerations alone but from the point of view of total system functioning, i.e., how it works together with the user. Reasoning mechanisms are generally very reliable on the level of components and elementary functions, and the probability of a failure here can be kept very low. Analyses of information system failures generally show that they arise not from failures in the technological components but from failures in how they are put together. Because the architecture throughout is driven by technological considerations, the reasoning mechanism becomes unnecessarily complex.

The alternative is to distinguish between the internal architecture (on the component level) and the external architecture as it appears to the user. The latter should not be driven by technology but, rather, should reflect user characteristics and needs. This approach will not only greatly facilitate information system failure analysis but also gradually reduce the need for it.

REFERENCES

Assemat, C. and A. Bonnet. (1986). Conceptualisation, coherence et validation des bases de connaisances. *The 6th International Workshop on Expert Systems and Their Applications*, Avignon, France, 28-30 April.

Coombs, M. J. and J. G. Stell. (1984). *A model for debugging PROLOG by symbolic execution: The separation of specification and procedure.* Glasgow: University of Strathclyde, Department of Computer Science.

Hollnagel, E. (1986). Cognitive system performance analysis. In E. Hollnagel, G. Mancini and D. D. Woods (Eds.). *Intelligent decision support in process environments.* Heidelberg: Springer-Verlag, 1986.

Hollnagel, E., G. Mancini, and D. D.Woods. (Eds.). *Intelligent Decision Support in Process Environments.* Heidelberg: Springer-Verlag, 1986.

Janis, I. (1986). Sources of error in the management of international crises. In this volume.

Wagenaar, W. A. (1986). *The Cause of Impossible Accidents.* The University of Amsterdam, The Sixth Duijker Lecture.

Wiener, E. L. (1986) Fallible humans and vulnerable systems: Lessons learned from aviation. In this volume.

FAULT MANAGEMENT, KNOWLEDGE SUPPORT, AND RESPONSIBILITY IN MAN-MACHINE SYSTEMS

Gunnar Johannsen

University of Kassel
Kassel, F. R. Germany

INTRODUCTION

The investigation of system failures can be based on fault management procedures. Different cognitive levels of human behaviour may be involved in each particular failure situation. Appropriate degrees of automation and knowledge support as well as suitable displays and graphics technologies will minimize the probability of failures and can increase the possibility for system recovery. Further, the issue of responsibility contributes to the avoidance and analysis of failures.

The present paper deals with these issues, referring to previous papers of the author (Johannson, 1983; Johannson et al., 1982-1986).

FAULT MANAGEMENT SITUATIONS

The possibility that a chain of incidents may lead to an accident or system shutdown has to be reduced. This problem can be tackled by designing an appropriate allocation of functions between human operators and computers. The solution has to consider the value of human insight and intuition as a substantial prerequisite for successful system operation and fault management. The two main problems are to provide the human with useful information and to aid him or her when difficult decisions are to be made.

Fault management situations may emerge in all kinds of technological, economic, environmental, health care, and national systems, among others. The technological systems include industrial plants, vehicles, and data communication systems. The fault management process is characterized by unforeseen deviations from normal operational conditions and by all efforts to return to a good or at least acceptable state. The process is often handled by humans but, depending on the system, may also be accomplished partially or fully automatically.

The fault management process is subdivided into four phases, namely, detection, diagnosis, correction, and compensation of failures. These are defined as follows:

(1) Detection is the act of discovering any kind of undesired deviation (whether suddenly occurring or slowly increasing) from normal systems operations.
(2) Diagnosis is the act of identifying the causes of malfunctions and, hence, identifying the location and the type of failure.
(3) Correction is the act of returning to normal systems states.
(4) Sometimes, only the preliminary goal of acceptable system\ states with degraded performance is reachable in cases where the systems operation has to be continued simultaneously. This phase is called compensation of failures.

The overall purpose of fault management is to achieve safe continuing operations by either preventive or corrective actions. In the first case, reduced performance, which eventually may lead to failures, requires intervention. In the second case, faults have already occurred which need to be detected, diagnosed, and corrected or compensated in time to avoid or limit any

NATO ASI Series, Vol. F32
Information Systems: Failure Analysis
Edited by J. A. Wise and A. Debons

system damage or breakdown. The concepts for dealing with these problems may be classified as follows:

1. Redundancy management
2. System recovery
3. Designing for transparency
4. Plausibility testing
5. Fast-time model predictions
6. Shifting control responsibility
7. Online maintenance

These concepts overlap each other to a certain extent. They all contribute to attaining a high coverage of failures. Coverage is defined as the conditional probability that, given the existence of a system failure, the system is able to recover and continue operation.

The process of redundancy management has the purpose of making systems fault-tolerant. In a decentralized computer control system, dynamic functional redundancy may be implemented in order to reduce the number of visible system failures. During normal operation, this redundancy is used to improve performance. In cases of faults, the respective class of failures is detected and the structure of the computer system is changed to allow continued process operation with only slightly reduced efficiency.

System recovery is a similar but more global concept of fault management. A distinction is made between automatic and manually controlled recovery algorithms, i.e., a computer or a human operator may be ultimately responsible for system recovery. The algorithms may be further subdivided into (1) full recovery, i.e., either returning to a known good state that existed prior to the fault; (2) degraded recovery of failsoft operation, i.e., returning to a fault-free state with reduced performance; and (3) safe shut-down or fail-safe operation, which is an orderly shutdown procedure avoiding damages and allowing an uncomplicated restart.

Another important concept is concerned with the design for transparency, which offers insight to the human or the computer on a supervisory level. Interactive fault-management systems should show a transparent information and decision structure. Failures have to be made transparent to system operations by, e.g., appropriate data base organization and display representation.

The computer and the human operator may check each other by means of plausibility testing and fast-time model predictions in order to ensure effective supervision of normal operation and reliable fault diagnosis in case of abnormalities. The same means can be used to ease the process of choosing between alternatives in fault correction. Plausibility testing implies that human operator inputs are screened by the computer for some obvious types of mistakes, and vice versa. More or less sketchy or precise internal models of the overall man-machine system can be used for fast-time predictions in fault diagnosis and correction by comparing expected and measured outputs of the system. In highly complex systems, it seems to be reasonable to implement some kind of this predictive capability on the lower level of intelligent subsystems to avoid the transfer of a huge amount of information to the higher level. In that case, the higher-level process computer will only make gross trend analyses.

There exists a variety of alternatives for fault correction in decentralized systems, ranging from accepting decreased performance to shifting control responsibility to other levels, either computer or human. For example, if one of several microcomputers working as lower-level decentralized controllers and decision-makers has failed, it can be bypassed by direct process computer control. In that case, the process computer may forward the appropriate information via an interface, e.g., a computer terminal, to the maintenance personnel. As the tasks of the microcomputer on the lower level have been taken over by the process computer, online maintenance without interrupting system operation would be possible.

The complete fault management process has reasonably high importance for the successful operation of many man-machine systems. Nevertheless, the fault management phases are almost always investigated separately This is achieved with different intensity of research effort. Many results exist for the phase detection and failure diagnosis, whereas only preliminary results are available for the phase correction and compensation (Johannsen and Borys, 1986; Rouse, 1983).

COGNITIVE LEVELS OF HUMAN BEHAVIOR

When dealing with the actual problems during particular fault management phases, the human operator can choose among three different cognitive levels: skill-, rule-, or knowledge-based behaviour. Generally, one can imagine that all three cognitive levels may be appropriate in the three phases of fault detection, diagnosis, and correction. This combination of three levels in three phases results in nine possible categories of human operator behaviour, which were described in Johannsen (1983).

The concept of differentiating among the three cognitive levels of behaviour was introduced by Rasmussen (1983). This concept seems to be applicable for all human task categories, particularly controlling and problem solving, including fault management.

Skill-based behaviour is well learned, highly automated behaviour using more or less subconscious routines. The human extracts stimulus input events via a feature formation from the sensory inputs, which are sampled according to a learned mode of perception. The input events act as time-space signals that evoke automated sensory-motor patterns, either for controlling a system or for solving problems within a system.

Rule-based behaviour applies in familiar situation that are controlled by stored rules available at the next higher cognitive level for coordination of subroutines on the skill-based level. Rule-based performance is based on the recognition of situations that can be associated with rules for actions, either learned or derived from instructions. The recognized input information consists of states, events, or tasks. The functional properties of the controlled or supervised systems are embedded in the rules that relate states, events, or tasks to actions. The activity itself is limited to the initiation and coordination of skilled routines. The overall appearance of rule-based behaviour is that of the stereotyped performance that predominates in many work situations.

Knowledge-based behaviour on the highest cognitive level is requested when situations are unfamiliar and proper rules are not available. The situation has first to be identified. Decisions have then to be made based on mental models, i.e., on knowledge of the functional properties of the system and on the priorities of various goals. This goal-controlled process leads to planning procedures, i.e., choosing among alternatives or generating new ones by associating with experience.

Typical descriptions of skill-based behaviour can be achieved through control theoretic models, sampling, and queueing theory. Rule-based behaviour can be described by decisions tables, associative nets, fuzzy sets, and natural language models. For knowledge-based behaviour, heuristic problem solving strategies and artificial intelligence techniques apply (Rasmussen, 1983).

DISPLAYS, SUPPORT SYSTEMS, AND RESPONSIBILITY

A new quality of automation can be based on the rapid development in displays, graphics, and related computer technology as well as in expert systems and decision/knowledge support (Johannsen, 1983; Johannsen et al., 1983, 1986). These new technologies offer a large number of options that have not yet been fully exploited for operational systems.

Without any support, the systems designer may find himself in a very time-consuming, ill-defined job, and the final user, the operator, may suffer under a design solution that is less than satisfactory for operational usage. It is more and more recognized that future automation will require user considerations and participation in selecting among the multitude of possible alternatives. Otherwise, a well-tuned dynamic function allocation between human operators and technological systems cannot be achieved, and expected improvements over traditional automation technology will not occur.

As an example for user participation, a combined task and knowledge analysis procedure was developed and applied in coal-fired power plants (Borys et al., 1986). The procedure was heavily based on interviews with engineers and operators. The task analysis yielded the detailed structure of job assignments and tasks of human operators, which resulted in a task network. The knowledge analysis added information about needed, available, and newly acquired knowledge for each node in the network. This also included results about knowledge sources and mental models used by the operators. Failure situations were particularly considered in the analysis, and a failure-oriented questionnaire was designed for this purpose.

The failure-oriented questionnaire is divided into four sections. The first three sections relate to the fault management phases and the fourth (Section D) deals with statistical data and overall systems performance and availability. The three questions of Sections A refer to a brief description of the failure situation, the symptoms, and an assessment of the different phases of the failure situation; the latter is done by using subjective scales. Sections B and C have six subquestions each, concerning the process of diagnosis and localization (B) as well as countermeasures (C).

The task and knowledge analysis was performed as part of the GRADIENT Project on Graphics and Knowledge Based Dialogue for Dynamic Systems, which is one of the joint academic/industrial cooperations in the ESPRIT-Programme (European Strategic Programme for R&D in Information Technology) of the European community. The final goal of the project is to develop intelligent design tools and a set of cooperating expert systems for supporting control room operators as well as display designers dealing with graphical dialogue in the area of process supervision and control. The graphical and knowledge support will be provided especially during fault investigation, emergency containment, and system modification. The type of support has to be related to the three cognitive levels of human behaviour.

The recent development of new technologies, when appropriately combined with knowledge about perceptional and cognitive aspects of human functioning, will lead to minimizing the probability of failures and, if such failures are unavoidable, can increase the possibility for system recovery.

CONCLUSIONS

In all discussion of fault management, the issue of responsibility has to be addressed. Although, responsibilities can be shared between humans and intelligent computers, the final responsibility will always lie with one or several humans -- operators, designers, managers, etc. Several of the recent major disasters with complex technological systems evidently occurred simply because people who were responsible either did not fully accept their responsibility or ignored important safety margins. Probably, good old traditions in science and engineering, such as thoroughness and thoughtfulness, need to be reinforced. Furthermore, people with responsible tasks have to resist, and must have the power to resist, undue political, economical, and managerial forces and ambitions.

REFERENCES

Borys, B. B., G. Johannsen, H. G. Hansel, and J. Schmidt. (1986). Task and knowledge analysis in coal-fired power plants. *Proc. IEEE Internatl. Conf. Systems, Man, Cybernetics*. Atlanta, GA, October.

Johannsen, G. (1982). Human-computer interaction in decentralized control and fault management of dynamic systems. *Proc. IFAC 8th Triennial World Congress*. Kyoto, August 1981. Oxford: Pergamon Press.

Johannsen, G. (1983). Categories of human operator behavior in fault management situations. *Proc. IEEE Internatl. Conf. Systems, Man, Cybernetics*. Bombay and New Delhi, December 1983/January 1984, pp. 884-889.

Johannsen, G. and B. B. Borys. (1986). Investigation of display contents and decision support in a rule-based fault correction task. In G. Johannsen, G. Mancini, and L. Mortensson (Eds.). *Analysis, Design, and Evaluation of Man-Machine Systems*. (Proc. 2nd IFAC/IFIP/IFORS/IEA Conf., Varese, September 1985). Oxford: Pergamon Press.

Johannsen, G., B. B. Borys, and L. Fejes. (1986). Ergonomic knowledge support in graphical interface design for industrial process operators. *Proc. 2nd Symp. Human Interface*. Tokyo, October.

Johannsen, G., J.E. Rijnsdorp, and A.P. Sage. (1983). Human system interface concerns in support system design. *Automatic. Special Issue on Control Frontiers in Knowledge Based and Man-Machine Systems*, 19, pp. 595-603.

Johannsen, G., J. E. Rejnsdorp, and H. Tamura. (1986). Matching user needs and technologies of displays and graphics. In G. Johannsen, G. Mancini, and L. Mortenson (Eds.). *Analysis, Design, and Evaluation of Man-Machine Systems* (Proc. 2nd IFAC/IFIP/IFORS/IEA Conf., Varese, September 1985). Oxford: Pergamon Press.

Rasmussen, J. (1983). Skills, rules, and knowledge: signals, signs and symbols, and other distinctions in human performance models. *IEE Trans. Systems, Man, Cybernetics*, SMC - 13, pp. 257-266.

Rouse, W.B. (1983). Models of human problems solving: detection, diagnosis, and compensation for system failures. *Automatic: Special Issue on Control Frontiers in Knowledge Based and Man-Machine Systems*, 19, pp. 613-625.

AN INTERACTIONIST'S VIEW OF SYSTEM PATHOLOGY

James Reason

University of Manchester
Manchester, U.K.

THE INTERACTIONIST VIEW

This paper argue that serious failures of a complex, centralized system arise from the interaction between two factors:

- The properties of the system as a whole (the systems domain).
- The information-handling characteristics of the human element (the human domain).

The task of improving the reliability of complex systems draws specialists from many different disciplines. Each is trained to focus on one particular aspect of the safety problem. What gets lost in the interstices of these various concerns is a global appreciation of the relationship between the system and human domains. The prevention of serious breakdowns in the future, as well as the ability to learn the right lessons from past failures, depends critically upon a better understanding of this interaction.

The illustrative material will be drawn from the nuclear power industry. This is appropriate not only because many nuclear power plant emergencies have been well-documented and extensively analyzed, but also because these relatively opaque and still partially understood systems raise important informational issues.

A MEDICAL METAPHOR FOR A SYSTEM FAILURE

At any point in time, all man-made systems -- particularly complex, tightly-coupled, highly interactive ones like nuclear power plants, chemical process plants and sophisticated weapons systems (see Perrow, 1984) -- are subject to minor failures. These can be either technical or human.

Such trivial failures are analogous to resident pathogens in the human body. For the most part, they are checked by built in protective measures (comparable to the body's auto-immune system), or detected and corrected by human operators.

On occasions, however, several of these small failures concatenate in unforeseen and unseen ways to thwart the protective devices, and produce intrinsically unpredictable and largely unavoidable disasters (see Bignall, Peters and Pym, 1977; Turner, 1978; Rolt, 1978; Reason, 1986; Wagenaar, 1986; Kasputin, 1986).

ACTIVE AND PASSIVE HUMAN FAILURES

From a systems perspective, human failures fall into two broad classes (see Rasmussen and Pedersen, 1982):

- Active Failures: Operator errors that initiate a major system breakdown, or that are committed during attempts to recover from such a breakdown.

NATO ASI Series, Vol. F32
Information Systems: Failure Analysis
Edited by J. A. Wise and A. Debons

• Passive Failures: Errors in design, installation, construction, planning or maintenance that create some latent failure condition (pathogen) within the system.

From a human perspective, these active and passive failures take fairly distinct forms. Detailed analyses of actual and simulated nuclear power plant emergencies (Pew, Miller and Feeher, 1981; Woods, 1982; 1984; Reason and Embrey, 1985) show that active failures repeatedly implicate two basic error tendencies: frequency-gambling and confirmation bias. Most prominent among the passive human failures, however, are the omission of necessary actions, particularly during maintenance activities (Rasmussen, 1980; INPO, 1984, 1985). Each of these two sets of error types will be discussed later. First, it is necessary to outline some of the basic characteristics of human information-processing.

THE CHARACTER OF HUMAN INFORMATION PROCESSING

Perhaps the most conspicuous achievement of human cognition, relative to other information-handling devices, is its ability to simplify complex informational tasks. It does this by employing, whenever possible, automatic rather than conscious modes of processing. Human beings are strongly disposed to exploit the effortless, parallel and pre-programed operations of highly specialized low-level processors and their associated heuristics (intuitive rules of thumb), and to avoid higher-level manipulations requiring the slow, laborious, resource-limited involvement of conscious processing.

Human beings owe their inclusion in hazardous systems largely to this unique knowledge-based ability to carry out "on-line" reasoning in novel situations. Ironically -- and notwithstanding the crew of Apollo 13 and other exceptionally talented jury-riggers -- they are not especially good at it; at least not in the conditions usually prevailing during most system emergencies.

ACTIVE ERRORS DURING NUCLEAR POWER PLANT EMERGENCIES

Below are listed some of the errors committed by control room crews during four post-Three Mile Island emergencies. The points to be made by these 'error inventories' are, first, that control room errors are the norm rather than the exception; and, second, that they tend to occur multiply in interactive combinations.

Oyster Creek

On 2 May 1979, a loss of feedwater emergency occurred in a General Electric Boiling Water Reactor (BWR) at Oyster Creek (see Pew et al., 1981). Three causally linked errors were made by the crew during recovery operations:

(a) A technician, carrying out a routine check, produced a hydraulic disturbance. The reactor tripped on a spurious high pressure signal. Four automatic recirculation pumps also tripped.

(b) An operator, intending to close pump discharge valves A and E, inadvertently closed B and C also. All natural circulation to the core area was shut off.

(c) The operators took the annulus level (160.8 inches) to indicate the level within the shroud. The actual shroud level was 56 inches above the fuel elements (due to valve closing error). Although the triple low-level alarm sounded 3 minutes into the event, and continued to sound, the error was not discovered for 30 minutes.

North Anna

On 25 September, 1979, a Westinghouse Pressurized Water Reactor (PWR) suffered excessive cool-down following automatic safety injection (see Pew et al., 1981).

(a) An operator, seeking to stop pressure loss through a stuck-open steam dump valve, closed the Non-Return Valve (NRV), then intended to close the Main Steam Injection Valve (MSIV). But, distracted by the automatic safety injection, he omitted to do so, and the Main Steam Injection Valve stayed open for six minutes, exacerbating the recovery problem. Procedures require that the Main Steam Injection Valve should have been closed before the Non-Return Valve.

(b) An operator failed to realign the charging pump orifices prior to initiating normal let-down. This caused the release of radio-active gas and water.

Ginna

On 25 January, 1982, a Westinghouse Pressurized Water Reactor experienced a steam generator tube rupture with "textbook" symptoms (see Woods, 1982):

(a) The shift supervisor delayed closing the Main Steam Injection Valve for eight minutes after isolating the faulted steam generator. Contrary to procedure, the cooldown began before the Main Steam Injection Valve was closed.

(b) Misinterpreting ambiguous operating instructions, the operators prevented the steam generator Pressure Operated Relief Valve (PORV) from opening.

(c) The operators, intending to depressurise the Reactor Coolant System, adopted the wrong Pressure Operated Relief Valve strategy, cycling it open and shut. The Pressure Operated Relief Valve stuck open on the fourth cycle.

(d) The operators delayed terminating the safety injections. The issue was debated for 30 minutes.

Davis-Besse

On 9 June, 1985, a Babcock and Wilcox Pressurized Water Reactor experienced a total loss of main and auxiliary feedwater (NUREG, 1985).

(a) An operator, wishing to initiate the Steam and Feedwater Rupture Control System manually, inadvertently pressed the wrong two buttons.

(b) The shift supervisor failed to initiate Make-Up/High Pressure Injection cooling at the point required by procedure. He did not realise that the steam operators had boiled dry.

(c) Two equipment operators were sent to reset the over-speed trip for each Auxiliary Feedwater Pump manually. Both failed to operate the handwheels correctly. Neither valve was left fully open as intended.

(d) An operator failed to appreciate that the Pressure Operated Relief Valve was stuck open.

(e) An operator wrongly sought to override or reset the auto close signal to No. 2 start up feed control valve.

Error Types

These errors fall into the two main classes listed below (see Norman, 1981; Reason and Mycielaska, 1982):

- Failures of Execution (slips and lapses), where the intention was correct but the actions did not proceed as planned, e.g., Oyster Creek (a and b), North Anna (a and b), Davis-Besse (a and c).

- Failures of Assessment and Planning (mistakes), where the actions may have gone as intended, but the plan was inadequate to achieve its purpose, e.g., Oyster Creek (c), Ginna (a,b, c and d), Davis-Besse (b, d and e).

The two basic error types differ markedly in their ease of detection and recovery. Woods (1984), reviewing data derived from 99 test scenarios using 23 crews in eight different events, noted that whereas half of the execution failures were detected by the crews themselves, none of the diagnostic errors were noticed by their perpetrators. Corrections of assessment errors occurred only through the intervention of some external agent (as at Three Mile Island and Oyster Creek). Nearly three-quarters of these errors remained undetected. Woods concluded that the commonest cause of mistakes was fixation

Crew Performance During Emergencies

The evidence from both actual and simulated emergencies indicates that crews are reasonably accurate in their initial-state diagnoses (Woods, 1984). What they fail to appreciate is how this initial condition changes during the course of the emergency. Instead of adjusting their assessments to match the changing conditions of the plant, they remain fixated upon their original diagnosis. Their subsequent behaviour is characterized by a succession of "key-hole" views, in which attention is captured by one focus of concern after another. Once seized upon, these "hot issues" come to claim a disproportionate amount of the crew's limited attention span. This was especially true of the Ginna Crew, who stumbled from one narrow focus to another until their attention was finally claimed by an overriding concern with the steam bubble in the reactor vessel head.

THE COGNITIVE ORIGINS OF FAILURE

Recurrent errors forms are rooted in necessary and adaptive properties of human cognition. The predictable varieties of human error are inextricable bound up with things at which the cognitive system excels, and especially with its facility for employing unconscious rules of thumb to deal with uncertainty. But each simplifying device is bought at a price: over-utilization. One heuristic, whose influence is evident in a wide variety of error forms, is frequency-gambling.

Frequency-Gambling

The frequency-gambling heuristic is essentially a guessing rule. It dictates that whenever cognitive operations are under-specified, the cognitive system should default to some contextually-appropriate, high-frequency response. Observations of a wide range of cognitive activities indicate that errors take forms that recognizably spring from an individual's repertoire of well-established reactions to a particular set of environmental circumstances.

Cognitive under-specification can arise for a variety of reasons: degraded or incomplete sensory inputs, patchy knowledge, insufficient attention to routine action, failures of prospective memory, fragmentary retrieval cues, stressors that diminish higher-level control, and so on. Notwithstanding these possible varieties of under-specification, their consequences are remarkably uniform: what emerge are perceptions, diagnoses, plans, thoughts, recollections, and actions that are typical of previous transactions with that particular situation. The more often a particular cognitive routine is successful in a given context, the more likely it is to reappear in conditions of incomplete specifications.

Correct performance is any sphere of activity is achieved by activating right "knowledge packages" (processors) in the right order at the right time. Cognitive control hinges critically upon the timely retrieval of these "processing packages" from long-term memory. Some of the necessary retrieval cues are supplied by intention, others by environmental signals, and still others by "descriptions" passed on from other active processors in the same task-linked sequence (Norman and Bobrow, 1979).

The "calling conditions" so provided are usually sufficient to specify the correct processor. Occasionally, however, these cues are incomplete and a number of partially matched " packages" become activated. Since the limited conscious "workspace" can accommodate only one such knowledge package at a time, some selection has to be made from among these partially matched candidates. This selection is automatically biased to favour high-frequency candidates. There is considerable evidence to show that event frequency information is encoded automatically (Hasher and Zacks, 1984) and preserved in some privileged analogue form in long-term memory (Hintzman, 1976).

These two retrieval mechanisms -- (a) the matching of "calling conditions" to the attributes of stored knowledge items, and (b) frequency gambling on partially matched candidates -- operate in a parallel, distributed, and largely automatic fashion. They are the primitives of the knowledge retrieval system. Only the conscious "workspace" can override the default options selected by this parallel search. But its capacity to do this is severely diminished in conditions of stress and high workload.

Recent data (Reason, 1986b) indicate that an individual's reliance upon frequency-gambling diminishes rapidly with increasing expertise. As the use of frequency-gambling decreases, dependence upon pattern-matching and inferential "gap-filling" correspondingly increases.

Confirmation Bias

Whereas frequency-gambling often determines which of a number of candidate hypotheses gets selected, confirmation bias works upon the criteria that allow a current hypothesis to be relinquished in the face of contradictory evidence. Confirmation bias has its roots in what Bartlett (1932) termed "effort after meaning." Cognition abhors informational anarchy. In the face of ambiguity, it rapidly favours one of the available interpretations, and is then loath to part with it.

Several studies indicate that preliminary hypotheses, formed on the basis of early, impoverished data, interfere with the later interpretation of better and more abundant data (see Greenwald, Leippe, Pratkanis and Baumgardner, 1986). In brief, preliminary hunches have a decided advantage in the judgment process. The various mechanisms underlying this phenomenon have been discussed at length elsewhere (see Nisbett and Ross, 1980).

LATENT FAILURES

The clear conclusion from a number of recent nuclear power plant error surveys (Rasmussen, 1980 INPO, 1984, 1985) is that maintenance-related omissions are the single largest category of latent failures, or "resident pathogens" This is in close accord with the relative incidence of different error types in everyday life (Reason and Mycielska, 1982), where the omission of planned actions (most often associated with failures of prospective memory) constitutes the commonest form of absent-minded lapse.

OMISSIONS: A CAUSAL TAXONOMY

In attempting to classify omissions, it is useful to divide the preparation and execution of an action sequence into three stages: (a) plan formulation; (b) plan storage; and (c) plan execution. Progression from formulation to execution is associated with a shift from higher to lower level processors.

Omissions can have their origins at any one of these stages. Failures in plan formulation are likely to involve the omission of much larger "chunks" of necessary actions than those at either the storage or execution stages. Various types of cognitive failure are possible at each of the three stages:

• Formulation stage

(a) Inappropriate goal selected (i.e., solving the wrong problem).
(b) Actions selected are inadequate to achieve goal: necessary actions are omitted from the plan, or the consequences (or side-effects) of actions are misjudged.

• Storage Stage

(a) Encoding failure: plan items inadequately committed to memory.
(b) Storage failure: plan item lost from memory.
(c) Retrieval failure: plan items not recalled at the proper time.

•Execution Stage

(a) Strong habit exclusion: actions are diverted along some familiar but intended pathway.
(b) Anticipatory leap: often cued by signals relevant to some future action.
(c) Place-losing: position in the action sequence is misidentified as being later than actual position.
(d) Premature exit: the next planned task is begun before the previous one is completed.

TASK FACTORS CONTRIBUTING TO OMISSIONS

The evidence from several studies (see Reason and Mycielska, 1982) indicate the influence of both general and specific task factors upon the incidence of everyday slips of action, among which omissions represent a major proportion.

General Factors

Slips of action of all kinds are most likely to occur in familiar surroundings during the performance of tasks requiring little by way of moment-to-moment conscious guidance. In addition, their occurrence is commonly associated with states of preoccupation or distraction in which the greater part of the limited attentional resource is captured by something other than the job in hand (Reason and Mycielska, 1982).

Specific Factors

There are a number of more specific task factors that are likely to increase the probability of making an omission error. These are listed below.

• The larger the number of individual steps in an action sequence, the greater the probability that one or more of them will be omitted.

- The greater the informational loading of a particular procedural step, the more likely it is that items within that step will be omitted.

- Procedural steps that are obviously cued by preceding actions, or that do not follow in a direct linear sequence from them, are highly likely to be omitted.

- When instructions are given verbally and there are more than five simple steps, items in the middle of the list of instructions are more likely to be omitted than those at either the beginning or the end.

- When instructions are given in a written form, isolated items at the end of a sequence (e.g., replacing caps or bushes, removing tools, etc.) have a reasonably high probability of being omitted.

- Necessary steps in an action sequence are more likely to be omitted during reassembly than during the original disassembly. In dismantling a piece of equipment, the order in which the steps should be carried out is naturally cued by the construction of the item. The same is not necessarily true of reassembly, where there are usually many more degrees of freedom.

- In a well-practiced, highly automatic task, unexpected interruptions are frequently associated with omission errors, either because some unrelated action is unconsciously "counted in" as part of the task sequence, or because the interruption causes the individual to "lose his place" on resumption of the task, i.e., he believe he was further along in the task prior to the interruption than he actually was. Such tasks are also especially prone to premature exits -- moving on to the next task before the previous one is completed, thus omitting some necessary final steps. This is particularly likely when the individual is working under time pressure, or when the next job is close at hand.

Applications

These observations have a number of practical applications. In the short term, they make it possible to identify in advance those steps in written maintenance procedures that are most likely to be omitted. Consider, for example, the following job description covering the valve inspection component of preventive maintenance on a compressor: "Check and clean suction and pressure valves. Replace defective valves. Replace packings. Clean valve chambers" (Kelly, 1984). The step most likely to be omitted here is the replacement of the packings. Having identified the most probable omissions, it is possible to provide the maintenance technician with a set of procedures, stored in a cheap lapheld computer, that not only give him step-by-step guidance in what has to be done, but also prompt him to check that easily omitted steps have been completed.

At best, however, this is a stop-gap solution applicable to existing plants. A more effective preventive is to eliminate likely errors at the design stage (see Lewis and Norman, 1986). Poor design plays a major part in creating unnecessary omission errors. A commonplace example is the SLR camera with automatic speed and exposure setting. Cameras are frequently replaced in their cases with the electronics left on. It is very easy to forget to turn off the switch located on the top of the camera. The opportunity for making this omission could easily have been removed by placing the switch beneath the wind-on lever. To replace the camera in its case, the wind-on lever must first be pressed in flush with the camera casing. A switch located beneath it would automatically be turned off. The key to "designing out" omissions is to minimize the number of functionally isolated manipulations that must be performed during routine operation or maintenance.

THE HUMAN COMPUTER MISMATCH

Whereas the incidence of maintenance-related omissions can probably be substantially reduced by the provisions of memory aids and by more "error-sensitive" equipment design, it is very difficult to see what can be done about the active failures These arise from a fundamental incompatibility between the properties of human cognition and the role it is required to play in the nuclear power plant control room.

Human cognition is supremely good at modeling the regularities of its previous dealings with specific environments, and then converting these stored representations into preprogrammed routines for guiding future action. It does this in the reasonable expectation that the event frequencies and recurrence of the past provide a fair guide to the likelihoods of the future And, in the course of normal everyday life, that is generally the case. But it is manifestly not true for nuclear power plant control rooms.

For the most part, the human controllers of these and comparable systems act in a supervisory capacity. That is, they are involved in "initiating, monitoring and adjusting processes in systems that are otherwise automatically controlled" (Sheridan and Hennessy, 1984). The primary emphasis here is upon "adjustment" in the event of system maloperation. They are there, in fact, because the system designers cannot foresee all the possible scenarios of failure. In short, the main job of the human operator is to handle the unexpected.

Such a role would be psychologically more acceptable if operators were experts at recognizing and coping with a wide range of failures types. But in the main they are not. Their actual expertise lies in tweaking the plant while it is operating within safe limits.

The problem is further compounded by the unpredictable course of nuclear power plant emergencies. They may start in a conventional way, but they rarely proceed along anticipatible lines. As case studies repeatedly show, it is in the nature of complex, tightly-coupled, highly interactive, largely opaque, and partially understood installations to spring nasty surprises. This means that even if it were possible to spend most of an operator's shift drilling him in the diagnostic and recovery lessons of previous emergencies, there is still no guarantee that these pre-established routines would be relevant to future incidents. Because emergencies invariably contain unexpected elements and thus demand resource-limited, knowledge-based processing, they are ideal for breeding erroneous frequency-gambles and for exaggerating the human predilection to seek out and find support for early hypotheses.

A PESSIMISTIC CONCLUSION

The unwelcome conclusion from this analysis is that human-system mismatches in present nuclear power plants (and comparable systems) make "active" operator errors inevitable. This suggests that for the short term we should focus our attention upon another question: If errors are unavoidable, how can their adverse consequences best be contained? For existing installations, error correction and containment seem to offer a better chance of avoiding disaster than error prevention. But however we cope with our immediate difficulties, it is necessary to go back to the drawing board and create a new generation of systems with the basic properties of human cognition firmly in mind from the outset. Design is too important to be left to the designers.

REFERENCES

Bartlett, F. E. (1932). *Remembering*. Cambridge: Cambridge University Press.

Bignall, V., Petrers, G. and Pym, C. (1977). *Catastrophic Failures*. Milton Keynes: Open University Press.

Greenwald, A. G., Leippe, M. R., Pratkanis, A. R. and Baumgardner, M. H. (1986). Under what conditions does theory obstruct research progress? *Psychological Review*, 93: 216-229.

Hasher, L. and Zacks, R. T. (1984). Automatic processing of fundamental information: The case of frequency of occurrence. *American Psychologist*, 39: 1372-138

Hintzman, D. C. (1976). Repetition and memory. In G. H. Bower (ed), *The Psychology of Learning and Motivation , (Vol 10)*. New York: Academic Press.

INPO (1984). *An Analysis of Root Causes in 1983 Significant Event Reports*. INPO 84-027, Atlanta, Georgia: Institute of Nuclear Power Operations.

INPO (1985). *A Maintenance Analysis of Safety Significance Events*. Nuclear Utility Management and Human Resources Committee, Maintenance Working Group, Atlanta, GA: Institute of Nuclear Power Operations.

Kasputin, R. (1986). Application of team concept/system approach to investigation of major mishaps. This volume.

Kelly, A. (1984). *Maintenance Planning and Control*. London: Butterworths.

Lewis, C. and Norman, D. E. (1986). Designing for error. In D.A. Norman and S.W. Draper (eds) *User Centered System Design*. Hillsdale, NJ: Erlbaum.

Nisbett, R. and Ross, L. (1980). *Human Inference: Strategies and Shortcomings of Social Judgement*. Englewood Cliffs, NJ: Prentice-Hall.

Norman, D. A. (1981). Categorization of action slips. *Psychological Review*, 88: 1-15.

Norman, D. E. and Bobrow, D. G. (1979). Descriptions: An intermediate stage in memory retrieval. *Cognitive Psychology*, 11: 107-123.

Pew, R. W., Miller, D. C. and Feeher, C. E. (1981). *Evaluation of Proposed Control Room Improvements through Analysis of Critical Operators Decisions*. NP-1982, Research Project 891, Cambridge, Mass.: Bolt, Baranek and Newman, Inc.

Perrow, C. (1984). *Normal Accidents: Living with High-Risk Technologies*. New York: Basic Books.

Rasmussen, J. (1980). What can be learned from human error reports? In K. Duncan, M. Gruneberg, and D. Walli (eds), *Changes in Working Life*. London: Wiley.

Rasmussen, J. and Pedrsen, O. M. (1982). *Formalized Search Strategies for Human Risk Contributions: A Framework for Further Development*. Riso-M-2351, Roskilde, Denmark: Riso National Laboratory.

Reason, J. T. (1986a). Catastrophic combinations of trivial errors. In T.Cox (ed.), *The Psychology of Occupational Safety and Accidents*. To be published by Taylor and Francis (London).

Reason, J. T. (1986b). Cognitive under-specification: Its varieties and consequences. In B. Baars (ed.), *The Psychology of Error: A Window on the Mind*. To be published by Plenum (New York).

Reason, J. T. and Embrey, D. E. (1985). *Human Factors Principles Relevant to the Modelling of Human Errors in Abnormal Conditions of Nuclear and Major Hazardous Installations.* Prepared for the European Atomic Energy Community under contract EC1 1164-B7221-84-UK, Parbold, UK: Human Reliability Associates.

Rolt, L. T. C. (1978). *Red for Danger.* London: Pan Books.

Sheridan, T.B. and Hennessy, R.T. (1984). *Research and Modeling of Supervisory Control Behavior.* Washington, DC: National Academy Press.

Turner, B. A. (1978). *Man-Made Disaster.* London: Wykeham Publications.

Wagenaar, W. (1986). *The Cause of Impossible Accidents.* The Sixth Duijker Lecture, Amsterdam, 18 March.

Woods, D. D. (1982). *Operator Decision Behavior during the Steam Generator Tube Rupture at the Ginna Nuclear Power Station.* Research Report 82-1C57-CONRM-R2, Pittsburgh, PA: Westinghouse R&D Center.

Woods, D. D. (1984). *Some Results on Operator Performance in Emergency Events.* Institute of Chemical Engineers Symposium Series No. 0: 21-13.

MENTAL MODELS AND FAILURES IN HUMAN-MACHINE SYSTEMS

Gerrit C. Van der Veer

Free University
Amsterdam, The Netherlands

INTRODUCTION

This article concerns the analysis of failures and failure prevention from the viewpoint of cognitive ergonomics, or the cognitive psychology of human-computer interaction. Causes of failures and the ways to prevent them will be reviewed with the help of theoretical notions on human behavior in human-machine communication processes.

A human-machine situation can be conceived as a two partner system, in which the partners are the human user or operator, and the machine. Both partners are to a certain extent in control of the interaction, both may initiate communication acts or communication sequences. Both partners take part in the interaction on the base of a model of the partner-system.

The computer reacts to human behaviour on the base of a model of the user, implemented by the designer either implicitly or explicitly. This model may be adaptive (in the sense of an expert system) in adjusting itself to individual differences between human users and to changing behaviour or level of expertise of the user.

The user interacts on the basis of a mental model of the system, establishing both his planning and his expectations, and his interpretation of outcomes and failures, on the mental model. The human mental model originates from metacommunication concerning the system, and evolves in interaction with the user-interface. A mental model is adequate if it is consistent with the conceptual model of the user-interface (the chief observable and inferable structure and characteristics of the system). Metacommunication inside and outside the system can be designed in order to improve mental models, taking individual differences in styles, strategies, and expertise into account.

The discrepancies between the mental model and the conceptual model of the system, and between the system's model of the user and thc actual characteristics of the user, may be investigated with the help of cognitive psychological research methods, to be illustrated.

MODELS OF SYSTEMS

Norman (1983) presents a clear terminology for mental models in the context of human-machine interaction. Three of his concepts are of crucial importance in our analysis: "system," "conceptual model," and "mental model."

System

A system in the sense of Norman consists of the combination of all hardware and software to which a user is interacting. As far as the user is concerned all parts belong together, and for most users a lot of system components may be unknown or their function may be obscure. But even for most designers, large parts of a system may be black boxes. As these parts play a role in the integrated whole, this implies the danger of a design that leaves some interactions between different components unnoticed.

NATO ASI Series, Vol. F32
Information Systems: Failure Analysis
Edited by J. A. Wise and A. Debons

For the user, however, only a very special aspect of the system is visible. The input and output behaviour of the user workstation, the terminal, both in the sense of hardware and in the sense of the relations between user's action and system's reaction, are the real point of contact between user and machine. This side of the system is called the user interface. In modern systems it may be especially constructed for its function of locus of interaction, being a separately defined layer between the system as designed for a set of tasks and the human partner. In this way, the designer may try to cope both with expected human behaviour and expected system reactions, and moreover may try to take measures for cases of unexpected actions from both sides. In those cases, it might be of importance to notice at once that something unpredicted has occurred, alarming both the user and the relevant other components of the machine.

Conceptual Model

The user interface should be designed with a model of the user in mind. It may be seen as a special layer between the task dedicated system and a human user. The term "conceptual model" may refer to the design of the user interface. This model or design defines the interaction between two partners (user and machine) in terms of levels of tasks, semantics, syntax, and key strokes, a distinction that is derived from Moran (1981).

Task Level

At this level of human-machine interaction the task domain is specified for which the system is applicable. The decomposition of tasks into subtasks and the delegation of some of the subtasks to the machine are described on this level of the user interface. Other subtasks ask for human decisions or the application of other tools or methods.

Semantics

The objects, the attributes of objects, the tools (special kinds of objects dedicated to a class of subtasks), the relations between objects, and the actions that operate on objects -- all of these concepts are the units which the user and the system have to manipulate in the process of delegation of tasks by the user to the system.

Syntax

Commands that are issued by the user in the process of interaction about semantic units, and system messages about the same kind of manipulation, ask for a specific format. Commands request the application of certain conventions for parameters. These rules are described at syntax level, as are the interpretation rules of the information sequences and structures that may be issued by the system. Two systems that are equivalent at the semantic level may have a different set of syntactic rules.

Key-stroke Level

In fact, the description of the interaction at this level not only concerns the physical characteristics of human input to the machine in its diversity, but also deals with the perceptual and physical aspects of system behavior, like audible signals, colour, movement, temporal patterns, and visual structure. The actual sequence of "key-strokes" may vary between different work-stations, even for machines that are equivalent or identical at other levels.

Mental Model

The mental model is the representation of the user's knowledge of the system and of the possibilities of interaction. The user interface is the only part of the system that the user is always aware of, both in its perceptible actions and reactions to human behavior, and in the inferred mechanisms that are responsible for the relations between user actions and system actions.

The mental representation of the user interface or "virtual machine" is the internal frame of reference that is used for planning activities (task delegation, manipulation of semantic units, formulation of structures of information, sequences of physical actions) and for interpreting system behavior (related to task accomplishment, meaning of message units, decomposition of information structures, analysis of signals). Mental models may be described at the same levels we used in analysing the conceptual model.

In empirical studies (Van der Veer, Felt, Van Muylwijk and Van Biene, in press) ,we found that mental models of different users may vary both in preferred mode of representation (visual-spatial, verbal, iconic) and in the amount of correct description at the different levels. Some users of a certain system prefer visual-spatial representation at task and semantic level; others show only verbal-iconic representations at key-stroke level. The behavior of users based on these different mental models may vary enormously. Some know very well what subtasks to delegate, and how to manipulate the objects and relations in a system, although they lack the skill to actually operate the machine. Other users perform very well at the terminal, provided others have decomposed the task and prescribed the application of tools and operations.

THE DEVELOPMENT OF MENTAL MODELS: METACOMMUNICATION

Mental models develop as the result of a learning process. To start with, the user has to apply an analogy to known situations, systems, or structures. This first metaphor will subsequently be reshaped, extended, discarded, or replaced as more information about the system (in fact, about the user interface) is collected. All information that the user needs to build or reshape his mental model has to be communicated to him, either by explicit teaching, by learning and exploration activities initiated by the user himself, or by incidental learning. Exchange of information in this sense will be called metacommunication: It consists of communication (between the user and any source of information) concerning communication (between the user and the system, about the task the system is to be used for).

The learning process about the system will never end. Even expert users will continually encounter situations in which their mental model does not fit the actual interaction with the system. The possibilities of metacommunication will remain important after the phase of initial learning, and different sources will have to be provided, related to the kind of problem and to the possibilities of the actual situation.

An important distinction is the source of metacommunication. This may be located either outside or within the system. The latter situation provides very quick information to the user. This is important in cases of accidents where speed of recovery is relevant. Some topics of metacommunication, however, are more easily dealt with by sources outside the machine, providing more flexibility or more opportunity for the user to collect information and exchange his vague notions for expert opinions or detailed background information in a way best suited to his personal style.

Metacommunication Outside the System

The first phase of the formation of a mental model very often is a case of acquisition of some feasible metaphor related to the task knowledge the user already has available. It is hardly ever useful, or even possible, to incorporate this part of the learning process into the normal human-computer interaction situation. Special learning devices may be developed, even including a simulated version of the system to be learned. Most of the knowledge about task level, however, may very efficiently be transferred by means of traditional learning situations. The same is true for general semantic knowledge, especially for systems for which the user does not have an adequate metaphor available (e.g., in the form of a mental model of an analogous system he has learned before). Explicit teaching, documentation, and expert users are the common sources for this kind of information.

Teaching

Introduction to new systems or to new applications of a system has to be planned with the explicit aim of inducing an adequate mental model. Education on human-machine interaction systems will have to be based on a complete and carefully constructed conceptual model of the system that covers the user interface and its behaviour for the target task domain. The conceptual model often has to be adapted to the user group, as different groups of users of the same machine or system may perform different sub-tasks, or have to interact with the system in a different way.

Teaching complex systems will inevitably take the form of presenting one or more metaphors that describe the task delegation and the system functions for the subtasks on the semantic level. A teacher should be aware of the application of metaphors and construct them purposefully, pointing to the aspects of the metaphor that apply and to those that do not. Presentation of partly invalid analogies will lead to invalid mental models, although that might only be discovered in exceptional circumstances. The application of metaphors therefore is a very delicate matter that can best be done in human-to-human interaction, where the teacher continually has to check the understanding of his students.

Manuals and Documentation

Written documentation is helpful for communication on all levels of the user interface. The semantic structure and the syntax may be presented verbally or graphically, which for the semantics will imply a metaphorical representation. Here again the author has to be aware of this, taking care not create implicit assumptions that are inconsistent with the actual user interface.

Semantic and syntactic descriptions should be treated separately to avoid mixing them up in the mental model. Since documentation is often consulted after the primary learning phase, users should be able to retrieve any information later on with the help of feasible indexing systems. For semantic concepts, this might include a menu-like retrieval system that is targeted from description of sub-tasks for syntactic information, the entries have to be chosen from a semantic description.

Consultation of Experts

If a user discovers that his mental model does not completely fit the interface, or is lacking information to continue work, the documentation often is not consulted or is found to be of no use. Users tend to apply to official or unofficial experts. In that way, they avoid the formulation of precise questions needed to use the manuals. Human experts are trusted to help them correct their mental model, allowing them to talk in vague metaphors. In case of emergency, however, a real expert might not be at hand. Expert knowledge should always be immediately available, either in the form of clear and well structured documentation or in the form of expertise implemented in the user interface.

<u>Metacommunication Within the System</u>

When a user is working in interaction with a system, information is being transferred between the two partners all the time. This information is to a large extent covering messages about the tasks that are being delegated to the machine. Since all information has to be transferred with the help of some structure of signals, the perception of these signals will give the user information about the interaction that is going on, and will suggest to him either that his mental model is adequate or that it is not. Apart from this implicit form of metacommunication, both the user and the system may take the initiative to start a discourse on the user interface, the mental model of it, and the interaction between user and machine.

Implicit Metacommunication

Any message exchanged will have the form of tokens. These may have many different appearances, ranging from verbal units, icons, graphical representations, sound-signals, to direct manipulation. Every representation will give rise to associations with available knowledge of the system, with semantic connotations, and with mental models and metaphors. The design of tokens no awareness of this mechanism of association may easily give rise to erroneous mental models.

Modern application software is a rich source of examples of invalid metaphors incorporated into the interaction tokens. The object stored on the Macintosh "clipboard" is not "unclipped" in the action of "pasting," but copied. The message "deleted" in the UNIX mail system is only marked "to_be_deleted," which is effectuated in case of one way of quitting the mail system, but not in other legal exit operation. The command "print" in the database system QUEST does not activate the printer, but selects rows from a datatable that subsequently may be printed with the command "outfile_lpr".

Users will never escape the implicit influence of the associations these tokens induce. Designer are aware of the mechanism, on the other hand, will explicitly apply this in designing tokens that contain hints to the semantic meaning of objects, actions, relations, and attributes. Iconic symbols have been shown to be very powerful means to this end (Van der Veer, in press).

Explicit Metacommunication

Either the human partner or the system may, in the course of interaction, discover that something is wrong. The user may figure out that his planning or prediction failed, since he turned out to be unable to delegate the intended subtask. He will conclude that his mental model needs adjustment at the semantic level (if he is aware of this concept), or at least he will feel the need for information about how to let the machine perform the intended subtasks. There are other situations in which the user feels the need for immediate information. If his mental model is too fragmentary at the syntax level, he is unable to actualize his plans although the semantic level of the mental model may be complete and valid.

To many users these two occasions are not naturally different, but the kind of help they need is at the level at which their mental model is incomplete or erroneous. Metacommunication within the system has to guide the user to the relevant information. He should be able either to ask for ways to issue a command in order to accomplish the manipulation of an object for which the semantic knowledge is available, or he should be provided the possibility to locate information on the semantic structure of the system associated with the delegation of some sub-task. On line information of this kind may have the form of a help system, structured around clear representations of the conceptual model.

The user interface, being the other partner in the interaction, may also discover that something is wrong. If a message from the user is syntactically incorrect, the reaction of the system (the error message) has to point to exactly that fact, offering the possibility of direct help at that level. If a message is incomplete, the problem might be at the semantic level. If the user interface is constructed with some model of the user and the intended kind of interaction in mind, it is even possible to detect situations where a message from the user, although complete and semantically correct, does not make sense at the task level. Insofar as this knowledge may be built in the system, warnings may be issued, leaving it to the user to decide whether he insists in having such actions performed (for reasons not to be understood by the system), or is happy to correct his semantic error.

THE USER INTERFACE: REQUIREMENTS FOR A CONCEPTUAL MODEL

The user interface is the location of metacommunication within the system. The design of metacommunication is an integrated part of the interface definition. Both the implicit metacommunication and the explicit forms will have to be taken into account. The value of on-line metacommunication is related to the fact that it may be immediately available on any occasion. On-line metacommunication at key-stroke level in possible only in a restricted sense. For special devices like function keys, help text on syntax may incorporate information on the application of the physical device. Another simple but adequate solution is the labelling of keys, meters, signals, and other devices in a clear and suggestive way, e.g., with iconic symbols.

Taking into account the wide diversity of systems and devices that have to be dealt with, it seems impossible to derive a complete and detailed set of guidelines for the development of conceptual models of a user interface. We will only mention some notions that are related to the mechanisms of development and adjustment of mental models during interaction with the system.

A Consistent Conceptual Model

The interaction between system and user should be designed with a semantic structure that is both related to the task and consistent with the "natural semantics" the user will expect from experience with the same kind of task outside the system. Objects, attributes, relations and actions should be designed analogous to their "natural" counterparts as far as these are available. The syntax has to be systematic and consistent with the semantic structure of the system. The choice of symbols, tokens and representations has to be both easy to remember, rich in (valid) semantic connotation, and clearly distinguishable from other signals.

Modes

Information processing systems often are designed around multiple modes. Monk (1986) shows the trade-off between the number of devices, displays, and keys against the number of different modes that have to be distinguished during the interaction. Either, systems of any complexity will suffer from the inpossibility for the user to attend to all different signals or devices, or the user will need clear and pregnant information about what mode is in operation at any moment. The span of attention and the mechanisms to draw attention have to be taken into account for the design decisions in this respect.

Metacommunication and Levels of the Interaction

Metacommunication has to be provided at the different levels of interaction. As the user often will not be aware of this distinction, the design of the user interface will have to provide adequate choices and clear descriptions of these options.

Control of Metacommunication

The user will be able to maintain a mental model only as far as the system is "visible". Any part that looks like a "black box" will result in a mental model of analogous quality, disabling the user either to plan or to interpret parts of the interaction. The user should therefore always have the opportunity to initiate metacommunication at any level and about any detail of the user interface he wishes. A special occasion exists when the user has lost track of the interaction. He may be unable to find out what next actions are possible, which of the partners in the interaction is expected to perform the next action, or what may have been the previous situation. If both the system and the user are thus waiting for the other partner to resolve the interactive deadlock, the user needs a simple and clear way to proceed, e.g., by pressing a clearly marked button to receive information about the status of the interaction.

The system, from its side, should initiate metacommunication when the actual situation asks for it: in case of incompleteness or errors at syntax or semantic level, and in cases of unexpected semantic constructions. This system-initiated metacommunication should, however, be designed in such a way that the user always may understand and interpret the system's initiative as a direct and sensible reaction to the previous interaction.

Another important kind of metacommunication from the side of the system is feedback on actions or commands issued by the user. Especially when the results are not directly visible at the interface, knowledge of the effect may reassure the user about his action, or let him know that his intended goal was not reached. That gives him a chance to try again or undo the last action. Although feedback has the appearance of a system controlled type of metacommunication, the user should have control at a higher level, e.g., by a general request to turn the feedback mode on or off.

Individual Differences Between Users

Users will differ in all kinds of relevant characteristics. From the point of view of the development of mental models, two distinctions seem to be particularly relevant. These concern the level of system expertise and the style of mental representation.

Expertise

Users who are novices in the use of a machine may be characterised by the lack of completeness and detail of their mental model. The semantic structure of this model lacks sufficient coherence, and the knowledge of the syntax consists of isolated fragments. These users are unable to manipulate large chunks of meaning in thinking about the system. They will easily have problems of memory and will fail to overview a complex situation in order to make the right decisions. The possibility to choose from meaningful options, from clearly labeled menus, will help them to survive.

Users with a certain level of expertise will not normally need these devices. They may apply to a complete and consistent mental representation, and issue their commands without any help. Having to wait for the menu to be displayed, or having to wait for a verbose interaction protocol to be completed, will hinder them and should be avoidable for any user when he so wishes.

Mode of Representation

Individual users will differ in their preference for an internal representation of the user interface. Some prefer visual-spatial models, whereas other use verbal predicates or apply motor memory of actions at key-stroke level. The representation of tokens in the interaction, and especially the metacommunication in the user interface, may be designed either in a strict way that leaves the user little choice, or may offer him several possibilities (This will in fact often make communication redundant, and ask for alternative units of metacommunication.)

FAILURE ANALYSIS

As we restrict ourselves to the user interface, failures in the cooperation of human-machine systems may be the result of two problem fields: either the interface does not cope with the psychological mechanisms of communication, or there are discrepancies between the mental model and the conceptual model of the interface.

The Model of the User

Failure in communication from the machine to the user may result if the interface is not designed to adjust to general psychological mechanisms. The team that is responsible for the development has to start with a clear model of the intended end user. In any case, this model should include a general description of the skills and educational background the operators will have to obtain.

Apart from these user characteristics, that to a certain extent may be defined at will, knowledge of human perception and information processing is indispensable for an optimal design. General psychological knowledge is especially important for the processes that are closely related to perception mechanisms, attention mechanisms, and working memory span.

Perception

Ergonomic science has much to say about physical guidelines for coping with human perception restrictions. Information about the size and distance of display devices, colours and contrasts, sound levels and noise, luminance of signals and background, is well known and may easily be taken into account.

Cognitive psychology offers contributions in the domain of recognition. Lindsay and Norman (1977) describe the mechanisms that are responsible for this phenomenon. Especially in situations of ambiguity or difficulty of detection, the process of attaching meaning to sensory input is the result of two processes. On the one hand, the physical characteristics, the structure of signal and background, will influence the perception ("data driven interpretation"). On the other hand, the available knowledge and the semantic connotations provided by the actual situation will guide the interpretation ("conceptually driven"). Signs and symbols, icons and verbal codes, have to be clearly discriminable from other messages that might be received in the same context in order to avoid misinterpretations in poor input circumstances (e.g., the verbal code TOUTO in Weiner, 1986). Analyses like these advocated by Weiner are indispensable for this aspect of the user interface.

Attention

Physical characteristics of the messages the interface presents to the user have to be chosen with knowledge of attention mechanisms in mind. Colour and sound will have to be applied carefully, as will multiple moving indicators and blinking cursors. An excess of these devices will distract or even actively frustrate the user's perception. Too much diversity in character size, display format, signal displays, and meters will require extra attention for the interpretation of the signal. The attention value of system components has to be investigated in an integrated situation. Separated components will show totally different results, invalid for real life situations.

Working Memory

Attention and the conscious processing of information are located in the human working memory, or "Short Term Memory." This memory unit may keep track of about seven different units of information at a time. In exchanging information, it is necessary to make use of units that compose a cluster or "chunk" of meaning. The name of a geographical location will represent a lot of semantic connotations, whereas the coordinates (actually two chunks in most cases) may, for most people, represent just a vague location at the map.

Choosing meaningful command-names, semantically loaded icons, relevant combinations of system parameters for a single meter, reduces short term memory load. Displaying large numbers of chunks, several screens and meters multiple windows on a single screen, will overburden the working memory and will result in a loss of possibly relevant information. Short term memory overflow is beyond human control, apart from deliberate rehearsal or taking notes as a substitute, but both methods ask for a temporal withdrawal from outside stimuli, a process that is often dangerous or impossible in an actual crisis. Analysis of the system in this respect will involve simulation of conditions of emergency, and measurement of memory overload or loss of attention to important chunks of information.

Mental Models

The second source of failure in human-machine communication results from discrepancies between the conceptual model of the interface, the virtual machine, and the mental model the user is applying. Analysis of this discrepancy asks for special methods. Mental models are cognitive psychological concepts, located inside the mind of the user. The distinctions we made earlier will be of help in designing ways of collecting differences in representation of mental models. Both the mode(s) of internal representation and the validity of the knowledge at the different conceptual levels of the user interface have to be established. The actual methods for collecting these data are strongly related to the type of interface and the modes of human-machine interaction that are possible.

General Representation of the Mental Model

Field studies on human-computer interaction (Van der Veer, Felt, Van Muylwijk and Van Biene, in press; Van der Veer, in press) show the possibility of provoking users to externalise their mental model of the system. A method we call "teach-back" consists of the request to the user to explain the delegation of some well chosen subtasks to the system. The user should be told he is absolutely free in his choice of mode or modes of representation. He may use natural language, drawings or any other kind of representation the situation allows, giving him the opportunity to stick as close as possible to his internal representations.

The kind of human to which this is to be explained has to be varied, from colleagues who are experts on comparable but different systems, to colleagues without any system expertise who are expert on the task. For some task domains, this last situation is unthinkable, so the questions will have to be adjusted. The general idea is to ask for a general representation of task delegation in situations that allow different levels of description: teach-back to system experts has been shown to lead in extreme cases to key-stroke representation; teach-back to novices to any system naturally asks for metaphorical representations. Striking differences among users have been found in the ability to represent the system, and in the same way, some representations turned out to be invalid in one or more relevant levels.

Representation of Crucial Functions or Key Concepts

Teach-back is a feasible method of collecting these representations. The questions have to be formulated in a rather specific way, emphasizing the need for completeness, but the user must be totally free to choose his own mode(s) of representation. After collecting impressions of the user's representations, one has to analyse these to find out whether certain representations are fit for the system, although only in extreme or rare circumstances leading to invalid predictions or to interpretations of cases that normally will not be met in "normal" practise, or learning situations.

Correctness of Knowledge about the System

In order to check the mental model on correctness, questions may be posed to a user that test misinterpretations on the semantic level (the meaning of objects, attributes, relations, actions, in relation to task decomposition). Separate questions can ask for syntactic knowledge, investigating the knowledge of the interaction grammar, the format of commands, the application of parameters, and the order of messages and signals in the interaction sequence. Knowledge at key-stroke level asks for yet another way of testing, preferably by doing, e.g., participating in a made-up interaction. Users of certain types of systems will be able to perform in realistic interaction in a correct way, even though they are unable to describe the sequence of physical events, to answer questions about it.

REFERENCES

Lindsay, P.H. and D. A. Norman, D.A. (1977). *Human Information Processing*. New York: Academic Press.

Monk, A. (1986). Mode errors: a user-centred analysis and some preventative measures using keying-contingent sound. *International Journal of Man-Machine Studies,* 24, 313-327.

Moran, T.P. (1981). The command language grammar: A reresentation for the user interface of interactive computer systems. *International Journal of Man- Machine Studies*, 15, 3-50.

Norman, D.A. (1983). Some observations on mental models. In A.L. Stevens and D. Gentner (eds.). *Mental Models*. Hillsdale, N.J.: Erlbaum.

Van der Veer, G.C. (in press). Computer systems and human learning. In F. Klix, Y. Waern and N. Streitz (eds.). *Man-Computer Interaction Research MACINTER II*. Amsterdam: North Holland.

Van der Veer, G.C., M.A.A. Felt, B. Van Muylwijk, and R. J. Van Biene. (in press) Learning an office system - A field study on the development of mental models. *International Journal of Psychology.*

TECHNOLOGICAL ISSUES
IN
FAILURE ANALYSIS

FAILURE OF PROCESS PLANT MONITORING SYSTEMS

Peter Andow

Loughborough University of Technology
Loughborough, Leicester, U.K.

INTRODUCTION

The chemical process industry produces a wide range of products for use in every sphere of modern society. In the past, there was very little public interest in this industry or its plants. In recent times, however, there has been much more awareness of the industry and, in particular, of the more spectacular failures of some of its plants. The problem of high hazard plants is now well known, but this is not the only constraint imposed by society. During the last 20 years or so, the following trends have developed:

- Reduction of general costs as competition has increased. This has tended to give us bigger plants.
- Reduction of pollution as awareness of the environment has increased.
- Reduction of energy consumption (and hence cost).

It should be noted that the net effect of these trends has been to push process plants into very tightly constrained areas of operation. We now have large plants (often with high material inventories) operating close to hazardous conditions. When a failure occurs, the consequences of that failure are therefore more severe not only in terms of financial loss but also in terms of danger to plant personnel and the general public.

The chemical industry has recognised these increased risks and has responded to them in two ways: 1) use of more automatic control systems, and 2) use of increasingly sophisticated protective systems.

A Typical Environment

In a typical process plant, a small operating team (based in a control room) will be responsible for a large area of the plant, with most of the routine control being carried out by one or more computers. In such a situation, it might be expected that the operators would normally have a very light work load, but in practice, the operators are often observed to be quite busy carrying out a wide range of disparate tasks:

- Continuous small changes to the process in response to changes in feed or product specification; changes on other plants in the same complex; changes in availability or quality of services.

- Coordination of maintenance activities, including issuing of permits to work, confirming test signals sent from the plant, etc.

- Coordination of plant operation with other operators out in the plant and in other related plants.

- Logging of information for production reports, etc.

Superimposed on these activities is a general duty to monitor the plant, to diagnose the cause of unusual behaviour, and to initiate corrective action, which may be a plant shutdown.

NATO ASI Series, Vol. F32
Information Systems: Failure Analysis
Edited by J. A. Wise and A. Debons

Alarms are used to attract the operator's attention to abnormal conditions. For many disturbances that are alarmed, the operator will be able to make a small adjustment to the process before any more serious condition occurs. For a more serious fault or one that the operator cannot diagnose correctly, the fault will escalate, usually causing more alarms, until a protective system causes automatic shutdown of the plant. The operator must then make sure that the protective system works correctly until the plant is returned to a safe state.

It is clear that alarms are very important in this activity since they are used to attract the operator's attention to abnormal conditions. In older plants, alarms are usually indicated by means of small back-lit glass fascia panels inscribed with warning messages. The glass is often colour coded to indicate the type of failure or its severity. The fascia panel will usually flash when the alarm condition is detected, and the operator normally acknowledges the alarm by means of an "accept" button. When the button is pressed, the flashing stops but the alarm panel continues to be lit until the alarm condition is no longer satisfied. (Note that one press of the button will stop all new alarms from flashing.)

The fascia panels usually contain 4 or 5 rows of alarm windows. Typically, each row will have 6 to 10 alarms. In many plants, there will be several panels of this size. In a nuclear power station, there will often be several thousand alarms, and many chemical plants include several hundred. Sometimes the first alarm active on a particular panel can be indicated, and sometimes there is a separate accept button for each panel or group of panels. An audible alarm is usually associated with any new (flashing) alarm. Generally, one audible tone is used to indicate an alarm, but sometimes different tones are used to indicate different alarm conditions.

A Modern Plant Environment

In more modern plants, the alarm fascia panels may be used for the most important alarms, but computer displays will normally be the primary means of indicating conditions. In these displays, alarms are often arranged in "pages," with each page roughly corresponding to a traditional fascia panel. Each page will normally use most of a VDU screen. One or two lines on the screen will often be reserved for information about new alarms on other pages or for general operator command echoing. In many plants, several VDU's may be used for simultaneous display of alarms. In some plants, the alarm display is integrated with the normal monitoring display, in which each variable uses one line of the display. A variable in alarm is often indicated by flashing or change of colour.

PROBLEM IDENTIFICATION

From the previous description of a typical process plant environment in terms of tasks and hardware, it might be reasonable to assume that disturbances or failures causing alarms are relatively rare. In practice, however, alarms often occur at a very high rate: in many plants, the rate is approximately one per minute. This rate may not seem high, but it entails a large number of alarms during a single operating shift.

This alarm rate is a major problem. The original intention of alarm systems was to attract the operator's attention to an abnormal event, but an event that occurs once a minute is by no means abnormal. The alarm frequency involves two distinct but interrelated problems:

1. Frequency Problem. The basic alarm frequency is so high that the operator is not in any sense "alarmed" by an event that occurs so often. Real alarm conditions are rare - many of the "alarms" are really better described as "minor deviations" from the expected condition. The real alarms are thus easily lost in the noise created by other alarm signals. (See Kragt and Bonten.)

2. Time Problem. The high arrival rate of new alarms means that the operator is left with very little time to think about the diagnosis of the latest alarm. The previously noted rate of about one per minute is only for normal operation. When a major fault occurs,

the arrival rate of new alarms can be much higher (e.g., several hundred alarms occurred at Three Mile Island in such quick succession that the alarm log printer was a long way behind with its messages). This lack of thinking time largely negates the value of a separate alarm indication.

In addition, the VDU display presents a special problem. With the traditional fascia panels, which are all on view at the same time, the presentation is essentially parallel. With the VDU, however, one page is presented at a time, or several VDU's give several pages from a (usually) larger number. The operator of a VDU based system must therefore actively hunt for the other relevant pages. This is referred to as the Keyhole Problem, since the operator views the plant through the "keyhole" of the display screen. The operator also has a Diagnosis Problem when presented with the alarms. Usually each alarm is associated directly with a very specific abnormal condition (e.g., Cooling Water Flow Low). The operator then has to perform diagnosis in a high-stress environment based on a pattern of such alarms.

MEANS FOR IMPROVEMENT

Process plants have evolved rapidly, and their short life cycle has given little learning time between one generation and the next. The alarm systems of process plants suffer from this general problem. They have grown rapidly from a few genuine alarms to systems with hundreds or thousands of alarm signals, but there has been very little critical evaluation of alarm systems.

The Frequency Problem

The frequency problem is regarded by the author as the basis of most alarm system problems. Other problems, to be covered in later sub-sections, are all reduced in severity by fewer alarm indications. The frequency problem is essentially derived from two basic facts:

- The number of "alarms" is related to the size of the plant, whereas the number of operators depends on the degree of automation.

- Modern "distributed systems" (using many embedded microprocessors) make it very easy -- indeed, too easy -- to specify extra alarm conditions.

Together, these two situations produce the "alarm inflation" problem that occurs in plants with a very large numbers of alarms and a small operating team. In many modern plants, there are large numbers of alarms that do not alarm the operator in any real sense at all. In fact, he would be more alarmed if alarms did not occur. Many alarms are really minor deviations that will be corrected by the automatic control systems. In most cases, it would seem that these signals could be removed from the monitoring system altogether. It often appears that some alarms are specified by the designer merely to be "on the safe side." In reality, however, each extra non-serious alarm actually degrades the credibility of all real alarm signals, since it manifests itself in the same way as a real alarm. Also, the significance of an alarm is not always consistent. In one specific circumstance, an alarm may be genuine, while in all other cases it is not serious.

The frequency problem can be resolved, at least in part, by asking designers to seriously consider the frequency of each alarm specified. This would seem to be a basic question but, in practice, it seems to be rarely considered. By asking the designer to consider the frequency of an alarm, we focus his attention on the total arrival rate of that signal, not just on that fraction of the arrivals associated with specific plant conditions. This approach has two distinct benefits:

1. When asked to consider the frequency, the designer may decide that the alarm will be too frequent or too general.

2. By defining an expected frequency, we give a value for comparison during later feedback. If too many alarms occur, they may help to focus attention on particular signals that occur more often than the designer anticipated.

The maximum expected overall alarm frequency should also be specified as part of the design. In practice, this is rarely done. If no maximum overall target exists, it is not surprising if we get a large number of alarms. Clearly, it is necessary to add up the predicted frequency of individual alarms for comparison with the expected overall alarm frequency. It is difficult to be precise about a desirable target figure, but once a minute (or more) must be far too high.

The Time Problem

The time problem is closely related to the frequency problem. The operator needs time to understand the meaning of each new alarm, which must be seen as part of an evolving pattern. Since it is rare for a plant to be running without a number of "standing alarms" - i.e., alarms that have been on display for some considerable time -- a new alarm must be considered in relation to many others.

At the simplest level, the thinking time can be improved by slightly more sophisticated alarm processing:

1. The raw alarm signal can be logged by the computer when it occurs but not be displayed unless it remains in alarm for a defined number of scans or for a pre-determined time period. This eliminates alarms caused by signal noise.

2. Several alarm signals related to the same basic cause may be grouped together. A single "group alarm" will then be displayed when any alarm in the group is active. Any further alarms in the same group may appear on the display but will not set off the audible alarm and may not flash.

3. A "first up" facility may be included so that the operator can easily find which alarm in a group was the first to occur. This can be implemented by means of a special button on the fascia panel which makes the "first up" alarm window flash when depressed.

4. A hysterysis band may be set up for alarms derived from analogue signals. Consider, for example, a high limit alarm.

Once a variable has gone into alarm, the alarm limit is effectively lowered. This means that the variable will not go out of alarm until a significant fall in the value occurs. If there is any noise on the signal, the alarm will remain active and will not continually go in and out of alarm. When the variable value does finally drop below the alarm limit, the alarm becomes inactive and the limit is reset to its initial value. These remedies are all relatively simple, but they can reduce considerably the number of new alarms displayed - even when the number of raw alarm signals remains high. This clearly helps to resolve the Time Problem. (See Visuri and Owre.)

Another aspect of this problem is that the operator does not usually know how long he has before a decision is needed. In some cases, rapid diagnosis is needed, while in other case many hours may elapse before any serious consequence occurs. This is clearly related to the design philosophy of the system. It would seem appropriate that systems be so designed that any alarm would always give some pre-defined time for diagnosis and corrective action. The amount of time might be indicated by, for instance, the colour coding of the alarm display.

The Keyhole Problem

The keyhole problem is clearly a design issue: the display system must support parallel presentation of most (or all) of the alarms needed for diagnosis. There is obviously a requirement for a defined minimum number of display devices, but this is not usually the real problem. The data needed for diagnosis needs to be organised carefully. This may require special display formats. The display system should be carefully defined so that the displays are suitable for diagnosis as well as for normal running.

In many plants, the operator is included in the control system primarily for handling of fault conditions, and yet the displays are rarely organised with this in mind. Keyholing is a problem not just when one alarm pattern is being investigated; it also affects what else can be done with the displays while a fault condition is being handled. The keying sequence to switch from one mode of display usage to another (and back again later) should be simple.

The Diagnosis Problem

Diagnosis is central to the alarm system. A whole range of factors are related to this problem:

1. The alarms are usually effects, while the operator is looking for causes.

2. The data relevant to a fault is normally available (in the computer or elsewhere), but is often buried in a mass of other data.

3. The sequence of fault indications may be confusing (e.g., A causes B, but the alarm for B occurs before the alarm for A).

4. The set of fault indications may be confusing. They may be slightly different than expected because of unusual plant conditions or failed instruments. It may be particularly difficult for the operator to recognise a low probability event with a fault pattern that is distinct from but similar to one experienced for a more frequent event.

5. Once the operator has diagnosed a fault incorrectly, he may find it difficult to change his mind, even though later fault symptoms conflict with the initial diagnosis (the "mind set" phenomenon).

6. The operator may have had little or no training in fault diagnosis, even though this is actually a major part of his job. (See Marshall et al.)

7. The displays may be unsuited to diagnosis. (See earlier remarks on Keyholing.)

8. The operator often has no job aids to help him - or if aids are provided, they may be out of date or incorrect.

Many of these problems can be remedied or at least mitigated by careful design. It is also possible to use "alarm analysis" or "disturbance analysis" job aids to help the operator with the diagnostic task. These aids attempt to match the fault symptoms against a pattern of expected indications. The result is presented to the operator via a VDU. The main obstacle to this type of aid is that the fault data is expensive to define. This is really another measure of the difficulty of the problem that the operator faces during diagnosis. These diagnostic aids also need careful design if they are to be used effectively. Such systems can be implemented using conventional software or by means of Expert Systems techniques. (See Andow.)

SYSTEMS ASPECTS

It should be clear that the author regards better design as the main means of improvement. HAZOP (Hazard and Operability) studies are now carried out on many new designs. This helps to ensure (among other things) that the operator will be able to observe a fault condition or its symptoms. This ability at least increases the probability that important signals are properly defined and available to the operator. The information used by the HAZOP team is also relevant to many of the other design problems noted earlier. One failing of HAZOP is that it often adds extra instruments to a design but does not really address the fault diagnosis issue directly. For example, if an instrument is added to monitor a particular fault, what other faults will also activate it? How can the operator distinguish between them?

Good design is not the only requirement. It is also important to monitor the system performance to ensure that design goals are realised. Key points to check are the following:

1. Are the alarms taken seriously by the operator? Does the operator say, for example, "one is always going off - it doesn't mean anything"? If alarms occur too often, they will not be effective when they do have a real meaning. How many alarms are "normally" on display?

2. What is the overall frequency of new alarms? What proportion do operators actually respond to?

3. Do the display system controls work effectively? How long are the key sequences to switch from one display mode to another and back again?

4. Are the display formats appropriate? Do the operators have to keep cross-checking from one display to another, perhaps in different parts of the control room?

It is also clearly important to ensure that measuring instruments and protective devices are regularly tested and in good working order.

FAILURE ANALYSIS

The previous sections have attempted to describe a typical process plant environment and to identify problems associated with fault analysis in that environment. In this section, a list of questions relevant to failure of a system are defined on the basis of the problems previously identified. The questions are organised in terms of the EATPUT model.

Event World

If the monitoring system is to be effective, the designer must have clearly defined those failures or disturbance events that are reasonably likely to occur (the so called "credible" events).

Q 1. Was there a systematic search carried out to determine all the likely failure or disturbance conditions (e.g., HAZOP or "What if" analysis)? (See Kletz.)

Q 2. Was there any system to feedback operating experience after plant startup in order to include events not previously covered by the monitoring system?

Acquisition

The pattern of signals from the acquisition devices must be capable of warning the operators of all credible events. (Note that some of the Reply received on credible failure events are the measuring devices themselves.)

Q 1. Was there a systematic search to determine if the sensors would be able to detect the credible failure events?

Q 2. Have reliability engineering techniques been used to ensure that the monitoring devices are reliable enough for the duty required? (See Henley and Kumamoto.)

Q 3. Was a Common Mode Failure analysis used to determine if any single failure could disable all channels of a redundant system or cause an equipment fault and a simultaneous fault in protective devices?

Transmission

Q 1. Was there any general requirement to ensure that transmission systems (such as pneumatic lines, electric cables, fiber optic cables, microwave links, radio and telephone links, etc.) are physically robust and reasonably protected from sources of noise?

Processing

The display devices must be selected and arranged to perform the functions necessary to detect and monitor the credible failure events.

Q 1. Were systematic task analysis methods used to define the display requirements for all modes of operation? (See Goodstein.)
Q 2. Was there any system to ensure that the display devices were adequate in both number and type for the required duty?
Q 3. Were any criteria defined for the overall performance of the entire fault-handling system (e.g., sensors, displays, operators, protective devices)?
Q 4. Were there defined maximum frequencies for a) all alarms and b) individual alarms?
Q 5. Were formal methods used to ensure that the alarm type (e.g., absolute, deviation, rate of change, etc.) was appropriate to the defined requirement?
Q 6. Were alarm limits carefully specified in terms of functional requirements?
Q 7. Was there a systematic method to define the importance of particular alarms?
Q 8. Were alarm displays arranged to use colour or position coding or other means to indicate alarm importance?
Q 9. Were individual displays or alarms properly labeled to indicate the signal source (e.g., is a valve position a control signal or a measurement signal)?

Utilization

Q 1. Was any method used to ensure that the operating team would be adequate, in terms of size and skills, to cope with the load imposed by the credible fault conditions?
Q 2. Was there a requirement to identify the time available for fault pattern interpretation?
Q 3. Were any methods used to ensure that the operator had sufficient facilities to identify the sequence of alarm indications (e.g., first up facilities, alarm logs)?
Q 4. Were any diagnostic job aids provided (e.g., diagnostic procedures or computer aids)?
Q 5. Were the operators trained in methods of diagnosis?

Transfer

Q 1. Was any method used to ensure that the operator was aware of the corrective actions that were possible?
Q 2. Were the controls arranged so that the operator could monitor the effects of his corrective actions?

CONCLUSIONS

This paper has tried to give an overview of some of the problems that exist in process plant monitoring systems. It should be emphasized that, by the nature of the topic, the paper has concentrated on the shortcomings in such systems. Many design engineers, however, take care to provide operators with a good task-oriented monitoring system. The good safety record of the chemical industry as a whole is testimony to the results of their efforts.

The questions outlined in the previous section cover a whole range of design and operational issues. The emphasis, however, is on the hardware and software components of one part of the overall monitoring system -- the process plant alarm system. The issues raised are not exhaustive - rather, they are intended as pointers to the relevant factors in monitoring system design and, hence, in failure analysis of monitoring systems.

REFERENCES

Andow, P. K. (1985). Fault diagnosis using intelligent knowledge-based systems. *Chem. Eng. Res. Des.*, 63.

Goodstein, L. P. (1982). Computer-based operating aids. *Proc. of Design '82*, I.Chem.E.

Henley, E. J. and H. Kumamoto. (1983). *Reliability Engineering and Risk Assessment.* Englewood Cliffs, NJ: Prentise-Hall.

Kletz, P. T.A. (1983). HAZOP and HAZAN. *I.Chem.E.*

Kragt, H. and J. Bonten. (1983). Evaluation of a conventional process alarm system. *IEEE. Trans. SMC-13*, 4.

Marshall, E. C., K. E. Scanlon, A. Shepherd and K. Duncan. (1981). Panel diagnosis training for continuous process installations. *The Chem. Engr.*

Visuri, P. and F. Owre. (1981). *A Candidate Approach to a Computer Based Alarm Handling System (HALO)*.. OECD Halden Reactor Project Report HWR-24.

USE OF SOFTWARE RELIABILITY TOOLS IN INFORMATION SYSTEM FAILURE ANALYSIS

Wolfgang D. Ehrenberger

Fachhochschule
Fulda, F. R. Germany

INTRODUCTION

Information systems contain more and more freely programmable processors. An increasing number of the tasks to be carried out are being done by software. Therefore, it is of great importance to discuss the problems of software reliability in connection with information systems.

In the terminology of the EATPUT model, only one item is connected with the topics to be discussed here: the item "processing." In the following discussion it is always assumed that the data of our information system have been successfully acquired and transmitted and are reliable. Should any error correction be necessary, it is being done either by hardware or by software. In the latter case, the corrective task itself forms part of the set of the software tasks and is therefore considered along the same lines as the other software.

The increasing role of software and software reliability in information systems is heavily influenced by two points:

- Due to the cost reduction of computer hardware, it is increasingly economical to have tasks executed by the computer.
- Software does not change its behavior in itself: it does not rust, and any changes in it that occur during operation come from outside, i.e., from the event world or from the hardware.

Because of this second property, software can principally be verified completely before operation starts. Ideally, no supervision of software behavior is necessary during operation.

Unfortunately, other difficulties arise. Since it is easy to implement specific tasks in software, the software packages get more and more tasks to fulfill and thereby become larger and larger. Since understanding of anything depends on its complexity and since software complexity naturally increases with the size of the code, it becomes increasingly difficult to understand the growing software packages. This, in turn, makes it more difficult to verify the correctness of the software part of an information system.

Many attempts have been made to overcome this difficulty. Development methods have been invented, and standards for software quality and software quality assurance have been set. In order to make things more economical, tools have been developed for the different phases of the software life cycle. Such tools help during requirements specification, software development, and software verification.

This paper concentrates on software verification tools. The related activities are known as testing, verification, and validation. Since any verification activity needs to be based on well known functional requirements, an example of a specification tool is also included. It is always assumed that the software being considered has an importance that requires developer-independent verification actions. The practical background for the following

NATO ASI Series, Vol. F32
Information Systems: Failure Analysis
Edited by J. A. Wise and A. Debons

discussion is the area of nuclear reactor safety.

PROBLEMS OF SOFTWARE RELIABILITY

Systematic Verification

Since software does not change its behavior by itself, it is possible to demonstrate its correctness by systematic means. Such mean are, in principle, a) mathematical proofs and b) tests whose completeness is demonstrated by logic.

Since the related mathematical and logical systems have attracted much interest and have been investigated for many years, the connected problems are very well known:

- In order to prove a program correct, it is necessary first to understand what it does. Such understanding is possible for small modules only and is sometimes difficult even there.

- Proofs need much effort.

- Proofs may contain errors.

In the case of exhaustive testing, the requirements for understanding are less severe, but extensive work is needed to determine what would provide a complete test. Also, such work can be completed successfully only if the programs are neither too complex nor too large.

At the present time, it is therefore impossible to verify the correctness of a large program system in any complete way. My experience in operating systems for very small computers can be considered extensive in the context of proving possibilities.

If complete testing is not feasible, other systematic means can be used. They will not provide complete verification, but they can give some relative confidence about program correctness. Such means may be the normal tests applied at the end of each development cycle, where the developer convinces himself that the program behaves as it should. Other systematic tests include the following:

- Execution of all explicitly specified requirements
- Crash test: all input conditions in extreme positions
- Execution of each statement at least once
- Execution of all edges of the control flow graph
- Execution of all interrupt sequences
- Execution of all possible data movements

For details, see Table E.4.2 in International Electrochemical Commission (1986).

Testing according to the intuition of the developer or according to the above list will not guarantee complete verification but only a certain unquantifiable software quality.

The different aspects of systematic software verification can be supported by automatic aids. The tools that exist for this can enlarge the area of application of a specific method and thereby increase the quality of the software.

Probabilistic Verification

Probabilistic verification efforts usually take a view of the software different from that used for systematic efforts. They do not consider internal details, but regard a software package as a black box. If this black box is tested with a certain number or sequence of input conditions, a specific failure probability can be attributed to it with a certain level of confidence. Such probabilities can also be derived from operating experience.

In this discussion, probabilistic verification questions do not play an important role because they cannot be supported by specific tools. Tool support is possible only in a very general way: by simulating the environment. Since probabilistic software verification usually requires a large number of test cases, the simulation of the event world is some help in providing the many test cases. Although simulation tools may be used for test case procurement, such tools are not software specific and therefore are not treated in the context of this paper.

PROBLEMS WITH SOFTWARE MAINTENANCE

Software failure detection and the removal of the underlying faults are important after the end of the development cycle. Sometimes it is not specific faults that are to be removed, but the software is to be adapted to new requirements. The common philosophy about treatment of such modifications advises going back to the earlier stages of the development. Depending on the size of the change in view, one should restart at software specification, software design, detailed design, or coding for the affected software parts. This will require an extensive documentation but no specific maintenance tool.

For this reason, no separate section about maintenance tools is included here. Maintenance uses just the ordinary tools that are being employed during normal development. Such use is certainly easier if the product to be maintained has been developed with tool support.

SOFTWARE SYSTEM FAILURE ANALYSIS

After failure of an information system, the question of finding the underlying fault may be very important. The system that is monitored by the information system might be allowed to resume operation only after fault detection and removal. Since information systems often contain software, the problem of whether or not they are fault free may be very important.

Since analysis of large software systems is very time consuming, any help in that work is welcome. The following section describes some of the automatic tools that can be used for analysis or assisting analysis. The post mortem analysis is essentially not different from the analysis carried out during the verification and the verification phase of the normal life cycle.

TOOLS

The following paragraphs include examples of existing or planned tools. The examples are necessarily incomplete, with respect both to their selection and their details. Some tools have properties that exist in other tools as well, although this is not mentioned explicitly. Basic properties are discussed in connection with one tool only.

Test Aids

Test aids are the oldest type of tools used in failure analysis. They are not yet common with high level languages, but are in widespread use with assembly languages. Typically, they allow three activities:

- To set break points at selected positions of the code.
- To input and output the contents of interesting variables at the break.
- To go on to the next break.

Depending on the tool, these actions can be done interactively or by means of a prepared supervision book.

One example of a test aid that is being developed for higher level languages is the Lilith debugger for Modula II programs. It becomes operable as soon as the test object breaks down. It then takes the memory dump as its basis and shows several windows on the CRT:

- Program window with the text of the code next to the breakdown position.
- Data window with the local variables of the procedure where the breakdown took place.
- Variable window, which gives insight into the structure of the arrays, records, and sets.
- Procedure window with the chain of calls of the procedures that were active at the moment of the breakdown.
- Dialogue window for the communication between debugger and user.
- Module window, which presents the modules that were loaded at the moment of the breakdown.
- Process window, showing the reason for the breakdown, e.g., index error or overflow.

Another example is the PEARL test system . Among other functions, it provides the following:

- The test object can be stopped at break points chosen by the user.
- The values of the task states and the synchronization or program variables can be displayed.
- The history of the program execution can be given.
- Deadlocks can be investigated.
- Task states synchronization states or program variables can be changed.
- The program execution can be traced by three different levels of tracing: procedure calls and task status changes, labels that have been passed and branching conditions that have been passed, and printing of each code line.

In the PEARL system, breakpoints can be installed in the following way:

```
BP1: BREAK WHEN INT1;
       BP2: BREAK ON OUTPUT DAT1;
       DISPLAY Steuerwert IN MOD1;
       CANCEL BP2;
       END;
      CANCEL BP1;
END;
```

When the condition INT1 (interrupt 1) becomes true, the outer break point BP1 inserts the inner break point BP2 and cancels itself. When an output to DAT1 is made, BP2 becomes active. It displays the value of the variable "Steuerwert" and cancels itself as well.

It is possible, of course, to debug any program without tools of the kind described. However, for programs that use real time features, such as interrupts, this is very time-consuming and difficult. Among the different types of information systems, those that take advantage of real time features will profit more from the tools mentioned above than the rest. Whether interrupts and other time-critical features are used in an information system depends on the required response times. The shorter such times are, the more important the tools will be.

Static Analysers

Static analysers qualify for verification purposes after debugging. They may be employed by an independent assessor. They are characterised by their static investigation of the test object, without executing it. Among the various types of static analyzers, the simplest forms are those connected with translators, cross reference lists, lists of procedure calls, lists of data usage in subroutines and the like. Several types are discussed below. Real tools usually own more than one of the described characteristics.

Proving Aids

During execution of any program, its data space is modified. The program is correct if the context of its data space is correct at execution end, depending on its content before execution. In the ideal case, the relationship between the two contents is specified in mathematical logic.

Proving aids make use of the logic relationship between the conditions on the data before and after the execution of each statement. They can derive preconditions and postconditions as required. Normally, the postcondition is given by the user, and the tool derives the condition that is true before statement execution, i.e., the precondition. Usually, the logic of Hoare forms the basis of the tool. One example is the product PASQUALE.

Normally, such tools require much human interaction. In particular, the human user must have understood certain program parts beforehand in order to derive some of the conditions that can be proven with the help of the tool. This applies, for example, to loop invariants. For problems existing with fault detection in information systems, such tools can be of limited value only.

Presentation of the Calling Trees and Data Dependencies

If a program is very large and if it consists of many individual subprograms - as it should - it becomes quite difficult to understand all the different calling relationships among the individual subroutines and functions. Tools like LAXYM (Mitsubishi,1982) provide some insight into the calling hierarchy by depicting the calling tree. See Appendix A as an example with a small program. Such trees are particularly helpful when during program modification the question arises of which influences must be considered if some of the subprograms are to be changed. For that purpose, the tool displays the usage of the COMMONs as well.

Presentation of Control Flow Graphs and Data Graphs

For software failure analysis it is important to understand the control flow and the data flow of a program. Some tools, such as the PEARL analyser of Puhr-Westerheide, provide related graphs. The control flow graphs show the possible sequences of code execution. Nodes represent statements; edges represent possible control movements. Backward movements, such as the latching branches in loops, are depicted with two lines. The numbers in the nodes represent the line numbers in the source code. See Figures 1 and 2.

The data flow graphs show the influences that exist on a specific datum. The datum to be influenced is chosen as the top node. In the nodes, the names of the data are shown, along with the code lines from which the data have received their assigned values. See Figure 3.

A more primitive form of such graphs is depicted in Appendix B and C, which show the tool analysis assembly code. On the left side next to the code are shown the possible control flow movements. The right side contains the data movements. The registers of the processor are shown, and at the far right the memory words, indications of indirect computation, change to the memory area, and calls of external devices. See Appendix B and C from TUV

Norrdduetschland (1984). The example is from the verification of the user programs of the control rod motion computer of the German boiling water reactor plants.

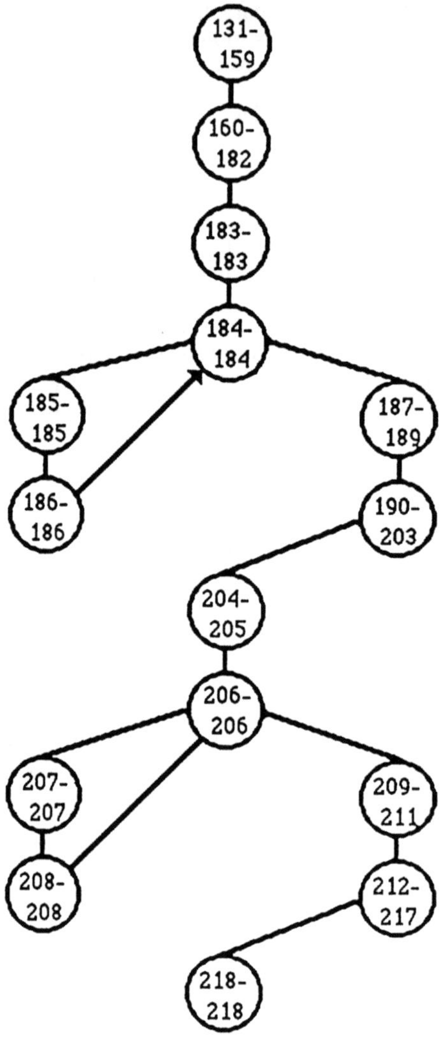

Figure 1. Control flow graph from a PEARL task (from Puhr-Westerheide, 1983)

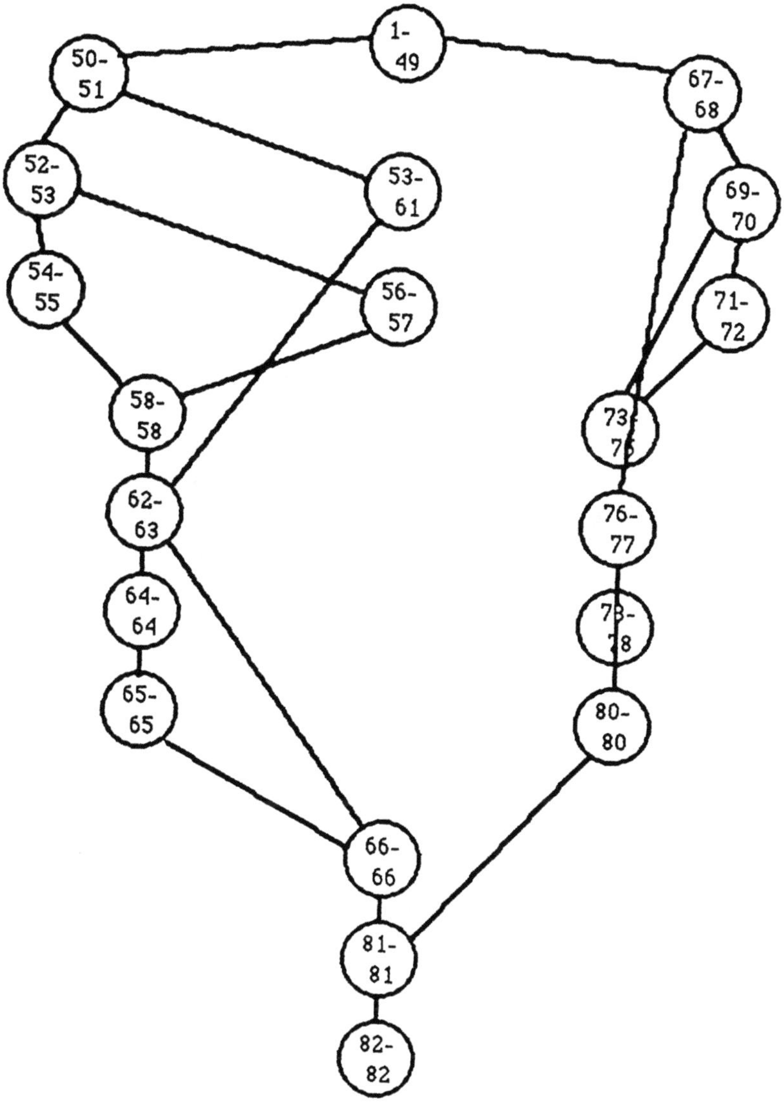

Figure 2. Control flow graph of a PEARL task (from Puhr-Westerheide, 1983)

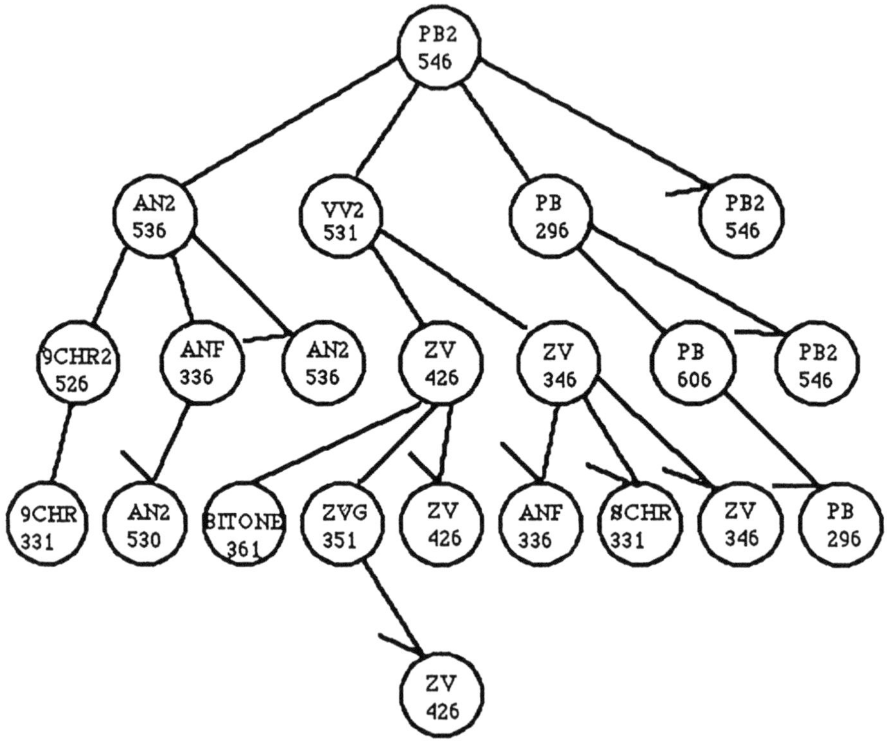

Figure 3. Data tree (from Puhr-Westerheide, 1983).

Tools of the kind described consider only sequential program behaviour. With respect to failure analysis in information systems, they are important as far as programs faults are to be detected in sequentially working program parts, i.e., in parts not dependent on interrupts, timing questions, or any sort of parallelism.

Verification Aid on the Basis of Machine Code

An important problem of software verification and reliability is the reliability of the translator. Since the tools thus far described work on the basis of the source code, they combine the advantage of easy understanding of the analysis results with the disadvantage of being useless if the source code is incorrectly translated. In order to avoid the difficult problem of translator correctness, a tool is currently being developed that starts its analysis from the machine code. The machine code is then transferred into a common assembly language called CAL from which the graphs are derived. The advantage of this approach is that different machine codes can be treated in the same way from the CAL on. The structure analyser works like the PEARL analyser. For details see GRS (1985). See also Figure 4.

The system is intended for the analysis of safety related software such as that for nuclear power plants. Since it is still under development, nothing can be said yet about its limitations and drawbacks.

Symbolic Execution Tools

In between the static analysers and the dynamic analysers are tools for symbolic execution.

Such tools do not really cause the investigated software to run, as the dynamic analysers do, but they are not confined to mere static analysis. They execute the test object symbolically by assigning symbolic values to input data and by going through the code step by step. Figures 5

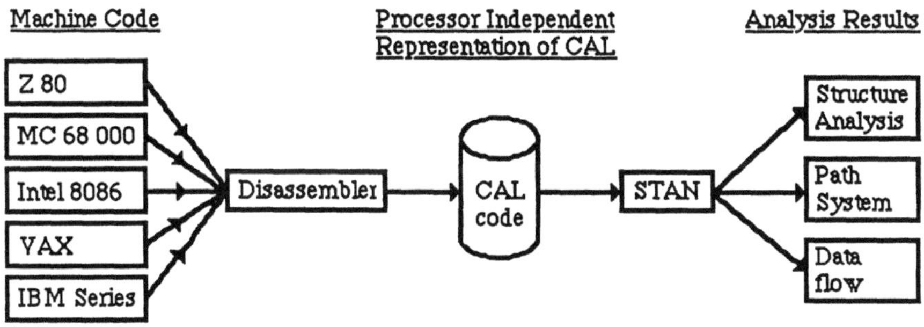

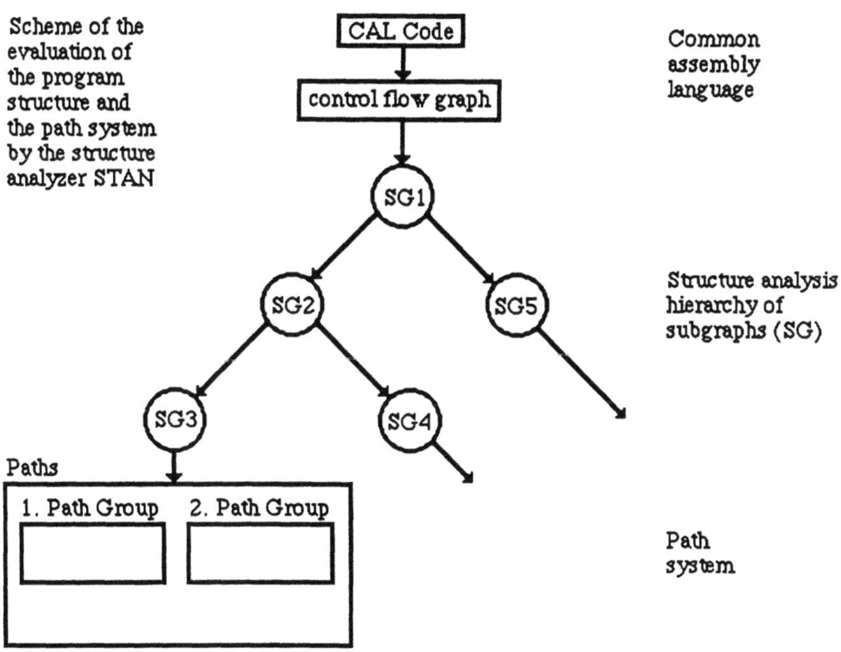

Figure 4. Schemes of SOSAT.
(SOftware SAfety Tools) (from GRS,1985)

and 6 from Walter (1986) give one example. Where possible, the expressions that occur during the symbolic execution are simplified. At each branching instruction, two sub-branches must be followed. For each of these two branches, the branching condition is transferred to a logic predicate. A path is defined through the AND relation of all its branch predicates. The splitting at each branching instruction can lead to very large symbolic execution trees for a fairly small program.

```
Procedure symbolic (A: integer; var Z; integer);
var B, C : integer
begin
    B : = 2 + A
    C : = B+ A
    if C< A then
        begin
        Z : = Z + 1
        if  C > B then
            Z : = Z + 1
        else
            Z : = Z + 2
        end
        else
            Z : = B + 1
        end;
```

Figure 5 . Example of Pascal procedure.

For programs with a realistic length, the tree with its path conditions is too complex to be meaningful or understandable. The method is useful for modules only. It has some similarity to the proving methods mentioned earlier.

<u>Dynamic Analysers</u>

During dynamic analysis, the program to be verified is not only investigated but also executed. The principle is to first instrument the program with specific monitors, translate it, and then run it on the target machine. The monitors gather specific data during program execution, such as which parts are being touched and at which time certain parts are executed.

Following a program run, this information is sorted out and displayed in table form. The principal features are shown by an example with the tool TFORT from L. Gallner (1986). Appendix D shows first the source program, then the instrumented program with its monitors, and then the tables that come out after execution with test data.

Among the most widely used dynamic analysers is the tool RXVP of the General Research Corporation. It provides both static and dynamic analysis and can be used for several languages. A particular feature is the self-metric component that provides several types of data:

- Number of calls for each subprogram.
- Number of executions for each logic decision and how often the true branch and the false branch were taken.
- For each variable on the left side of an assignment, the first, last, maximum and minimum values.

For details see General Research Corporation (1985). Due to copyright restrictions, it is not possible to give illustrations here.

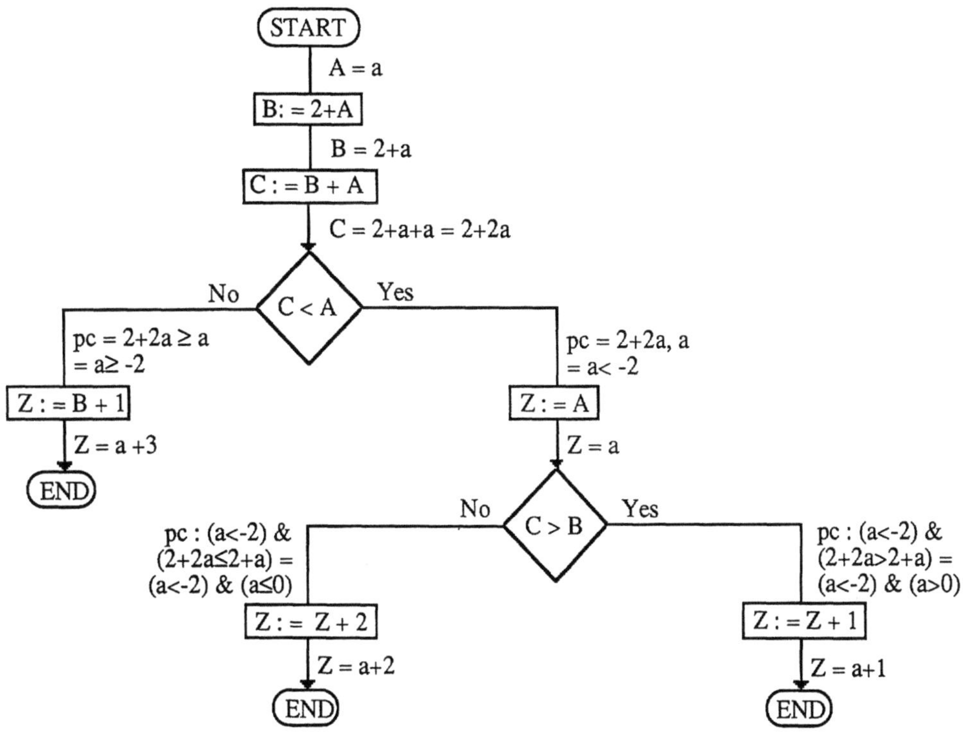

Figure 6. Symbolic execution tree from the procedure of Figure 5. The boxes contain program instructions. The symbolic values and path conditions are given at the edges.

Specification Tools

An important point during program verification is to determine against which requirements an analysis result must be checked. Since only precisely known functional requirements can be a solid basis for verification, tools that help to describe the functional requirements play an important role. Appendix E includes an example of the use of a specification tool, taken from TÜV Norrddeutschland (1984). The specified tasks are for the control rod motion computer for the German boiling water reactor plants.

In the graphs, the edges represent data flows or transmissions that have to be explained for a specific request by the tool, as shown in Figure 1 of Appendix E. The nodes of the graphs represent actions that have to be described separately. The boxes stand for the connections to the event world. The meaning of the data is described in a data dictionary.

The system works top down. On each lower level, a node is spread out to a new sheet. The tool, called porMod, has been used with satisfying results in the above-mentioned project of the control rod motion computer.

ASSESSMENT

At present, many software projects are still being carried out without the aid of any tool. This may finally prove not to be cost efficient. In particular, if formal licensing is required and many documents about the program are needed, tool usage may be of great help.

On the other hand, some existing tools suffer from severe shortcomings:

- Lack of operating experience with the tool.
- Awkward handling.
- Difficulty with the installation on the hardware that is available for the project.
- Applicability for small test objects only.
- Restriction to specific dialects of the source languages.

The most important point, however, is that some of the tool-supported principles do not work for software of the size of real programs.

Since software is taking an increasing share of the costs during information system projects, undoubtedly more effort will be made to overcome the problems. Future tool sets will probably contain several, if not all, of the properties described in this paper. These sets will be designed to cooperate with the translators for each language and will accompany the whole life cycle of a program. New features can be expected from large data bases and the techniques of artificial intelligence.

During failure analysis of information systems, the analyst tries first to find out whether the failure cause was technical or human. If the cause was human, he will further investigate whether the failure is due to a computer and if so, whether it was in the hardware or the software. In the case of software failure, the analyst will search for the fault. In this search, the tools described above will be of great help. The procedure will be top down and will include a specific sequence of different tools of a tool set. The more global tools, i.e., the system oriented ones, will be used first.

As for the details, very few suggestions can be made here, as anybody knows who has ever personally searched for faults in programs. The details of software failure analysis u require much intuition and are essentially based on the experience of the analyst. Tool support, however, will become more and more important because it increases the overview of the individual and his capabilities for drawing conclusions.

REFERENCES

Adrion, W.R., M. A. Branstad, and J. C. Cherniavsky. (1982). Validation, verification and testing of computer software. *Computing Surveys*, 14(2), 159-192.

Alberts, D. S. (1978). The economics of software quality assurance. in E. Miller. and W. E. Howden, (eds) *Tutorial: Software Testing and Validation Techniques*, Second Edition. Piscataway, NJ: IEEE, 143-156.

Ammann, W. (Hrsg.). (1984). Fchtagung Programmier-sprachen und Programmentwicklung der GI. *Informatik-Fachberichte*, 77, Berlin : Springer.

Balzert, H. (1985). Systematischer Modultest in Software-Engineering-Environment-System PLASMA. *Elecktronische Rechenanlagen*, 27.Jg.m Heft 2, 75-88.

Balzert H. (1982). *Die Entwicklung von Software-Systemen Prinzipien, Methoden, Sprachen, Werkzeuge*. Mannheim-Wien-Zurich

Berg, H.K., W. E. Boebert, W. R. Franta, and T. G. Moher. (1982). *Formal Methods of Program Verification and Specification*. Englewood Cliffs,, NJ: Prentice Hall.

Bergeretti, J.F. and B. A. Carre. (1985). Information-flow and data-flow analysis of whole-programms. *ACM Transactions on Programming Languages and Systems,* 7(1), 37-61.

Chandrasekaran, B. and S. Radicchi. (1981). *Computer Program Testing*. Amsterdam-New York: North Holland.

Clarke, L.A. (1976). A system to generate test data and symbolically execute programs. *IEEE Transactions of Software Engineering*, SE-2 (3), 215-222.

Clarke L. A. and Richardson, D.J. (1981). Symbolic Evaluation Methods Implementations and Applications, in Chandrasekaran, B. and Radicchi, S. (1981). *Computer Program Testing*. Amsterdam: North Holland, pp.65-102.

Clarke, L. A. and D. J. Richardson. (1984). Symbolic evaluation: An aid to testing and verification, in: H. L. Hausen. (ed). (1984). *Software Validation: Proceedings of the Symposium on Software Validation*. Amsterdam: North-Holland, pp.141-167.

Darringer, J.A. (1979). The use of symbolic execution in program testing. *Software Testing*, 2, 67-84.

Donaho, J.D. (1984). Maintenance tools and techniques description and technology review. In *A Review of Maintenance Technology* , NTIS: AD A-082 985.

Ehrenberger, W. (1985). Vorlesungsunterlagen: "Softwarezuverlassigkeit."

Floyd, C.: and Kopetz, H. (Hrsg.). (1981). Software engineering entwurf und spezifikation, *Tagung II/1980 des German Chapter of the ACM*, vom 12. bis 16. 9. 1980 Berlin, Stuttgart 1981.

Gallner, L. (1986). *Erstellung der dynamischen Testhilfe TFORT5*. Internal Report from the Company for Reactor Safety.

Geiger, W. (1980). Experience with an automated test and documentation system for Fortran programs. *Practice in Software Adaption and Maintenance: Proceedings of the SAM Workshop in Berlin in April 1979,* GRZ Berlin : North Holland.

Geissmann, L. (1985). Der Lilith-Debugger: Ein modernes Werkzeug zur Fehlersuch in Modula-2 Programmen. *Elecktronische Rechenanlagen*, 27 Jg., Heft 2, 95-105.

General Research Corporation. (1985). *RXVP80 The verification and validation System for FORTRAN Version 4.0, User's Manual.*

Gmeiner, L. and Voges, U. (1982). Erfahrungen mit dem Einsatz automatischer Werkzeuge Fur die Qualitatssicherung wahren Entwurf und Test. in H. Sneed and W. R. Wiehle. (1982). *Software Qualitatssicherung*, Tagun I des German Chapter of the ACM am 25./26. 3. 1982 in Neubiberg bei Munchen, pp. 235-252.

GRS (Company for Reactor Safety). (1985). Jahresbericht 1985.

Hantler, S.L. and J. C. King. (1978). An introduction to proving correctness of programs, in

E. Miller and W. E. Howden. (1978). *Tutorial: Software Testing and Validation Techniques, Second Edition*, Piscataway, NJ: IEEE, 57-79.

Hausen, H.L. (Hrsg.). (1984). *Software Validation. Proceedings of the Symposium on Software Validation.* Amsterdam: North-Holland.

Hausen, H.L.; Mullerburg, M. (1984). An introduction to quality assurance and control of software, in H. L. Hausen. (ed). (1984). *Software Validation. Proceedings of the Symposium on software Validation.* Amsterdam: North-Holland. pp. 3-8.

Hoare, C.A.R. (1986). An axiomatic basis for computer programming. *Communications of ACM*, 12(10).

Hoare, C.A. R. and N. Wirth.(1973). An axiomatic definition of the programming language Pascal. *Acta Informatica*, 2.

Hohlfeld, B. (1984). Programmverfikation in lauffahigen Pascal Programmen, In W. Ammann. (Hrsg.). (1984). Fchtagung Programmier-sprachen und Programmentwicklung der GI. *Informatik-Fachberichte*, 77, Berlin : Springer.

Hohlfeld, B. and A. Kley. (1985). Formale Methoden in der Software-Entwicklung. *Kolloguium: Software-Entwicklungs-Systeme und Werkzeuge*, Ostfildern 2, Technische Adkademie Esslngen, September.

Howden, W.E. (1977). Symbolic testing and the DISSECT symbolic evaluation system. *IEEE Transactions on Software Engineering*, SE-3 (4), 280-285.

Howden, W.E. (1976). A survey of dynamic analysis methods, in E. Miller and W. E. Howden. (1976). *Tutorial: Software Testing & Validation Techniques, Second Edition.* Piscataway, NJ: IEEE. pp. 185-206.

Huang, J.C. (1975). An approach to program testing. *ACM Computing Surveys*, September, pp. 113-128.

IEC TC 45A, *WG3 Software for Computers in the Safety Systems of Nuclear Power Stations.* International Electrotechnical Commission, Standard Published in Geneva, 1986

Kimm, R., W. Koch, W. Simonsmeier, and F. L. Tontsch, *Einfuhrung in Software Engineering.* New York: Berlin.

Majoros, M.(1982). SOFTEST: A System for the Automatic Verification of PL/1, COBOL, and Assembler Programs, in H. Sneed, and H. R. Wiehle, (1982) *Software Qualitatssicherung.* Tagun I des German Chapter of the ACM am 25./26. 3. 1982 in Neubiberg bei Munchen. Stuttgart. pp. 203-218.

Miller E. and W. E. Howden. (1976). *Tutorial: Software Testing & Validation Techniques, Second Edition.* Piscataway, NJ: IEEE.

Mitsubishi Research Institute. (1982). LAXYM - Large Code System Maintenance Report

Myers, G.J. (1979). *The Art of Software Testing.* New York.

Osterweil, L.J. and L. D. Fosdick. (1984). DAVE a validation error detection and documentation system for Fortran programs, in Hausen, H.L. (Hrsg.). (1984). *Software Validation. Proceedings of the Symposium on Software Validation.* Amsterdam: North-Holland. pp. 73-103.

Panzl, D.J. (1976). Automatic software test drivers, in E. Miller and W. E. Howden (1976). *Tutorial: Software Testing andValidation Techniques, Second Edition.* Piscataway, NJ: IEEE. pp. 266-271.

Pearl-Testsystem, Benutzerhandbuch der Ausbaustufe 2, Luneburg, Werum Datenverarbeitungssysteme GmbH, Erscheinungsdatum bekannt.

Puhr-Westerheide, P. (1983). *Verfahren der Software-Produktion und Software-Qualitatssicherung.* Report from the Company for Reactor Safety, Nr GRS-A-856 July.

Puhr-Westerheide, P. and J. Martz. (1982). *Der Pearl- Analysator, ein Software-Werkzeug zur analyse von Pearl - Programmen.* Garching, Gesellschaft fur Reasktorsicherheit.

Quirk, W. J. (ed). 1985. *Verification and Validation of Real-Time Software.* New York: Verlag.

Ramamoorthy, C. V. and S. B. F. Ho. (1978). Testing large software with automated software evaluation systems, in E. Miller and W. E. Howden. (1976). *Tutorial: Software Testing and Validation Techniques, Second Edition.* Piscataway, NJ: IEEE. pp. 157-169.

Rubey, R.J., Dana, J.A. and Biche, P. W. (1983). Quantitative Aspects of Software Validation, in E. Miller and W. E. Howden. (1976). *Tutorial: Software Testing and Validation Techniques, Second Edition.* Piscataway, NJ: IEEE. pp. 314-319.

Saib, S.H.: RXVP Today and Tomorrow. in H. L. Hausen. (Hrsg.) (1984). *Software Validation. Proceedings of the Symposium on Software Validation.* Amsterdam: North-Holland. pp. 103-127.

Sneed, H. and W. R. Wiehle, H.R. (1982). *Software Qualitatssicherung.* Tagun I des German Chapter of the ACM am 25./26. 3., Neubiberg bei Munchen. Stuttgart.

Stoffel, C. and G. Goller, G. (1982). KTDS- Ein integrierter Satz von Werkzeugen Fur Analyse und Test und Warung von Software, in Sneed, H. and Wiehle, H.R. (1982). *Software Qualitatssicherung.* Tagun I des German Chapter of the ACM am 25./26. 3., Neubiberg bei Munchen. Stuttgart. pp. 203-218.

Stucki, L.G. (1977). New directions in automated tools for improving software quality in: *Current Trends in programming Methologie II.* Englewood Cliffs,NJ: Prentice Hall, pp. 80-111.

Taylor, R.N. An integrated verification and testing environment. *Software: Practice and Experience,* 13, pp. 697-713.

TUV Norrddeutschland. (1984). *Sicherheitsanalyse der Programme der Steuerstabfahrrechner.* AbschluBbericht Report from TUV Norrddeutschland, Hamburg, Dez.

Voges, U., L. V. Gmeiner, and A. Mayrhausen,. (1980). SADAT: An automated testing tool. *IEEE Transactions on Software Engineering,* SE-6(3), 286-290.

Walter, K. H. (1986). *Bestehende Verifikationswerkzeuge - Uberblick and Beurteilung.* Unpublished Diploma Thesis at the Fachhochschule Fulda.

APPENDIX A
Output of the tool LAXYM

```
--VARTHRU 8001 --  **  STRUCTURE MAP  **

          1              2                 3                 4                 5                 6                 7

1 MAIN            ---PRONUM
 M     3  WD      I
                  I
2                 ---REED
                  I
                  I
3                 ---ENTBLA
                  I
                  I
4                 ---STATPUF ---REED   ===SAME
                  I S  4  D         I    -                  2- 2
                  I                 I
5                 I                 ---RITE
                  I                      -
                  I
6                 ---OPHILF
                  I     -
                  I
7                 ---EINLES
                  I S  1  D
                  I
8                 ---DAHLAE
                  I     -
                  I
P                 ---SATWRI
                  I S  S  WD
                  I
10                I KSATKA
                  I     -
                  I
11                ---RITE            ==SAME
                  I     -                 5- 3
                  I
12                ---SEGMET
                  I     -

--VARTHRU 8001 --  *** CB-SP MAP  ***
                                               (SUBPROGRAMS)
                                    0   0   0   0   0
                                    0   0   0   0   0
                                    1   2   3   4   5

                                    E   E   M   S   S
                                    I   N   A   A   A
                                    N   T   I   T   T
              (COMMONS)             L   B   N   P   W
     ID          NAME               E   L       U   R
HEX DEC     S                       A       F   I
                                    ========================================

$01      1      /DAPC      /I               +
$02      2      /SATZC     /I               +   +   +
$03      3      /ZENPUF    /I       +   +   +       +
$04      4      /SATBUF    /I               -   +
```

APPENDIX B
Example of Program Execution Diagram
Assembly language AKRA of AEG (from TUV, 1984)

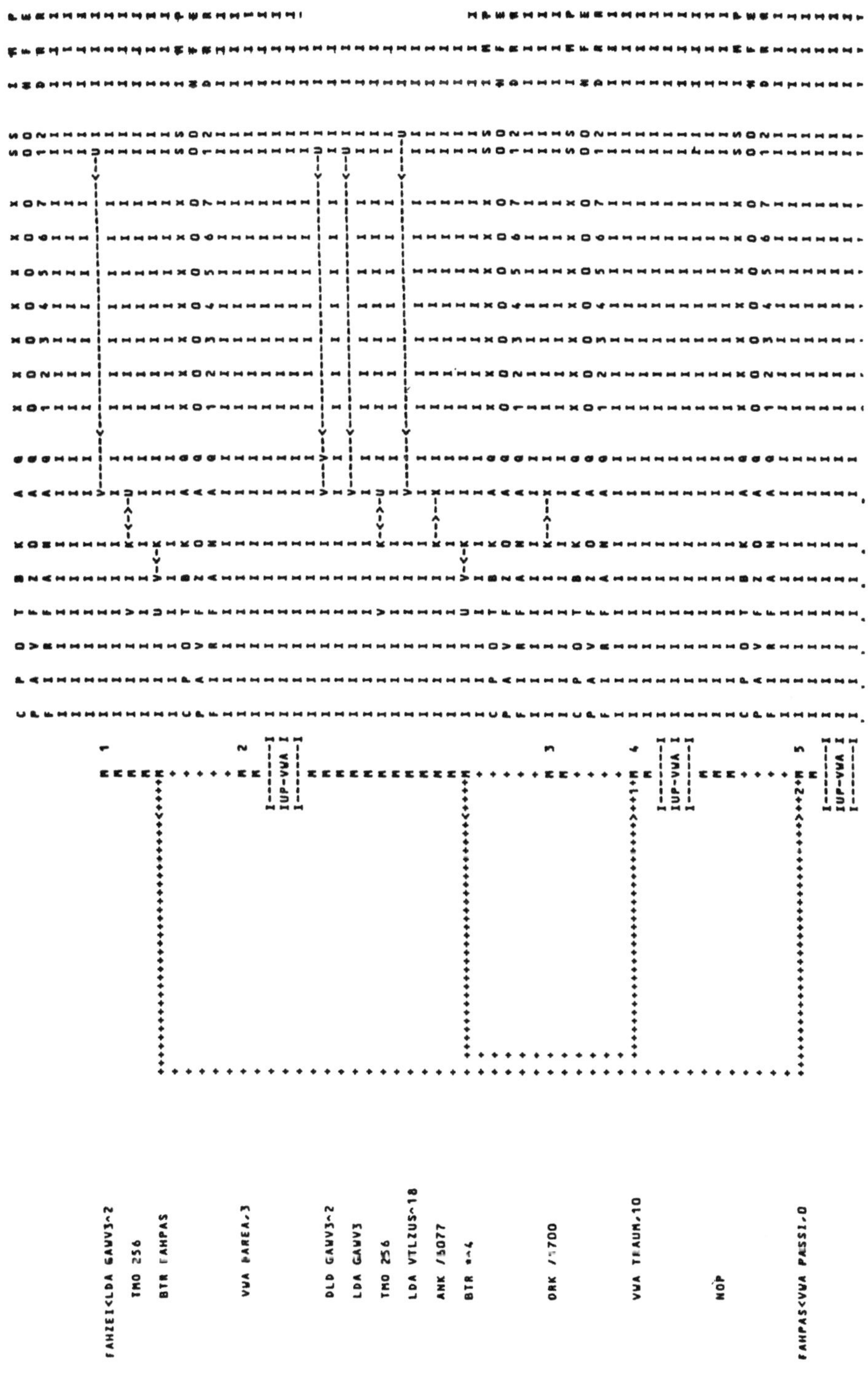

APPENDIX C
Explanation of the symbolic of a program execution diagram (from TUV, 1984)

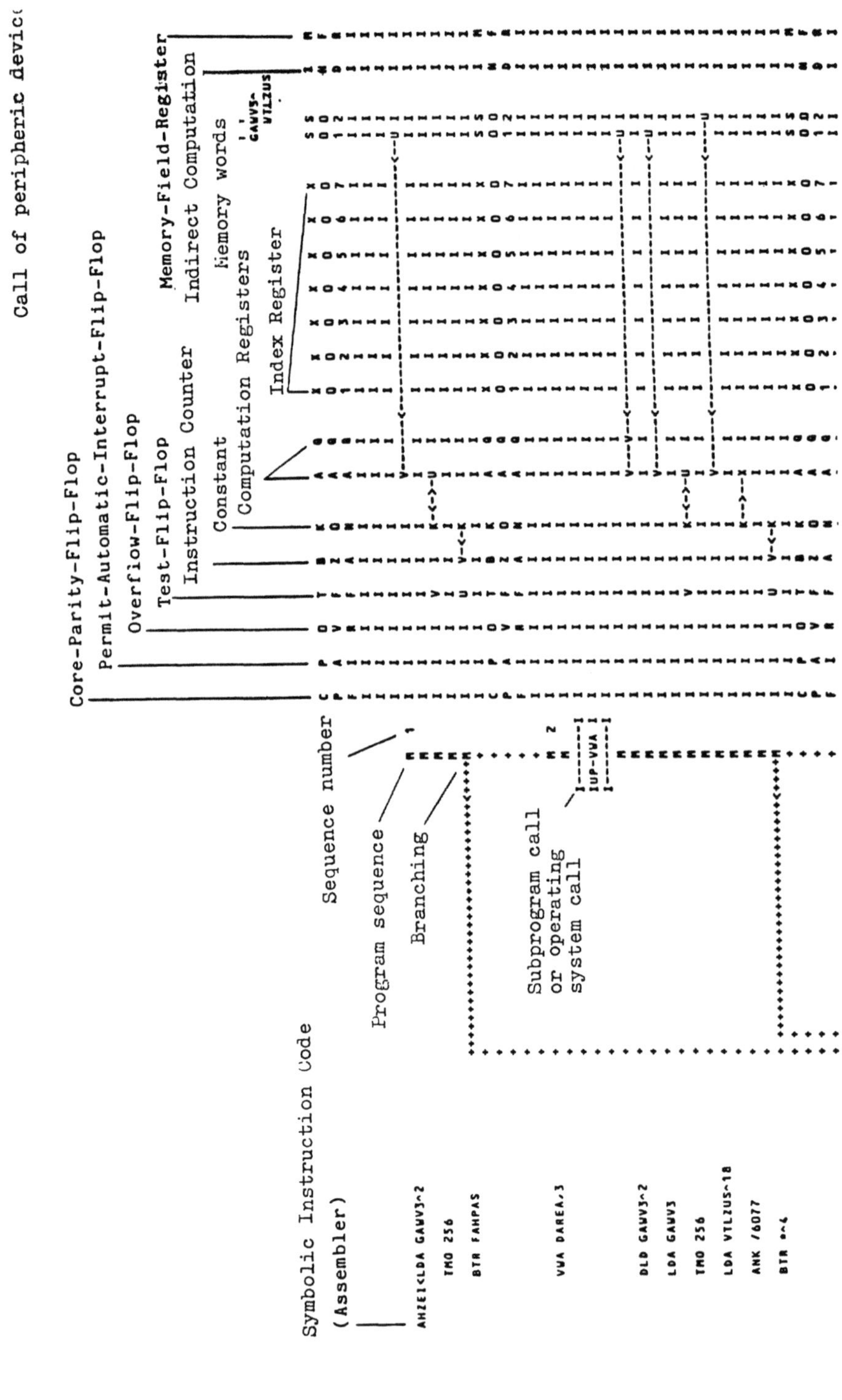

APPENDIX D
Example of FORTRAN program, the instrumented program and the report from the the tool after execution.

ANHANG:

DEMONSTRATIONSPROGRAMM

```
C       DEMONSTRATIONS PROGRAMM
C       FUER PROGRAMMTESTER 'TFORT'
C
        COMMON EPS,ZAHL,INDEX,IFAKUL,LIMIT
C
        LIMIT = 20
        EPS = 1.E-05
C
        CALL DRUCKE(1)
C
        ZAHL =.2.
        READ (5,*) INDEX
        IFAKUL = INDEX
C
        DO 10 WHILE (ABS(2.71828 - ZAHL) .GE. EPS)
           IF (INDEX.LT.LIMIT) THEN
              CALL RECHNE
           ELSE
              CALL FEHLER
           ENDIF
 10     CONTINUE
C
        CALL DRUCKE(2)
        STOP
        END
C-------------------------------------------------------------------------
        SUBROUTINE RECHNE
C
        COMMON EPS,ZAHL,INDEX,IFAKUL,LIMIT
C
        INDEX = INDEX + 1
        IFAKUL = IFAKUL * INDEX
        ZAHL = ZAHL + 1./FLOAT(IFAKUL)
C
        RETURN
        END
C-------------------------------------------------------------------------
        SUBROUTINE DRUCKE(N)
C
        COMMON EPS,ZAHL,INDEX,IFAKUL,LIMIT
C
```

```
      IF (N .EQ. 1)
     *   WRITE(6,1000)
      IF (N .EQ. 2)
     *   WRITE(6,2000) ZAHL,INDEX ,EPS
C
      RETURN
C
 1000 FORMAT('1******  DEMO PROGRAMM FUER TFORT: START ******' )
 2000 FORMAT('0 WERTE DER  VARIABLEN: '/,
     1       '0ZAHL = ',F16.8,' INDEX = ',I12,' EPS = ',1PE12.5/,
     2       '-***** DEMO PROGRAMM FUER TFORT: NORMAL END *****')
      END
C-----------------------------------------------------------------
      SUBROUTINE FEHLER
C
      COMMON EPS,ZAHL,INDEX,IFAKUL,LIMIT
C
      E = 2.71828
C
      WRITE(6,1000) LIMIT,EPS,ZAHL,E
      STOP 111
      RETURN
C
 1000 FORMAT('0******  DEMO PROGRAMM FUER TFORT: ABNORMAL END'/,
     1       '0LIMIT = ',I12,' EPSILON = ',1PE12.5,' Z = ',E12.5,
     2       ' E = ',E12.5)
      END
```

INSTRUMENTIERTES PROGRAMM

```
C     DEMONSTRATIONS PROGRAMM
C     FUER PROGRAMMTESTER 'TFORT'
C
      COMMON EPS,ZAHL,INDEX,IFAKUL,LIMIT
C
      COMMON/TFORT/ NPASS(50000),OTIM(1000),GTIM(1000)
      REAL*8 OTIM,GTIM,$TIME$
      INTEGER NPASS
                         CALL OE02B
                         GTIM (     1) = GTIM (     1) - $TIME$(0)
                         OTIM (     1) = OTIM (     1) - $TIME$(0)
                         NPASS(     1) = NPASS(     1) + 1
      LIMIT = 20
      EPS = 1.E-05
C
                         OTIM(     1) = OTIM(     1) + $TIME$(0)
      CALL DRUCKE(1)
C
                         OTIM(     1) = OTIM(     1) - $TIME$(0)
      ZAHL = 2.
      READ (5,*) INDEX
      IFAKUL = INDEX
C
      DO 10 WHILE (ABS(2.71828 - ZAHL) .GE. EPS)
                         NPASS(     2) = NPASS(     2) + 1
         IF (INDEX.LT.LIMIT) THEN
                         NPASS(     3) = NPASS(     3) + 1
                         OTIM(     1) = OTIM(     1) + $TIME$(0)
            CALL RECHNE
                         OTIM(     1) = OTIM(     1) - $TIME$(0)
         ELSE
                         NPASS(     4) = NPASS(     4) + 1
                         OTIM(     1) = OTIM(     1) + $TIME$(0)
            CALL FEHLER
                         OTIM(     1) = OTIM(     1) - $TIME$(0)
         ENDIF
                         NPASS(     5) = NPASS(     5) + 1
 10   CONTINUE
                         NPASS(     6) = NPASS(     6) + 1
C
                         OTIM(     1) = OTIM(     1) + $TIME$(0)
      CALL DRUCKE(2)
                         OTIM(     1) = OTIM(     1) - $TIME$(0)
                         OTIM (     1) = OTIM (     1) + $TIME$(0)
                         GTIM (     1) = GTIM (     1) + $TIME$(0)

                         CALL OE02C
      STOP
      END
```

```
C--------------------------------------------------------------- ---
      SUBROUTINE RECHNE
C
      COMMON EPS,ZAHL,INDEX,IFAKUL,LIMIT
C
      COMMON/TFORT/ NPASS(50000),OTIM(1000),GTIM(1000)
      REAL*8 OTIM,GTIM,$TIME$
      INTEGER NPASS
                              GTIM (     2) = GTIM (     2) - $TIME$(0)
                              OTIM (     2) = OTIM (     2) - $TIME$(0)
                              NPASS(     7) = NPASS(     7) + 1
      INDEX = INDEX + 1
      IFAKUL = IFAKUL * INDEX
      ZAHL = ZAHL + 1./FLOAT(IFAKUL)
C
                              GTIM (     2) = GTIM (     2) + $TIME$(0)
                              OTIM (     2) = OTIM (     2) + $TIME$(0)
      RETURN
      END
C---------------------------------------------------- -------------
      SUBROUTINE DRUCKE(N)
C
      COMMON EPS,ZAHL,INDEX,IFAKUL,LIMIT
C
      COMMON/TFORT/ NPASS(50000),OTIM(1000),GTIM(1000)
      REAL*8 OTIM,GTIM,$TIME$
      INTEGER NPASS
                              GTIM (     3) = GTIM (     3) - $TIME$(0)
                              OTIM (     3) = OTIM (     3) - $TIME$(0)
                              NPASS(     8) = NPASS(     8) + 1
      IF (N .EQ. 1) THEN
                              NPASS(     9) = NPASS(     9) + 1
         WRITE(6,1000)
      ENDIF
                              NPASS(    10) = NPASS(    10) + 1
      IF (N .EQ. 2) THEN
                              NPASS(    11) = NPASS(    11) + 1
         WRITE(6,2000) ZAHL,INDEX ,EPS
C
      ENDIF
                              NPASS(    12) = NPASS(    12) + 1
                              GTIM (     3) = GTIM (     3) + $TIME$(0)
                              OTIM (     3) = OTIM (     3) + $TIME$(0)
      RETURN
C

1000 FORMAT('1******  DEMO PROGRAMM FUER TFORT: START ******' )
2000 FORMAT('0 WERTE DER  VARIABLEN: '/,
    1      '0ZAHL = ',F16.8,' INDEX = ',I12,' EPS = ',1PE12.5/,
    2      '-***** DEMO PROGRAMM FUER TFORT: NORMAL END *****')
     END
```

****** COVERAGE REPORT ****** DATE 85.11.21 TIME 12:59:
DATE AND TIME OF EXECUTION OF THE INSTRUMENTED PROGRAM: DATE: 85-
SAMPLE(S) 1
SORTED ACCORDING NAMES

NAME	# OF CALLS	TESTED %	# OF PATHS	# EXEC. PATHS	COUNTER # OF MAX.	VALUE OF MAX
DRUCKE	2	100.0	5	5	8	
FEHLER	0	0.0	1	0	0	
MAIN	1	83.3	6	5	2	
RECHNE	7	100.0	1	1	7	

SORTED ACCORDING NUMBER OF CALLS

NAME	# OF CALLS	TESTED %	# OF PATHS	# EXEC. PATHS	COUNTER # OF MAX.	VALUE OF MAX
RECHNE	7	100.0	1	1	7	
DRUCKE	2	100.0	5	5	8	
MAIN	1	83.3	6	5	2	
FEHLER	0	0.0	1	0	0	

SORTED ACCORDING SUMMARY TIME

NAME	# OF CALLS	TESTED %	# OF PATHS	# EXEC. PATHS	COUNTER # OF MAX.	VALUE OF MAX
MAIN	1	83.3	6	5	2	
DRUCKE	2	100.0	5	5	8	
FEHLER	0	0.0	1	0	0	
RECHNE	7	100.0	1	1	7	

SORTED ACCORDING ONLY TIME

NAME	# OF CALLS	TESTED %	# OF PATHS	# EXEC. PATHS	COUNTER # OF MAX.	VALUE OF MAX
DRUCKE	2	100.0	5	5	8	
MAIN	1	83.3	6	5	2	
FEHLER	0	0.0	1	0	0	
RECHNE	7	100.0	1	1	7	

TOTAL TEST COVERAGE IS 11/ 13 = 84.6 %

*** COVERAGE REPORT END ***

APPENDIX E
Examples of the use of an automatic specifiaction aid.

```
****************************************************************
*  proMod V1.1e          08-17-84     11:20:57      -PAGE   5 -   *
*  Lic_No 110                                                    *
*  TUEV_Nord        PROJECT : stfrsa    Analysis Report (RA&D)   *
****************************************************************
                                                Translation

0 Steuerstabsteuerung                           control rod control

NODES         1 Fahrrechner;                    1 motion computer

TERMINATORS 1 Stellungsmess,                    1 position measurement
            2 Stabsteuerung,                    2 rod control
            3 Watch_Dog,                        3 watch dog
            4 Warte,                            4 control room
            5 Bediener;                         5 operator

DATAFLOWS AD_fertig FROM Stellungsmess TO Fahrrechner,
          Wartenbefehle FROM Warte TO Fahrrechner,
          Bedienereingaben FROM Bediener TO Fahrrechner,
          Anwahltest FROM Stabsteuerung TO Fahrrechner,
          Stellungen FROM Stellungsmess TO Fahrrechner,
          Wartenmeldungen FROM Fahrrechner TO Warte,
          Bedienerausgaben FROM Fahrrechner TO Bediener,
          An_Abwahl FROM Fahrrechner TO Stabsteuerung,
          AD_Steuerung FROM Fahrrechner TO Stellungsmess,
          Uhr_aufziehen FROM Fahrrechner TO Watch_Dog,
          Adressen FROM Fahrrechner TO Stellungsmess
```

Translation

AD_ ready FROM position measurement TO motion computer, control room orders FROM control room TO motion computer, operator orders FROM operator TO motion computer, selection test FROM rod control TO motion computer, positions FROM position measurement TO motion computer, control room messages FROM motion computer TO control room, operator output FROM motion computer TO operator, selection _discard FROM motion computer TO rod control,AD_ control FROM motion computer TO position measurement, set watch FROM motion computer TO watch dog addresses FROM motion computer TO position measurement.

Figure B1 First sheet for specification with a tool: first node, terminators and data flow.

```
             Copyright(c) for proMod, GEI Aachen 1983
*************************************************************
*  proMod V1.1e         08-17-84     11:20:57      -PAGE   6 -   *
*  Lic_No 110                                                     *
*  TUEV_Nord        PROJECT : stfrsa     Analysis Report (RA&D)   *
*************************************************************


1 Fahrrechner


Fahrrechner hat 4 Hauptaufgaben:
  Ueberwachen der Funktion der Steuerstabsteue-
  rung durch
    zyklisches Einlesen aller Stabstellungen
    und Fuehren eines aktuellen Stabstellungs-
    bildes in allen Betriebsarten incl. ausga-
    ben von Adressen, AD_Steuerung (AD : Ana-
    log_Digital_Wandler) und verarbeiten der
    Rueckmeldung AD_fertig,
    Pruefen der Stellungsmesselektronik bei jedem
    Stabstellungslesedurchlauf mit Hilfe von
    Teststellungen,
    Zyklisches Vergleichen der eingelesenen
    Stabstellungen mit der aus Fahrprogramm
    und aktuellem Fahrschritt folgenden
    Sollstellung in den Rechnerbetriebsarten,
    Feststellen von Stellungsfehlern, loeschen
    der gerade gueltige Rechnerbetriebsart bei
    Stellungsfehlern und Ausloesen von
    Wartenmeldungen und Bedienerausgaben

  Steuern der Stabanwahl in den Rechnerbetriebs-
  arten mit
    Auswaehlen der zu fahrenden Steuerstaebe bei
    anliegendem Fahrbefehl entsprechend
    Fahrprogramm und Fahrschritt,
    An_Abwahl der ausgewaehlten Steuerstaebe,
    Pruefung der Anwahlelektronik und der aus-
    gegebenen Anwahl ueber Anwahltest.

  Allgemeine Aufgaben, das sind:
    Empfangen, Pruefen und Interpretieren aller
    Wartenbefehle,
    Erzeugen aller Wartenmeldungen,
    Empfangen, Pruefen und Interpretieren aller
    Bedienereingaben,
    Erzeugen aller Bedienerausgaben.

  Selbsttest durch
    Programmablaufkontrolle,
    CPU-Test,
    Speichertest,
    Peripherietest.
    (Programmablaufkontrolle und Peripherietest sind
     nicht als gesonderte Module von Selbsttest zu
     realisieren. Sie sind als Kontrollroutinen in
     die Module von Ueberwachung und Steuerung zu
```

Figure B2. Specification with a tool: description of the functions of the first node.

Translation of the text of Figure 2B.

1 Motion computer

Motion computer has 4 main tasks:

a) Supervision of the function of the rod control through cyclic reading of all rod positions and recording an actual rod position map in all operation modes, including output of addresses; analog/digital (A/D) control and processing of the message A/D ready; electronic checking of the position measurement during each run of rod position reading and means of test positions; cyclic comparison of the read rod positions with the intended position derived from the motion program and the actual motion step in all operation modes; statement of position faults; reset of the actual computer operation mode in case of position faults; and release of control room messages and outputs to the operator.

b) Control of rod positions in the computer operation modes with selection of the rods to bemoved in case of pending motion order according to motion program and motion step, select discard of the selected control rods, check of the selection electronic and the selection issued via selection test.

c) General tasks: reception, check and interpretation of all control room orders, generation of all control room messages, reception, check and interpretation of all operator inputs, generation of all operator outputs.

d) Self-check through program run supervision, CPU test, memory test, and test of the periphery. Program run supervision and test of the periphery are not to be implemented as separate modules of this task. They are to be integrated as check routines in the modules of supervision and control.

```
              Copyright(c) for proMod, GEI Aachen 1983
*****************************************************************
*  proMod V1.1e           08-17-84     11:20:57     -PAGE  26 -   *
*  Lic_No 110                                                     *
*  TUEV_Nord        PROJECT : stfrsa    Analysis Report (RA&D)   *
*****************************************************************


                         DATA_DICTIONARY
                         ***************


Abwahl

@

 Mit Abwahl sperrt Sollist bei festgestellten
 Stellungsfehlern die weitere Ausfuehrung von
 Fahrbefehlen in einer Rechnerbetriebsart.
@

    USED IN DFD: Steuerung,  Prozessueberwachung,  Fahrrechner
    USED IN MSP: Steuerung,  Sollist,  Prozessueberwachung



Adressen

@

 Mit Adressen wird der Analogmultiplexer des Digital_
 Analog_Wändlers auf den angesprochenen Steuerstab oder
 die festverdrahtete Testspannung geschaltet.
@
    TYPE   Wort;
    STRUCT   Stabadressen
           + Testadressen

    USED IN DD:  Stellungsmess
    USED IN DFD: Fahrrechner,  Steuerstabsteuerung
    USED IN MSP: Fahrrechner
```

Figure B3. Description of the meanings of the data used and the places where they occur.

Translation of the text of Figure B3

Data Dictionary

Discard: If position faults have been detected, the intended position bars any further execution of motion orders by means of discard in a specific computer control mode.

USED IN DFD: control, process supervision, motion computer

USED IN MSP: control intended position, process control

Addresses: By means of addresses, the analog multiplexer of the AD converter is switched to the related control rod or the hard-wired test voltage.

TYPE WORD
STRUCT rod addresses
+ test addresses

USED IN DD: position measurement
USED IN DFD: motion computer, rod control
USED IN MSP: comtion computer

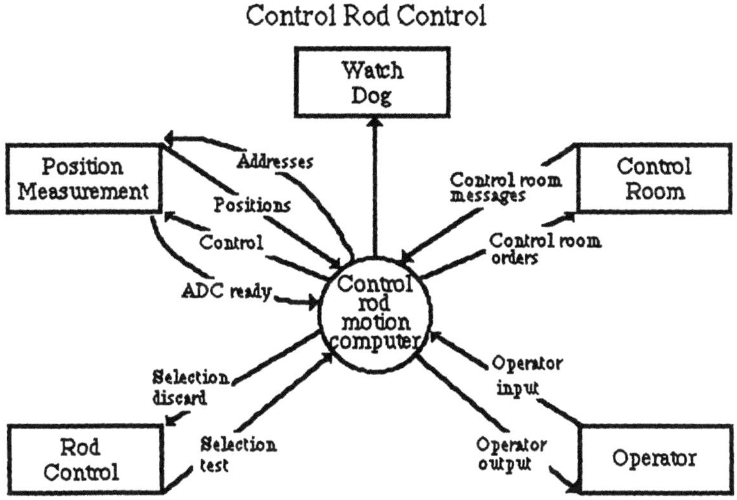

Figure B4. System description: connections of the first node to the terminators.

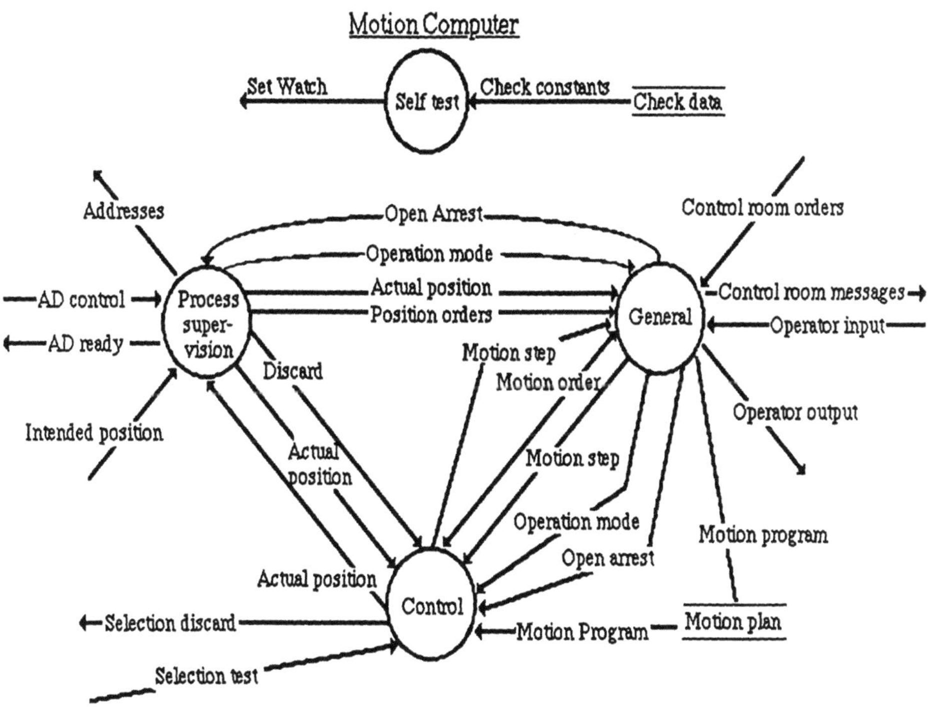

Figure B5. First step of detail: the first node is split into 4 more detailed ones.

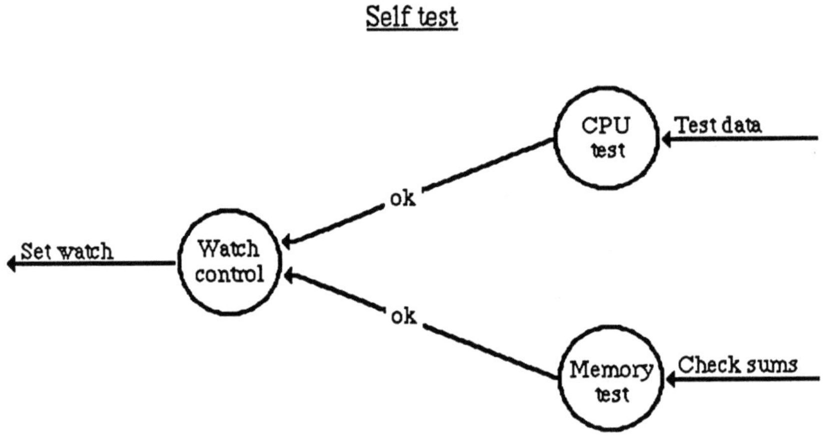

Figure B6. Second step of detail: one of the nodes of the second step is further detailed

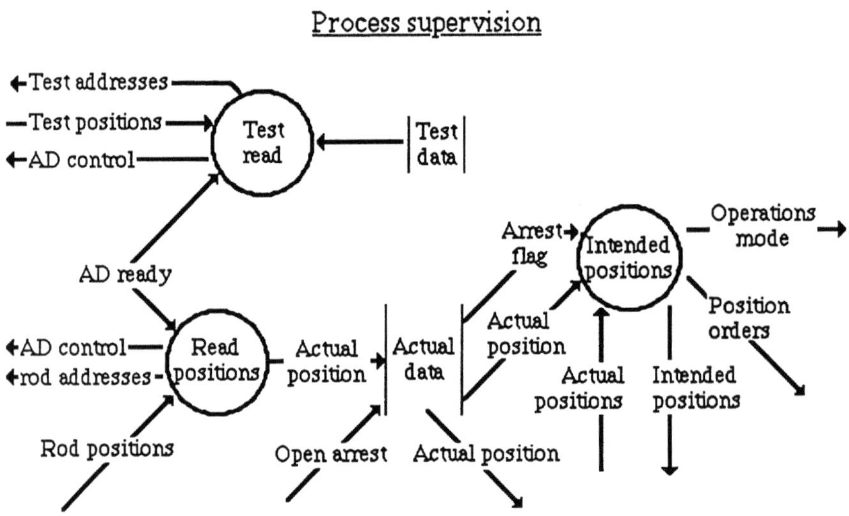

Figure B7. Second step of detail: one of the nodes of the first step is split.

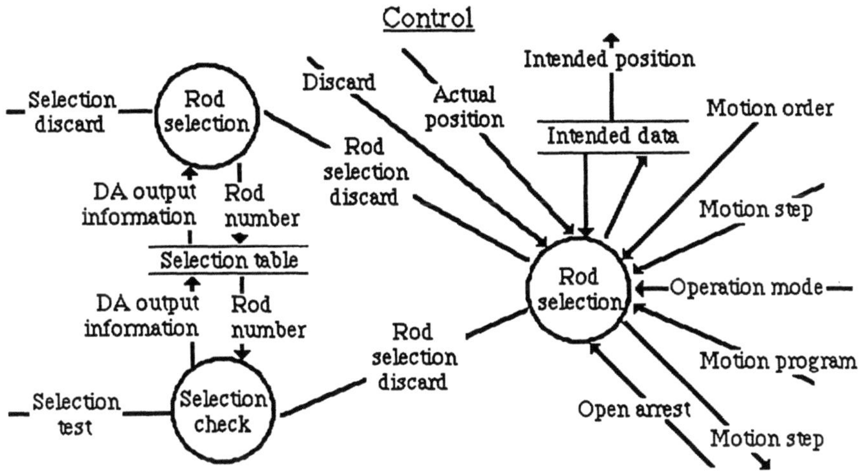

Figure B8. Second step of detail: one of the nodes of the first step is split.

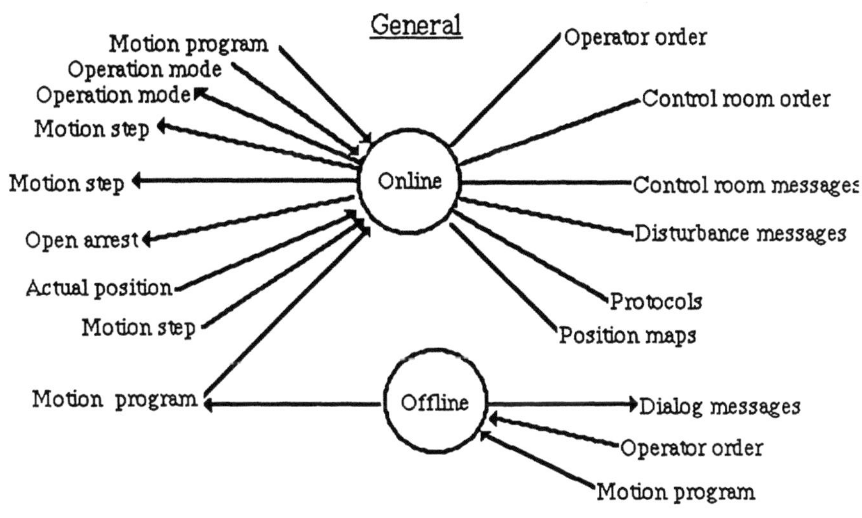

Figure B9. Second step of detail: one of the nodes of the first step is split.

FAILURE ANALYSIS OF INFORMATION SYSTEMS WITH A CYBERNETIC SYSTEM PERSPECTIVE

Halim Ergunalp

Telecam A.S.
Turkey

PRIMARY FOCUS: STRUCTURE

More important than the information systems structure as it operates and fails are the design criteria and characteristics that encapsulate potential weaknesses for possible failures. This is one of the points that comes to mind when one approaches the problem solely from the designers' point of view. However, failure is not a singular event that stems from the design efficiency, but a systemic interactive phenomenon that is generated in time by people, technology, and procedures.

For a start, it may be helpful to analyze the EATPUT model of Information Systems from the point of view of distribution of sources with weighted averages that generate the structural components of the model, as depicted in the matrix below.

FAILURE ANALYSIS OF INFORMATION SYSTEMS
WITH CYBERNETIC SYSTEM PERSPECTIVE

Information System	Source generating the structural component		
Component	People	Technology	Procedure
Event World	0.4	0.4	0.2
Acquisition	0.2	0.7	0.1
Transmission	0.3	0.6	0.1
Processing	0.2	0.7	0.1
Utilization	0.5	0.2	0.3
Transfer	0.6	0.1	0.3

Such a matrix would be unique to a specific information system operating within and for another system that is larger and that can control it.

It may be difficult, or even impossible regarding some components, to set the weighted average figures of composite probabilities of occurrence (in percent of time and cases) for different sources. But by proper assessment of source distribution at all levels of the information system in question, a global categorical framework for approaching failure analysis can be formed.

To take an example from the depicted sample matrix above: when processing is independently considered, it is seen that at this level the information system is heavily technology-dependent. Also, when the whole system is considered, by the multiplication of all factors at each level under the technology column, it is seen that technology prevails again. The global categorical framework may provide the primary focusing needed to begin a failure analysis, letting us know where to look first and perhaps what to dwell on most.

NATO ASI Series, Vol. F32
Information Systems: Failure Analysis
Edited by J. A. Wise and A. Debons

SECONDARY FOCUS: LINKAGES

An information system is embedded in a larger system, such as an organization of a specific purpose or a human society, and helps such systems in making decisions. In this respect, an information system's architecture has a specific relationship in space and time with the decision network of the large system.

An information system feeds all the decision-making points (decision nodes) at various levels of the decision network. The specific need of a decision node may be manifested as raw data, processed information, or distilled knowledge of experience, depending on the circumstances. It is the changing need of decision nodes regarding the form of information that sets the interface area of information system and decision network.

The interface area is full of relationships among various levels of information system, such as acquisition, transmission, processing, etc., and various decision nodes, such as a manager, a technician, a machine operator, a committee, a stereotype procedure, or a computer. It is a delicate region of connections and can be designated as the potential area of possible information system failures.

The vital element in the interface area is the linkage. Linkage is a bundle of connections among three components:

(a) the two components of information system, like the one between event world and acquisition, or between acquisition and transmission, etc.

(b) a component of the information system and a decision node

(c) the two distinct decision nodes of a decision network.

Linkages are dynamic entities, changing in time. They have their own form and capacity to relate two components or a component/node pair as the circumstances dictate. A linkage form may appear as a letter, informal chats, a series of formal meetings, a bunch of electrical cables, a series of reports, or a stream of fluid hitting a membrane. Linkages carry a things in various forms of complexity, such as data, information, and knowledge. And their operational capacity depends upon what they relate on both ends, that is upon the nature of two components, a component and a node, or two nodes.

Linkages, having an interactive dynamism, may sometimes fail to adapt to the changing circumstances. The lack of adaptation of a linkage stems from either one or both of the two linked sides (component/component, component/node, or node/node) that, in turn, fail to adapt to changing circumstances.

LINKAGE FAILURE

How do we observe a linkage failure? In which form of explicit manifestation does it occur? In general, we observe a linkage failure through (a) a misadaptive form of linkage that is out-of-date but persists for many situational reasons and/or (b) an insufficient capacity to meet the demands of at least one of the two linked sides.

Suppose for instance, that at particular time slice, the form and capacity of a linkage are set as three readings per day recorded from an instrument and one memo per week to be held to exchange views. However, a prolonged emergency, starting later, requires 24 readings per day of several instruments, instant group meetings on the spot of the staff concerned 15 times per week, and no memos at all. In such a situation, the linkage is severely strained and needs to be changed. If a proper change of required magnitude cannot be realized, it remains misadaptive. Information cannot flow as needed, and the linkage fails.

Linkages fail mainly because of the disharmony among people, technology, and procedures, which at a deeper level stems from the misfit among the three elements. Some sort of disharmony sets in when either one or all of the elements begin to be unable to cope with the external environment's demand for change. When this demand is sudden, disharmony is particularly severe and so is the manifest form of failure.

Linkages are information conduits, sometimes tangible, sometimes intangible. Therefore, when they fail, an information system cannot operate efficiently and effectively; decision nodes of the decision network cannot receive the information relevant to their needs to form the information system; hence, they are weakened. As an interactive outcome, decision quality and effectiveness deteriorate to an indecisive weakness position.

What makes linkages adaptive and flexible against outside perturbations is the existence of feedback loops, operating between the components of the information system and the decision network. Through those feedback loops, the linkages continuously adjust themselves from within continuously against the challenging demand of the environment for change. It is fundamentally essential that feedback loops should remain operative under all circumstances so that the information system itself and its interface area with the decision network maintain their dynamic and adaptive identity. Otherwise, starting from the back end of the decision node and moving up to the front end of the event world, a long array of consequences begin to emerge:

- Decision nodes cannot receive information in needed quantity, quality, and sequence from the utilization, processing, or transmission component of the information system. No decision or a wrong decision is made, and hence inaction or false action follows.
- Some decision nodes can be unproportionally less or more informed, leading to the network's malfunction.
- The transfer component cannot induce an appropriate output when required.
- The utilization component may arrive at a wrong synthesis.
- Processing, transmission, and acquisition from the event world may be incomplete.

At the end, the singular or integral outcome is observed as a manifest information system failure.

A FLOW SYSTEM SHOULD FLOW

An information system, by acquiring, transmitting, processing, and generating, utilizing, accumulating and transferring information, plays a vital role in the internal and external communication and control of the system in which it is embedded.

At its core, it is a flow system. It makes information flow from one point to another in various modes like EATPUT. It has a flow rhythm and a pulse.

With its flow rhythm and pulse, it makes information flow at different rates (units/time) in different modes, and it accumulates different amounts of information in different modes at different levels (units).

The rates and levels are the basic structural elements for a quantitative analysis of the dynamics of the flow system.

The quantitative analysis aims to show the dynamic terms and conditions that enhance the continuity of flow system. That is how is it that the flow system can allow flow. When the system cannot flow in continuity, a linkage is broken at some point and the system fails.

At each component phase of the EATPUT model, at what we call "the mode" of the flow system, information is carried at a different rate (R) and accumulated at a different level (L) as depicted in the figure below.

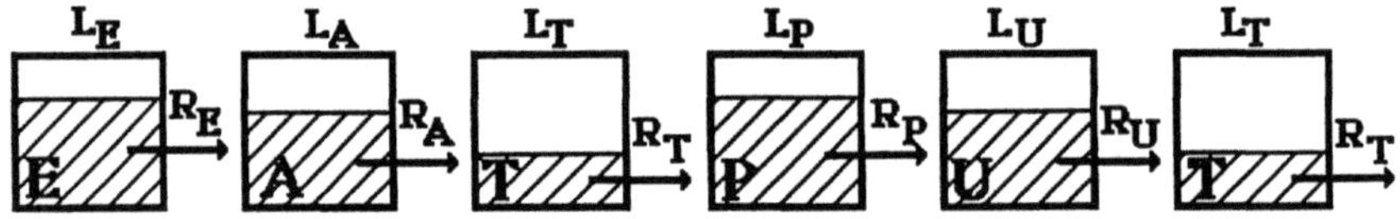

Taking any two adjacent modes at any time slice, the necessary and sufficient boundary conditions for continuous flow will be as follows:

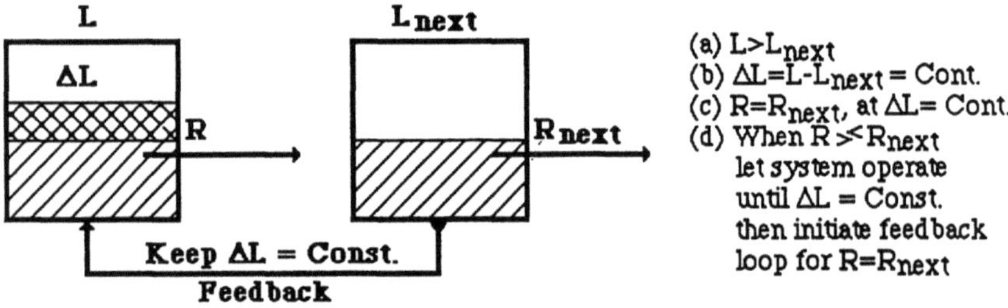

Therefore, the flow system should flow' can only be realized as follows:

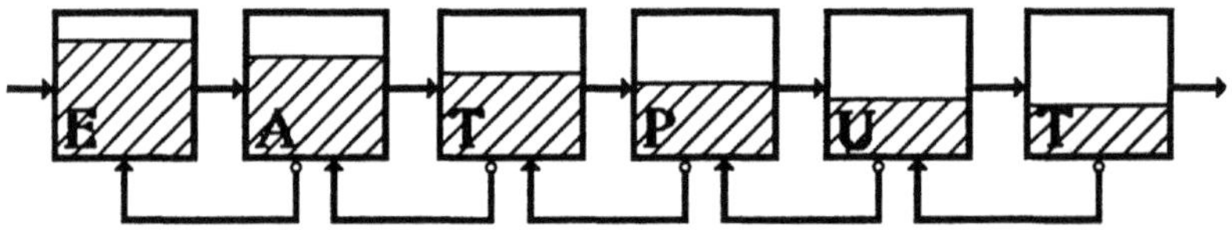

This is the dynamic equilibrium condition based on the quantity of information stored and the rate of information flow. As long as this equilibrium is maintained, information flow remains continuous and information system operates safely.

Any set of disturbances that causes a prolonged disequilibrium brings an inevitable information system failure. To avoid a failure, whatever the circumstances may be, information should therefore flow continully throughout the system. However, historical statistics of system failures indicate that sometimes sets of disturbances manage to get organized and upset the equilibrium. Surely, this is the universal trend of running into disintegration. But what is it in particular?

OBSTACLES TO FLOW

The specific source of disturbances is the misfit within the people/technology/procedure triangle. These three elements of information systems environment generally cannot get along well in the course of time and cannot form a harmonious trio. To investigate this question, we need to pose a series of subsequent questions and to collect accurate answers, which will be very much situation-dependent.

- Have people innovated the technology themselves in their own sociocultural environment, or was it imposed upon them from outside?
- Have people learned and associated with the technology easily, or do they have difficulties in coping with various aspects of it?
- Have people overadapted to technology and begun to develop 'operational blindness' by taking critical things for granted?
- Do people have an affinity for procedural thinking or do they prefer short-circuited jumps?
- Do people have specific biases, originating from their own socioculture, that work against certain interactive situations?
- Do procedures fit the complexity of the technology?
- Which one of the three elements is in control, and why? Keeping the previous questions in view, is this control appropriate?

The profile of the misfit discovered through a careful search can effectively reveal the intricacies of past failure as well as foretell the potential weakness of a running system.

ARTIFICIAL INTELLIGENCE TECHNIQUES FOR THE DISTRIBUTION OF CRITICAL INFORMATION

L.F. Pau

Technical University of Denmark
Lyngby, Denmark

INTRODUCTION

This paper is specifically concerned with real-time operation implementation opportunities for artificial intelligence techniques in information systems handling critical data. It is not within our scope to deal with knowledge based techniques in post-event failure analysis of these information systems; likewise, we are not concerned directly with related organizational issues.

MAIN APPLICATIONS

Such real-time operation implementation opportunities for artificial intelligence techniques (Barr and Fiegenbaum,1982) have been mostly investigated in relation to research or actual operation of certain networks:

- Command and control information networks, with heterogeneous users, nodes and transmission links (Athans, 1982; Frosher and Grishman, 1983; IEEE, 1985; Pau and Negret, 1986).
- Alarm and monitoring networks for technical facilities (Pau, 1980; Kleinrock, 1985; IEEE, 1985; Pau and Negret, 1986).
- Office automation networks (Pau, 1986a).
- Electronic funds transfer networks (Kleinrock, 1985 ; Pau and Negret, 1986).

MAIN METHODOLOGY

Following classical failure diagnosis methodology (Pau, 1981b; 1986b), I divide the information system diagnosis into the following steps and subcategories (as done in Pau and Nerget, 1985):

<u>Specification of Failure and Event Modes (E)</u>

- Hardware failures or degradation at information source or destination, or during transmission (nodes, links)
- Software failures, at same locations
- Procedural failures
  - Information input errors
  - Information transmission errors (flow control, routing, protocol, transaction)
  - Information output/display errors
  - Action errors
  - Security procedures
- Human failures
- Systems operation
- Decision-making failures

NATO ASI Series, Vol. F32
Information Systems: Failure Analysis
Edited by J. A. Wise and A. Debons

Failure Localizations (L)

In information systems, I suggest taking as generic failure localizations the seven layers of the OSI architecture for open communications networks (Heitmeyer and Wilson, 1980):

Layer 7: Application/mission
Layer 6: Presentation
Layer 5: Session
Layer 4: Transport
Layer 3: Network
Layer 2: Data links
Layer 1: Physical

Observables (Y)

Among the possible observables, I (Pau, 1981b; 1986b) include the following for each localization:

- Hardware status
- Software status
- Procedural status
- Information contents
- Information transmission and distribution delay

Operation Environment (O)

This covers exogenous conditions:

- Historic data (Pau, 1981b)
- Operational physical environment
- Operations mission specifications/requirements
- System design data

Diagnostic Tasks (D)

- Failure/degradation forecasting
- Failure/degradation detection
- Failure/degradation localization
- Failure/degradation diagnosis
- Failure/degradation maintenance/repair/reconfiguration

Diagnosing an information system is then specifying fully the mapping d (Pau, 1981b; 1986b): $d \varepsilon D$: $(Y \times L \times 0) \rightarrow E$.

The contributions of artificial intelligence (Barr and Feigenbaum, 1982) reside in four areas:

- Selection of knowledge representations able to handle jointly elements of Y x L x O.
- Elicitation/acquisition of additional knowledge about Y, L, O, E in the form of knowledge bases.
- Inferencing $d \varepsilon D$ and $e \varepsilon E$ from current information about Y x L x 0, supplemented by knowledge bases about the same.
- Offering interactive man-machine facilities in carrying out and explaining the diagnosis d, while reducing operator work-load.

The following sections highlight specific important examples of artificial intelligence contributions to information system diagnosis. Only the principles are given; see references for further details.

KNOWLEDGE-BASED FAILURE DETECTION AND DIAGNOSIS OF THE PHYSICAL LEVEL

Framework

E: Hardware failures and procedural failures
L: Physical level and data links (Layers 1 and 2)
Y: Hardware status, procedural status
O: Physical environment, mission requirements, design data
D: Detection, localization, diagnosis, repair

Approach

Knowledge-based failure detection and diagnostic systems (also called diagnostic expert systems) (Pau, 1986b) rely on distributed specialized knowledge bases about elements of E, L, Y, O, with possibly a common blackboard architecture, to perform the tasks $d \in D$ using heuristic search, test generation forward/backward/mixed chaining, constraint propagation, and diagnostic metarules.

Three major difficulties exist:

- Knowledge acquisition about failure/degradation modes (Pau, 1981b).
- Selection of useful sensor data by feature extraction (Pau, 1981b).
- Still today, insufficient validation of the completeness of the knowledge bases in view of safety, repeatability, and reduced diagnostic requirements.

Four major contributions are expected, or already experienced, from knowledge engineering techniques (Pau, 1986b):

- Generation of functional test sequences through metarules.
- Reduction of false alarms through built-in tests (BIT) with knowledge bases (Pau, 1980).
- Reduction of skills in test and maintenance tasks.
- Use of diverse observation types (see the section below on Information Fusion).

These contributions are likely to go into seven generic systems:

(a) Computer-aided preliminary design for testability (IEEE, 1985).
(b) Smart built-in test to identify intermittent faults, reduce false alarms, and carry out recalibration (Pau, 1981b).
(c) Smart system integrated test for system level testing while in operation (Pau, 1981b).
(d) Box maintenance knowledge-based system to provide offline test management with self-improvement of functional test (Pau, 1986b, IEEE, 1985).
(e) Automatic test program generation (Bell and LaPadula, 1973).
(f) Smart bench, which is a maintenance knowledge-based system developed for use with bench test equipment controlled by an engineering workstation (Pau, 1986b).

INTELLIGENT CONTENTS-BASED MESSAGE ROUTING IN INFORMATION NETWORKS

Framework

E: Procedural failures, decision-making failures
L: Application/mission, session (layers 5,7)
Y: Information contents, procedural status, information and distribution delay
O: Operational mission specifications/requirements
D: Failure detection, reconfiguration

Approach

Critical to information systems is proper information distribution in a safe, reliable, timely, and secure way (Frosher and Grishman, 1982; Heitmeyer and Wilson, 1980; Bell and LaPadula, 1973; Pau et al., 1979). In all networks, the purpose of the routing implemented at each node is to select the next destination for each message/packet datagram in such a way that it gets closer to its destination, while sub-optimizing a number of criteria (transmission delay, cost, flow, congestion, fault-tolerance, network management overheads, security) (Pau and Negret, 1985; Kleinrock, 1985; Heitmeyer and Wilson, 1980).

Common to state-of-the-art routing procedures is the fundamental assumption that message contents are not used in order to select a tradeoff among routing criteria applicable to that message.

Next, none of the routing protocols allow for the syntactic or semantic parsing (fragmentation) and concatenation (assembly) of messages, thus providing to the user smaller semantically meaningful entities (*Aviation Week*, 1985).

Extensive research and development have been carried out on knowledge-based routing for implementation in the node processors of application specific closed heterogeneous networks (data packets, and packet radio co-existing) (Frosher and Grishman, 1982; Pau and Negret, 1986). It is based on three sets of techniques (Pau and Negret, 1985):

1. Apply natural language analysis (Frosher and Grishman; 1982, Bolc, 1978; Pau, 1985) to each incoming packet (or full message) in order to assign semantic criticality measures to it: $x \stackrel{\Delta}{=}$ (urgency, message type, security level, required transmission reliability, length of message contents) (Pau and Negret, 1985).

2. Apply a multiple-criteria, adaptive, global topology update algorithm (Pau and Negret, 1985) to generate the best non-dominated routes toward all possible destination nodes.

3. Select the route by knowledge-based inference, to match message criticality measures with the attributes of the possible routes generated by (2).

After (1), (2), (3), the result of (3) is stored in the routing table of the node switch for execution on existing node hardware, in agreement with the applicable protocol. The same techniques make it theoretically possible to select the protocol and transmission medium according to message contents, although this has not yet been tested.

Some major difficulties remain :

(a) Lack of fast dedicated architectures for natural language processing (1) and inference (2) (3).

(b) Restriction to applications with limited vocabulary and constrained text.
(c) Spelling, phonetic errors, and other human factors.

Current implementations (Pau and Negret, 1985) represent 5000 lines of Prolog II, besides 5000 lines about language and protocol attributes.

KNOWLEDGE-BASED MESSAGE DISTRIBUTION

Framework

E: Systems operation, decision-making failures/errors
L: Presentation and session (Layers 5,6)
Y: Procedural status, information transmission, and distribution delay
O: Operational mission specifications/requirements
D: Failure localization and diagnostics

Approach

It is possible to encode procedural information about message distribution, e.g., in a rule-based way, to carry out verifications or actions on the basis of the semantically inferred criticality measures X of each message (Pau and Negret; 1985, Pau, 1986a) (section 4). Specialized rule sets cover the following:

(1) Message distribution lists on the basis of individuals/units indicated as destinators in the message, after information access control (urgency, security) (Frosher and Grishman, 1982; Heitmeyer and Wilson, 1980).
(2) Generation of acknowledgments and backup, based on message type or action mentioned in the message.
(3) Message collection and indexing.
(4) Approval procedures, based on contents, source, destination, security (Bell and LaPadula, 1973).
(5) Queueing rules, where message priority in buffer is reassigned based on level of urgency in X (Pau et al., 1979).

The major limitations are proper specification of the message distribution rules (Heitmeyer and Wilson, 1980), and relative efficiency of a knowledge-based implementation vs. a classical distribution software (eventually operator assisted). Requirements like the ones outlined are typical of C^3 networks, such as NMICOSS, HERMES, SIGMA.

KNOWLEDGE-BASED THREAT ANALYSIS OVER TIME

Framework

E: Procedural failures, decision-making failures
L: Application/mission (layer 7)
Y: Information contents
O: Operational physical environment, historic data, operational mission specifications/requirements
D: Forecasting, detection, diagnosis

Approach

By a proper data representation of asset types and of related threats represented by transactions between these assets, it is possible to generate threat analysis over time in the form of inferences by a kernel system (Pau, 1986c). By filling in, through dedicated knowledge acquisition editors, the application specific attributes, the kernel will carry out application specific threat analysis. The asset types that may be the object of a threat, and are required to carry a threat out, are the following:

1. Monetary assets
2. Physical assets
3. Policy assets
4. Service and information/intelligence assets

Actions and events over time are reflected as transactions between asset types, where the threat action using asset type i is needed to acquire asset type j (Pau, 1986c) as shown in Table 1.

A transaction is characterized by application-specific attributes:

- Settlement terms: distribution over time and type of asset type, in return for i.
- Logical constraints in transaction sequence (represented through scripts).
- Transaction costs expressed as a vector with one component by asset type.

Table 1. Transactions between Asset Types.

(i,j)

Asset Types i:	1	2	3	4
j: 1	Financial	Sale	Blackmail	Service/ Income
2	Investment	Barter/ Destruction	Political/ Influence	Media Influence
3	Aid	Control	Negotiation	Manipulation
4	Access/ Insurance	Collect data/ Service	Assistance	Intellectual

With application-specific rules and the transaction attributes the threat inference engine will enable a search for underlying transaction combinations in view of the observable attributes about the assets. The mixed chaining will then reveal the transaction types and order, thus also the type of threat developing over time.

This approach has been tested in the following areas: political campaign, currency rate adjustments, and threat analysis in a C^3 environment (Pau, 1986c). The difficulties encountered have been incomplete information about the attributes, and lack of incorporation of subjective judgemental values [Dempster] (Athans, 1982).

INFORMATION FUSION

Framework

E: Hardware, software, decision-making failures/errors
L: Physical, presentation, application (Layers 1,6,7)
Y: Hardware status, software status, information contents

O: Historic data, operational mission specifications
D: Detection, localization

Approach

Information fusion is essentially relying on hierarchical knowledge representation techniques in order to combine (through filiations among knowledge representation schemes) information appearing in such diverse forms as signals, images, text, procedures. The standard approaches (Pau, 1981a) are (1) registration, (2) overlay, (3) merging feature lists, and (4) merging syntactic grammars by combining their rewriting rules.

REFERENCES

Aviation Week & Space Technology. (1985). Command, control & communications. Dec. 8.

Athans, M. (1982). The expert team of experts approach to command and control organizations. *Control Systems Magazine.* Sept.

Barr, A. and E. Feigenbaum (Ed). (1982). *Handbook of Artificial Intelligence.* San Franscisco: William Kaufman.

Bell, D. E. and L. I. La Padula. (1973). *Secure Computrer Systems.* Vol I-II, MTR 2547, MITRE Corp., Bedford, November.

Bolc, L. (ed). (1978). *Natural Language Communication with Computers.* New York: Springer-Verlag.

Froscher, J. and R. Grishman. (1983). A linguistically motivated approach to automated analysis off military messages. *Proceedings 1983 American Artificial Intelligence Conference.* Rochester, MI.

Heitmeyer, C. L. and S. H. Wilson. (1980). Military message systems: current stats and future direction. *IEEE Trans. Communications*, COM-28(9), September.

IEEE. (1985). Special issue on artificial intelligence techniques. *IEEE J. Design and Test of Computers.* 2(4), August.

IEEE. (1985). Expert systems in government. *Proceedings IEEE 85 CH-2225-1.* Washington, D.C., October.

Kleinrock, L. (1985). Distributed systems. *IEEE Computer Magaizine.* November.

Pau, L. F. (1980). *Data Communications Network Monitoring for Failure Diagnosis in the Presence of Non-digital Links.* T.R., E.N.S.-D-80006 Telecommunications, Paris.

Pau, L. F. (1981a). Fusion of multisensor data in pattern recognition. In J. Kittler, K.S. Fu, L.F. Pau (eds), *Pattern Recognition Theory and Applications.* Boston: D. Reidel.

Pau, L. F. (1981b). *Failure Diagnosis and Performance Monitoring.* New York: Marcel Dekker.

Pau, L. F. (1985). Inference of the structure of economic reasoning from natural language analysis. *J. Decision Support Systems*, 1, 313-321.

Pau, L. F. (Ed) (1986a). *Artificial Intelligence in Economics an Management.* Amsterdam: North Holland.

Pau, L. F. (1986b). Survey of expert system for detection, test generation and maintenance. *Expert Systems J.*, 3(2), April.

Pau, L. F. (1986c). An expert system kernel for the analysis of strategies over time. In Pau, L.F. (Ed), *Artificial Intelligence in Economics and Management.* Amsterdam: North Holland, 1986.

Pau, L. F. & J. M. Negret. (1985). *SOFTM, A Software Maintenance Expert System.* Internal report, Battelle Memorial Institute, September.

Pau, L. F. & J. M. Negret. (1986). Intelligent contents based message routing in communications networks. *Proceedings Nordic Teletraffic Conference.* Lyngby, August.

Pau, L. F., B. Wafae, and A. Bousquet. (1979). Fault detection capability implications for flow control in data communication networks. *Proc. 1979 IEEE International Conference AUTOTESTCON.* Minneapolis, September.

FAILURE ANALYSIS OF INFORMATION SYSTEMS: AN OVERALL DISCUSSION AND A SIMPLE CALCULATIVE RATIONALITY MODEL

T. Yarman, Müjgan Kerman, and Faruk Yarman

Anatolia University
Eskisehir, Turkey

INTRODUCTION

The phrase "failure analysis of information systems" implies a large area of meanings and related action. In the first place, the term "failure analysis" indicates an analysis of how a given system has failed or could fail. Simple examples of system failure may be how, on September 25, 1978, a collision occurred between Pacific Southwest Airlines Flight 182, a Boeing 727, while descending for a landing at San Diego's Lindbergh Field, and a single engine Cessna 172 approaching the same airport; how TMI 2 and Chernobly 4 reactors produced their historical accidents; how the Challenger space shuttle was lost.

THE MEANING OF "FAILURE ANALYSIS"

"Failure analysis" for the accidents mentioned above means "the investigation and analysis of relevant clues to the accident." Although we define here a very clear mission, we believe that the term should specify post-mortem failure analysis, which includes various tasks:

- Search for the possible modes of failure of an information system.
- Collection and study of experiences crystallized out of investigations of accidents.
- Systems evaluation.

We would also like to add another dimension to the meaning of "failure analysis" by extending it to cover pre-mortem aspects of failure:

- Continuous effort toward perceiving, comprehending, and learning the nature of a working system that has been created, such as a nuclear reactor, a space shuttle, or the conjectured ultra complicated SDI software system.

- Analysis of past data related to the behavior, expected failure, or unexpected accident stored or displayed by components or man-machine interfaces of a given system with the hope of predicting and thus controlling the future of the system.

We may perhaps elaborate a little bit further on these dimensions. There is the question of identifying the channels of failure. This is not quite a risk assessment task. Rather, it entails conjecturing realistic scenarios or patterns of behavior of a system and checking whether the system works within the frame of each of these. If it does not, we improve the system by changing it or by adding safety features to it.

This dimension of failure analysis can be called "understanding, debugging and/or bettering" the system. Within this context, failure analysis and safety analysis overlap. The difference between the two may be considered to lie in the difference between two questions:

NATO ASI Series, Vol. F32
Information Systems: Failure Analysis
Edited by J. A. Wise and A. Debons

1. Should the system fail? and if so, how?
2. Is the system safe enough? how could it be made safer?

INHERENTLY IMMEASURABLE VULNERABILITY OF A SYSTEM

The inevitable and regrettable feature of failure analysis, or safety analysis, is the existence of an unpredictable pattern of failure or accident associated with the system. All tragic accidents have been "inconceivable" No one thought that the "invincible" Titanic, during her first trip, would collide with an iceberg and sink in a short time. No safety analyst would have conceived of a scenario for the accidents at TMI2 or Chernobyl 4.

In other words, it appears that there is no absolute way to ensure that a system will work the way it has been designed and expected to work. We may improve a system and decrease its degree of vulnerability without ever reaching a perfectly functioning system. Furthermore, it appears that while systems get more and more complicated, their degree of immeasurable vulnerability increases.

The two dimensions failure analysis are not totally unrelated, but they give rise to different sets of exercises. The first one of these was just explained. While the exercises of the first set are thus carried for a given system over postulated plausible incidences, those of the second set aim to quantify and evaluate the experience collected over not just one unit, but over many units of a given sort, or even further if usable, over great many units that may bear different overall functions. The second set of exercises is then carried out, component by component on a micro level; whereas exercises of the first set (based on postulated plausible incidence) are pursued on a macro system level. By the word "component" we mean either a material component or a man-machine interface.

The bookkeeping of components, either one by one on the inter-system level or statistically on an intra-system level, allows us to predict the behavior and the performance of a component of interest, thus providing us with the capability of making changes at either the manufacturing or the operational stages. We can call this dimension of "Failure Analysis" in short, "the close tracking of the components (including human factors) and the control of the system", mostly on a statistical basis.

RECAPITULATION

In summary, we are inclined to perceive the actions covered by the term "failure analysis along the three following, somewhat interrelated dimensions:

1. Analysis and understanding of a given accident (that has taken place).
2. Understanding, debugging and/or bettering a working system.
3. Close tracking of components, including man-machine interfaces, of a given system before or after failure in order to take essential measures and master the control of the system to avoid any potential accidents.

The work that is carried out with the aim of fulfilling the requirements of the second and third dimensions (defined above) of failure analysis shall obviously be very useful in conducting a failure analysis of a given accident (within the frame of the first dimension mentioned above).

Yet it seems impossible to nail down the failure patterns displayed by major accidents. Thus, the exertion undertaken through the failure analysis second dimension, though perhaps necessary, still looks insufficient for avoiding the occurrence of catastrophes. Getting to know a complex system that we have created unfortunately somehow requires that the system tragically fail. There is no end to learning about a system. The more cumbersome a system is, the more difficult it is to come to a full comprehension of it.

In this sense, most important inputs to a failure analysis, are the data related to all of the past failures and accidents that may be of interest. (As Napoleon Bonaparte pointed out, there is no better way to make up one's own strategy than to study strategies previously developed.)

Let us add that the second and third dimensions of failure analysis as we have defined them are interrelated within the frame of the interaction between micro and macro levels.

The failure analysis of an information system such as the TMI 2 reactor necessitates not only identifying the initial causes, steps, and mechanisms of the accident in question but also discovering how such causes took place.

We recall that one of the very first causes of the TMI 2 accident was that the workers, throughout the maintenance, had forgotten to reopen the front valves of the emergency pumps (which they had initially closed on purpose) of the secondary circuit of the reactor. As a result, these pumps were left functionless during an emergency call. In other words, the primary cause of the tragic TMI 2 accident was human distraction, which can take place anywhere and at any time.

The failure analysis of the TMI 2 accident thus requires not only the identification of the cause that had started it but also an intelligent study of how distraction or simple human failure operates and affects the health of the working system and how the system can thus become ineffective if not totally nullified. This kind of analysis looks yet extremely hard, if not impossible.

Here perhaps our second and third dimensions can complement a failure analysis undertaken along the first dimension of our definition. A microanalysis of the elements and characteristics of human failure, especially distraction at the man-machine interfaces, might indeed be the vital key to avoiding future technological disasters that might be initiated like the one at TMI.

In the Appendix to this paper, we present a simple "calculative rationality" model. This term, which has been borrowed from R. Westrum, refers to using the past to compute the future.

ANSWERS TO QUESTIONS OF INTEREST

The discussion presented above may be considered fairly self-evident. Nevertheless, we felt the need to classify the elements of failure analysis of information systems by following three related dimensions reason, because each calls for a different set of exercises and exhibits different characteristics. Answers to questions posed will therefore differ from one dimension to another.

In regard to our first dimension of failure analysis of information systems, there might not be just one universal approach to the causes of failure, although perhaps all of the possible or conjectured ways of finding answers to the failure in question can be described as a Lieutenant Colombo approach. The elements of this kind of approach are the following:

- Clues left after a failure (or crime).
- A very bright and advanced sense of observation.
- A keen aptitude for reasoning and assembling clues in order to draw the integrity of the real picture, yet unknown.
- Checks and tests to verify the appropriateness of the conclusions that were reached.

A team working with this methodology, of course, facilitates the task. Although experience and cognitive sharpness are primary, they are difficult to define and analyze.

In regards to second dimension of failure analysis, i.e., conceiving realistic scenarios about failing patterns and checking whether the system would ever fail through any of these, there seems just no way to scan exhaustively the spectra of possibilities for failure or accident. As a matter of fact, even very advanced studies generally disregard a potential cause of accident that eventually takes place. It seems that only through tragic happenings do we grow in experience and enlarge our vision of potential hazards. We thus feel very inadequate in our efforts to identify a methodology.

With the third dimension of failure analysis of information systems, it appears possible, if not to rely on it, to use it in improving statistical or forecasting methods to predict the future of a working element or a group of working elements, based on the tracing and study of their past behavior.

ACKNOWLEDGMENTS

We would like to thank Professor Cemal Yalinpala of Vanier College. Montreal, for reading this manuscript, and Dr. Zulal Balpinar of Anadolu University and our dear friend Leyla Sevinc for their help in getting this paper typed.

APPENDIX

A Simple Calculative Rationality Model

We borrow the expression "calculative rationality" from R.Westrum and use it to mean using the past to compute the future.

Here we sketch a simple calculative rationality model that we developed. Let us consider items of a given sort. These can be, for instance, airplane engines or a given component of these engines.

These items break down and have to be repaired. For simplicity, we suppose the number of the items in question to be a constant.

Let us note that the technique outlined below may be regarded as an example of the third dimension of failure analysis of information systems defined throughout the text of this paper.

Let N_i be the probability of an item to fail within the unit time interval throughout the ith period of study. Let HL(t) be the number of items in question available (in "good health") at time t. Let λ_i be the probability of failure of an item in question within the unit time interval throughout i^{th} period of study. Let BL(t) be the number of items that failed and are awaiting repair at time t.

Since the total number of our items was considered to be a constant, one can easily see that we can write the following differential equations for HL(t) and BL(t):

$$\frac{dHL(t)}{dt} = - \mu_i HL(t) + \lambda_i BL(t) \quad , \tag{1}$$

and

$$\frac{dBL(t)}{dt} = -\lambda_i BL(t) + \mu_i HL(t) \quad . \tag{2}$$

These equations are solved to find our unknowns:

$$\mu_i HL(t) = \frac{\lambda_i c_1}{\lambda_i+\mu_i} - \mu_i c_3 e^{-(\lambda_i+\mu_i)t} \tag{3}$$

$$\lambda_i BL(t) = \frac{\lambda_i c_1}{\lambda_i+\mu_i} + \lambda_i c_3 e^{-(\lambda_i+\mu_i)t} \tag{4}$$

c_1, c_3 are integration constants.

We keep track statistically of how the item fail and how we repair them. In other words, we record the "failure activity" displayed and "repair activity" given to the items. Thus, we will know how physically, for instance, the quantity $\mu_i HL(t)$, i.e., the number of items failing within the unit time interval, behave with time throughout i^{th} period of study. We would like to determine μ_i and λ_i based on our recorded data. For this purpose, we propose the following procedure:

At times $t_1, t_2, \ldots, t_n, \ldots, t_N$, we suppose we have recorded for $\mu HL(t_k)$ the values of $s_1, s_2, \ldots, s_n, \ldots, s_N$.

Let us make the following definitions:

$$x = \frac{\lambda_i c_1}{\lambda_i + \mu_i} , \quad (5)$$

$$y = \mu_i c_3 , \quad (6)$$

$$k = e^{-(\lambda_i + \mu_i)} . \quad (7)$$

We let the origin of our time scale be t_1, so that

$$t_1 = 0 . \quad (8)$$

Following Eq. (3) we can thus write

$$s_1 = x - y , \quad (9)$$

$$s_2 = x - yk , \quad (10)$$

$$\vdots$$

$$s_N = x - yk^{N-1} . \quad (12)$$

$$\vdots$$

$$s_n = x - yk^{n-1} , \quad (11)$$

At time t_n, we now define the error quantity ε_n to be

$$\varepsilon_n = s_n - x + yk^{n-1} . \quad (13)$$

We require that our theoretical curve $\mu_i HL(t)$ with respect to time as given by Eq. (3) coincide with s_1 and s_N at times respectively t_1 and t_N. Moreover, we require that the sum of errors vanish:

$$\sum_{n=2}^{N-1} \varepsilon_i = 0 , \quad (14)$$

or

$$\sum_{n=2}^{N-1} s_n - x + yk^{n-1} = 0 . \quad (15)$$

This can be shown to come to

$$\left(\sum_{n=2}^{N-1} s_n\right) - (N-2)x + yk \frac{k^{n-2}-1}{k-1} = 0 \quad , \tag{16}$$

Neglecting k^{n-2} in front of 1 and defining S to be

$$S = \sum_{n=2}^{N-1} s_n \quad , \tag{17}$$

Eq. (16) can be written as

$$k = \frac{s_{N(n-2)} - S}{s_N(n-1) - s_1 - S} \tag{18}$$

Eq. (18)_ with Eq. (7), furnishes the value of $(\lambda_i + \mu_i)$.

If, on the other hand, yk^{N-1} can be neglected in front of x, as expected for large t_n's, then following Eq. (12),

$$x = s_N; \tag{19}$$

and, following Eq. (9),

$$y = s_N - s_1 \quad . \tag{20}$$

Therefore, $\mu_i HL(t)$ of Eq. (3) can be written as

$$\mu_1 HL(t) = s_N - (s_N - s_1)\, e^{-(\lambda_i + \mu_i)t} \quad . \tag{21}$$

Now we would like to determine λ_i and μ_i over the period of study. For this purpose, we propose to minimize the sum of the squares of errors, as defined by Eq. (13), with respect to λ_i and μ_i, over the period of study

$$\frac{\partial}{\partial\lambda_i, \partial\mu_i} \sum_{n=2}^{N-1} \left(\frac{\lambda_i c_1}{\lambda_i + \mu_i} - \mu_i c_3 e^{-(\lambda_i + \mu_i)t_n} - s_n\right) = 0 \quad ; \tag{22}$$

or

$$\sum_{n=2}^{N-1} \left(\frac{\lambda_i c_1}{\lambda_i + \mu_i} \quad \frac{c_1(\lambda_i + \mu_i) - \lambda_i c_1}{(\lambda_i + \mu_i)^2} \quad + \quad \frac{\lambda_i c_1}{\lambda_i + \mu_i} \mu_i c_3 t_n e^{-(\lambda_i + \mu_i)t_n}\right.$$

$$-\mu_i c_3 e^{-(\lambda_i+\mu_i)t_n} \frac{c_1(\lambda_i+\mu_i)-\lambda_i c_1}{(\lambda_i+\mu_i)^2} - \mu_i^2 c_3^2 t_n e^{-2(\lambda_i+\mu_i)t_n}$$

$$-s_n \frac{c_1(\lambda_i+\mu_i)-\lambda_i c_1}{(\lambda_i+\mu_i)^2} - \mu_i c_3 t_n s_n e^{-(\lambda_i+\mu_i)t_n} \;) = 0\;, \qquad (23)$$

$$\sum_{n=2}^{N-1} \Bigg[- \frac{(\lambda_i c_1)^2}{(\lambda_i+\mu_i)^3} - \frac{\lambda_i c_1}{\lambda_i+\mu_i} c_3 e^{-(\lambda_i+\mu_i)t_n}$$

$$+ \frac{\lambda_i c_1}{\lambda_i+\mu_i} \mu_i c_3 t_n e^{-(\lambda_i+\mu_i)t_n} + \mu_i c_3 \frac{\lambda_i c_1}{(\lambda_i+\mu_i)^2} e^{-(\lambda_i+\mu_i)t_n}$$

$$+ \mu_i (c_3)^2 e^{-2(\lambda_i+\mu_i)t_n} - (\mu_i c_3)^2 e^{-(\lambda_i+\mu_i)t_n} t_n e^{-(\lambda_i+\mu_i)t_n}$$

$$+ \frac{\lambda_i c_1}{(\lambda_i+\mu_i)^2} s_n + s_n c_3 e^{-(\lambda_i+\mu_i)t_n} - s_n \mu_i c_3 t_n e^{-(\lambda_i+\mu_i)t_n} \Bigg] . \qquad (24)$$

The key point here is that we take ($\lambda_i + \mu_i$) as given through Eq. ('8).

The term T1

$$T1 = \sum_{n=2}^{N-1} t_n e^{-(\lambda_i+\mu_i)t_n} \qquad (25)$$

can be shown to be

$$T1 = \frac{k}{(1-k)^2} \;, \qquad (26)$$

with k defined via Eq. (7).

If ($\lambda_i + \mu_i$) can be assumed to be small enough, then T1 can be written as

$$T1 = \frac{1-(\lambda_i+\mu_i)}{(\lambda_i+\mu_i)^2}$$

The term T2 that appears in Eqs. (23) and (24),

$$T2 = \sum_{n=2}^{N-1} e^{-(\lambda_i+\mu_i)t_n} , \tag{27}$$

can be shown to be

$$T2 = \frac{k(1 - k^{N-2})}{1 - k} ; \tag{28}$$

but if k^{N-2} is small enough, then

$$T2 = \frac{k}{1 - k} ; \tag{29}$$

and if, $(\lambda_i + \mu_i)$ is small enough, then T2 becomes:

$$T2 = \frac{1 - (\lambda_i+\mu_i)}{(\lambda_i+\mu_i)} . \tag{30}$$

Note that the quantities λ_i,c1 and λ_i c3 are supposed to be known through the comparison of Eqs. (21) and (3). Thus:

$$\frac{\lambda_i c_1}{(\lambda_i+\mu_i)} = s_N, \tag{31}$$

$$\mu_i c_3 = S - s_1 . \tag{32}$$

With this material c_1 and c_3 can be easily extracted from Eqs. (23) and (24) respectively, to be

$$c_1 = \frac{(\lambda_i+\mu_i)(s_N-s_1)\left[4\bar{s}_n(1-\lambda_i-\mu_i) + (s_N-s_1)(1-2\lambda_i-2\mu_i) - 4s_N(1-\lambda_i-\mu_i)\right]}{4\mu_i \quad (\lambda_i+\mu_i)(N-3)\,\mu_i \quad (s_N-\bar{s}_n) - (s_N-s_1)(1-\lambda_i\,\mu_i)}, \tag{33}$$

and

$$c_3 = \frac{\left[-4(\lambda_i+\mu_i)^2 s_N(N-3)\right](s_N-\bar{s}_n) + \left[(\lambda_i+\mu_i)(s_N-s_1)(1-\lambda_i-\mu_i)\right](-8\dot{s}_N+4\bar{s}_n)+4(s_N-s_1)^2(1-\lambda_i-\mu_i)^2}{\left\{(\lambda_i+\mu_i)\left[-4(\lambda_i+\mu_i)\; s_N(1-\lambda_i-\mu_i)+(s_N-s_1)(1-2\lambda_i-2\mu_i)+4s_n(\lambda_i+\mu_i)(1-\lambda_i-\mu_i)\right]\right\}} \tag{34}$$

with of s_n, an average of s_n's are defined rather to shorten the expressions in question

Once c_1 and c_3 are determined, λ_i and μ_i can be found from Eqs. (31) and (32), knowing already $(\lambda_i + \mu_i)$. Whether the sum of $\lambda_i + \mu_i$ just found checks the quantity $(\lambda_i + \mu_i)$ previously used should be verified, and fortunately, it does to a satisfactory extent if the approximations assumed throughout so indeed hold.

This exercise ought to be repeated for each one of the periods of the study, and I pairs of (λ_i,μ_i) should be determined for each of the periods of study in consideration.

The way our parameters change from one period to another may be expected to give us how the failure and repair patterns progress, allowing us to make predictions for the future and take necessary steps to make interventions if necessary.

Below we give a numerical example concretizing our approach.

Numerical Example

A set of numbers related to a failure activity with respect with time is given in the following figure. For the sake of simplicity, we consider just one period of study covering the entire scale of time showing the future.

Thus k, following Eq. (18), is found to be

$$k = 0.\dot{9}. \tag{35}$$

$\lambda + \mu$ then comes to be

$$\lambda + \mu = 0.08 \tag{36}$$

Through Eqs. (19) and (20, Eq. (21) is read:

$$\mu HL(t) = 26 + 70\ e^{-0.08t} \quad . \tag{37}$$

This expression with respect to time is drawn on our original figure (see below). c_1 and c_3 are found from Eqs. (33) and (34) to be

$$c_1 \cong 110 \tag{38}$$

$$c_3 \cong -2030 \tag{39}$$

Thus λ and μ are found to be, respectively,

$$\lambda = 0.03 \tag{40}$$

$$\mu = 0.04 \tag{41}$$

The sum of these numbers checks out satisfactorily with the number $l + m$ given in Eq. (36).

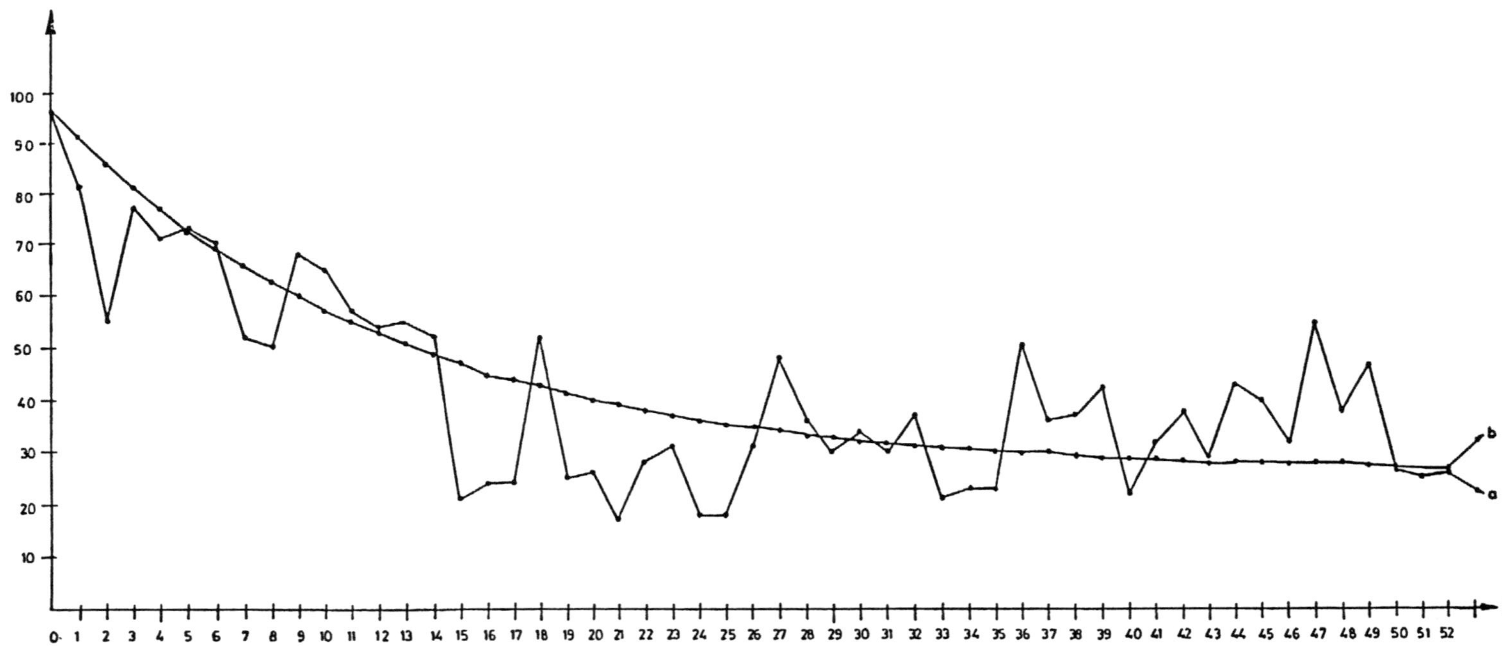

Figure : Activity of failure of the items in question versus time

(a) Physical Data (b) Predicted Behavior within the context of our approach

A PRACTICAL APPROACH TO AUTOMATED TESTING IN PC BOARDS

F. A. Yarman, B. S. Yarman, N. Nadar, and C Kurada

University of Anatolia
Eskisehir, Turkey

This paper reviews a system developed for automated testing, fault tracking, and isolation in possibly ill-defined circuits to be used in personal computers. To assess the factors relevant to failure analysis of information systems, the discussion focuses on system objectives and the approach used in reaching them.

The system consists of a hardware side, a microcomputer with the necessary interface, a software side, the unit under test, i.e., some printed circuit boards, and the best operator as the human element.

Based on the concept of fault tracking as an extended version of testing, and starting with the criteria for a well defined test, an attempt is made to set requirements for fault tracking in Printed Circuits (PC) boards of electrical networks. Moreover, thinking of these tests as a comparison of behavior between unit under test and a reference, a standard approach can be established for fault tracking in failed PC boards. This approach can be mapped to the EATPUT model by identifying the relevant steps.

THE SYSTEM

A hierarchical organization is set with the test operator at the bottom, the test engineer in the middle, and the "Test Spirit" at the top (Figure 1).

Starting from the bottom, the test operator has at his disposal the unit under test (UUT) and a reference unit (RU) to carry on the comparison. A test signal is generated by a personal computer (PC), and the responses of both UUT and RU are compared either automatically to avoid misreadings, or manually to let the test operator really feel that the test is running.

The measurements, provided as raw data, are processed by the PC within the framework of a test program (Level-1). It should be noted that the steps carried by the test program data acquisition and transmission routines include not only steps from the event world to the information set but also the following steps to decision (Level-2). This is simply a suggestion to the test operator about either a new position for the probe (a new position for sensing) or a new step to be taken (e.g., changing the settings, replacing a circuit element, etc.). In effect, even the implementation of the order (suggestion) can be achieved automatically. Therefore, in a fully automated test session, from sensing to transfer, there would be really no need for the test operator except for initial settings.

Unfortunately, unless all possible behaviors of UUT are predicted in Level-1 (Test Generation Level), such a level of automation would endanger the success of the test session. But it should be noticed that a communication medium between the test engineer (Level-1) and the test operator (Level-2) would ultimately lead to perfection if the process of learning is well set in Level-0 of the Test Spirit. To reach such perfection, the judgment faculty of the test operator is needed along with his fallible nature.

To achieve improvements, some options in running the test are provided to the test operator. Experience (learning) files are used in Level-1 by the test engineer to improve his test program. But theoretically, the opportunity for operator failure is still not eliminated since even

NATO ASI Series, Vol. F32
Information Systems: Failure Analysis
Edited by J. A. Wise and A. Debons

with no room for misjudgment, he can still provide misinformation. Nevertheless, the iterative approach seems to move toward a mishap-cleared system.

At the top of the hierarchy is Level-0, the Test Spirit. In our work, it is assumed that universal rules can be found at least within the event world of typical classes of PC boards. Based on this assumption, poorly skilled test engineers are allowed to work only in a Test Program Generator environment. Their decision (Level-1), therefore, always gears the bias of the Test Spirit.

Electrical networks are among the lucky systems where one can say the model (e.g., capacitance, inductance, resistance, etc.) is a fairly good approximation of the real element (e.g., capacitor, inductor, resistor, etc.) within the ranges of normal operating conditions. One should therefore expect that all possible behaviors could ultimately be obtained through analysis of the simulated circuit. It is well known that two sets of variables (node voltages and branch currents) are necessary and sufficient for a complete description of the system behavior. It should be noted that for the practical reasons our fault tracking routines concentrate on node voltages.

Although the UUT may contain undefined blocks, it is assumed that the node voltages (response of the UUT) for normal operation are known. In case of failure, their faulted values can be measured, and test procedures for functional testing can be provided (Kurada, 1986). The question can then be raised whether a complete set of "all credible failures" is considered. An appropriate answer for this question, we presume, should be given in terms of experience and past performance of the routines.

An alternative for the functional test routines is the in-circuit test routines. For bilateral checks of the two approaches, one could use circuit partitioning and network sensitivity analysis routine included in the latter package (Adar 1986).

The user-friendly Test Program Generator (Level-0) can be considered a product of the Test Spirit. The program provides a guide pointer, memory allocations, and addresses needed for preparations, settings, checks, steps, and substeps for test and fault tracking procedures. On the in-circuit side, one can find a circuit analysis program, fault tracking routines, and acceleration routines (discussed below).

THE APPROACH

In general, testing can be understood as a comparison of the expected behavior of the considered system and its observed behavior. A well defined test will therefore require the following:

1. A well defined system and its environment.
2. A test signal from the test environment.
3. A reference unit (a system assumingly unfailed).
4. An appropriate sensing (mostly measurements).
5. A set of adequately collected data.

The least efficient approach would then be to compare the raw data from both reference and suspected units, if feasible. In fact, however, such a brute force approach may necessitate unfeasibly large memories and execution time.

Difficulties are found, even in the first requirement, in defining the system well. In our case, the definition of the PC Board Test environment should unambiguously include the UUT and RU, the test equipment (roughly referred to as hardware), the test generating program, the test programs (referred to as software), and the test operator. Should possible sources of noise be included in the environment? What could they be? An implicit definition may risk clarity, while a precise and explicit one may render the list needlessly long. As a result, in failing to define the system well, one can hardly derive a complete set of predicted credible effects and failure patterns.

Similar shortcomings would exist also for the reference unit, especially those resulting from output fluctuations caused by random behavior of the circuit elements that remain within their tolerance limits. On the other hand, the effectiveness of collected data would be affected not only by its adequacy but also by the reliability of the sensing.

Theoretically, the concept of comparison with a hypothetical reference does not hold, since in reality it is not possible to carry measurements on a system without any interaction -- hence the necessity of symmetrical and compatible perturbation on both the UUT and RU. An additional worry is the possibility of harming the circuits with the test signal (which is not necessarily comparable to one under normal operating conditions). The risk suggests that a simulated circuit and its possible responses should be established during the test program generation.

A usual practice, defined during test program generation, is to obtain the response matrices of the unit under varying input conditions. When referring to tests, one usually thinks of the comparison of only the output signals in response to a given input chain on both the UUT and the RU. It should be noted that the response at any spot internal to the circuit is included in our concept of test.

At first glance, to consider the process of fault tracking as an extended version of testing may sound like an understatement. But we shall look at the problem of fault tracking in a more complex way.

Once the suspected unit is surely faulted, random probing at the internal nodes of the card with two assumptions will ultimately lead to two nodes that embrace the failed circuit element. The first assumption is that the failure of the sought circuit elements affects only the neighboring two nodes. Obviously, this assumption does not hold most of the time. In other words, the failure does not propagate. The second and conceptually less important assumption is that only one failed element exists and it is connected to two nodes. This assumption is made for simplicity, and can later be dropped.

With the two assumptions, the most primitive process of fault tracking is reduced to a sequence of certain number of tests. Furthermore, these assumptions can be released at the expense of additional requirements to those mentioned above within the framework of a test in general.

Since fault tracking is more complicated than just a brute force application of a sequence of tests, it is not surprising that certain requirements must be added in addition to the ones mentioned above:

6. Data processing routine to lead to effective information.
7. Some optional information checks against data collected.
8. Information utilization routines to lead to decision.
9. Decision implementation routines.
10. Controlled loops to repeat steps from sensing to transfer.

In the case of manual fault tracking, the adequacy and reliability of data collection depends heavily on the natural capacities and expertise of the test operator. The quality of the transition from the event world to data acquisition, without loss of valuable data and the omission of useless data, requires not only a thorough knowledge of the system but also fast access to that knowledge. Increasing the complexity of the network favors the use of computers rather than the human operator for both control and monitoring units and hence increases efficiency, effectiveness, and even reliability at the risk of the automated system falling short of predicting all credible cases. In a transition period from semi-manual to fully automated system, optional checks of information against raw data may be comforting, although time- consuming. In the long run, basing the overall system on a self-learning and self-improving scheme gives room for an increasingly reliable system.

A further step is the inclusion in the test program of some sort of acceleration procedures. The routing will seek answers to certain questions: Where in the failed circuit one should start to fault track? What is the most possible efficient way of analyzing the failure for the failed component cognition?

In answering both questions, we rely on our assumption that the responses of the failed circuit to selected test signals are of great help in the analysis of the failure. One answer to the first question could be to map the PC Board to a multi-dimensional array of errors, starting with the node that maps to the highest error. Although this approach may sound practical, it can be misleading when, for example, an error is amplified.

Another possible answer to the first question could be to map the circuit to a network, constructing a fault tree where each component is indicated by its probability of failure. This approach omits the cause- effect chain in the failure mechanism and also possible evaluations of diagnosis during the test session.

A third possible answer in accordance with the past experience could be to search for the old failures or to try to identify the diagnostic having the acquired knowledge of the system with previous failures.

Finally, we would like to mention an interesting approach that starts from the normal operating conditions, perturbs the system, and records the case histories to be identified. For this purpose, existing techniques of network sensitivity analysis would be of great help. The last three approaches mentioned above would also help to answer the second question on fault backtracking.

CONCLUSIONS

Fault tracking is closely correlated to testing, the former being an extended version of the latter through intelligent accelerating procedures.

For failed component cognition, our approach heavily relies on understanding not only the subsequent events but also the response of the failed system in selecting test signals. In both functional and in-circuit testing, a reference for the system and/or its behavior can certainly be derived from a real physical duplicate as well as from the simulation of the system under consideration.

Testing can be conceived as a hierarchy of three levels. At the highest level is the Test Spirit. Although this level may be thought of as simply a test generating program, it is based on the theoretical assumption that a universal strategy can be adopted for testing in general. It is expected that this strategy will be improved with increasing expertise.

At the intermediate level, a test program can be generated within the structure imposed by Test Spirit. Finally, at the lowest level, the test operator runs the test according to the test program.

Aiming at a learning system, there is room to record feedback information in all three levels so that the overall performance could be improved over time.

We believe that automation, although it lacks the judgment capability of an operator, will increase the efficiency, effectiveness, and reliability of a system. The lack of the operator's capability to judge is minimized by the flexibility inherent in our three-level approach.

ACKNOWLEDGMENTS

The authors would like to thank the Turkish Air Force, General Selahattin Kavustu, General Fazil Aydinmakina, Capt. Nuri Acvioglu, Dr. Erdener Ildiz, and many other officers for their constant support and cooperation.

Thanks are also due to Dr. Zülal Balpinar, Dr. Tolga Yarman, Ms. Müjgan Kerman, and Mr. Kenan Isik, without whose help this work could not have been completed.

REFERENCES

Adar N. (1986). *A Computer Program for Point-to-Point Fault Backtracking*. B.S. Thesis. Eskisehir: University of Anatolia. July.

Isik, K. (1986). *Function Generator Design for ATE*. B.S. Thesis. Eskisehir: University of Anatolia. July.

Kurada, C. (1986). *Fundamental Concepts in Automated Testing Equipments*. B.S. Thesis. Eskisehir: University of Anatolia. July.

Yarman, F. A. (1985). *Computer Aided Testing and Fault Tracking in Complex Logic Circuits*. Exclusive Communique presented to Turkish Air Force Officers. Eskisehir, November.

FAILURE ANALYSIS OF INFORMATION SYSTEMS IN SMALL MANUFACTURING ENTERPRISES: THE IMPORTANCE OF THE HUMAN INTERFACE

Ana Maria Ramalho Correia

Laboratorio Nacional de Engenharia e Technologia Industrial
Lisbon, Portugal

INTRODUCTION

This paper is based mainly on the experience of managing an Information Center in an R&D institution providing information services to industrial firms in Portugal. Generally speaking, in industrial organizations there are always two categories of information to be handled:

- Internally generated information
- Externally generated information

Internally generated information includes the following:

- Reports on original R & D data
- Marketing information (e.g., customers' requirements, preferences and expectations, sales analysis)
- Registries of correspondence
- Quality control reports
- Design of new products
  (Haygarth-Jackson,1980; Rzevski and Farrar, 1984; and Pratap-Lingham, 1983)

The variety of external information sources is extremely broad. It includes many different types of published material:

- Review journals
- Scientific and technological journals
- Trade journals
- Monographs/textbooks
- Directories
- Government reports
- Non-government reports
- Standards
- Patents

The information required by an industrial enterprise may be in the form of words or text, data or numbers, drawings, or graphics (e.g., chemical structures, engineering drawings) or a combination of any of these forms. Any industrial function (R & D, production, or marketing) requires a complex blend of all the information available (scientific, technical, commercial, economic, financial, political, social, environmental) (Haygarth-Jackson, 1980).

INFORMATION NEEDS OF SMALL COMPANIES

Recently, there has been an increasing interest in the study of information-seeking behavior and information needs of small manufacturing enterprises (SMEs) (Capital Planning

NATO ASI Series, Vol. F32
Information Systems: Failure Analysis
Edited by J. A. Wise and A. Debons

Information, 1982; Pipin, 1983; Sweeney, 1985; Hemer, 1984). This interest is related in part to the capacity that SMEs in the manufacturing and technical services sectors have demonstrated in their flexibility to meet market and technological change and also to the fact that they have become major sources of new employment (Sweeney, 1982).

In various countries throughout the world (e.g., The Netherlands, U.K., Denmark, Ireland), different services and schemes exist that aim to provide and promote the uses of information by SMEs (Sweeney, 1985). The basic objective of these schemes is usually to aid the transfer of the results of technological research to enterprises in order to stimulate industrial production, or to identify problems so as to suggest different production processes so that they can modernize and innovate quickly. Due to their size and shortages in qualified resources, most SME industrial enterprises have to rely on systems that are aimed at providing information services for industry.

One of the major problems faced by industrial enterprises, independent of their size, is that information is not properly managed within the organization, leading to failures. Recognition of the problems of inadequate information management and ineffective systems can be found in many papers. Vickery (1983), for example, presents an interesting general scheme of the information flows in industrial organizations. He identifies five main information flows:

1. Operational information that includes the order, production, delivery, and accounting documentation (mainly forms) and some routine interdepartmental communication (memoranda, correspondence, committee papers).

2. Management information (e.g., tabulated data or regular analysis of data on production, raw material costs, machine utilization, sales, cash flow, profitability) that are tapped from the operational information flow at certain points, substantially performed by Data ProcessingDepartment.

3. International technical information (e.g., laboratory reports, product information, health and safety, equipment evaluations) that flows between departments and also to the outside world (mainly customers). Two subsystems are particularly important to this process: (a) flow from the laboratories to marketing and sales, and (b) preparation of technical reports for release to remote factories.

4. Customer information (i.e., feedback from or about customers) that flows via the sales force to marketing and technical departments.

5. External technical and commercial information flowing largely via the information unit, where one exists, as well as directly into different parts of the organization.

It is also interesting to note that Launo (1984) in several papers describes the need to relate internal information to external information. He points out that there is a need for combining internal and external information and that this need has to be filled by in-between systems, i.e., by human communications.

It has also been reported in literature that there are considerable differences in the use of formal published sources of information by scientists and engineers, depending on the discipline or the activity (research, production, marketing). Allen (1977), who studies communication in R & D laboratories, reported that engineers spent three times more time on informal literature (e.g., unpublished reports) than on formal literature (e.g., books and periodicals). On the other hand, technologists and industrialists prefer direct and immediately usable information rather than references. Thus, small firms tend to use periodicals for specific rather than general information.

Further evidence of the different use of the certain information sources by industrial personnel can be found in a study by Rosenberg (1967). This author ranks the preferred information sources and research methods as follows:

1. Search personal library
2. Search in building where you work
3. Visit knowledgeable person nearby
4. Telephone a knowledgeable person
5. Use library not in your organization
6. Consult a reference librarian
7. Write a letter
8. Visit a knowledgeable person more than 20 miles away

It has also been reported that informal discussion through person-to-person contact is the preferred channel of communication for transfer of information, particularly for acquiring information among all people, whether academics, industrial scientists, technologists, or managers (Sweeney, 1980). Face-to-face communication is rich because it overcomes barriers that exist in written or telephone communication with a known or unknown person. This is the reason that, in a number of countries, industrial information services operating through Information or Liaison Officers are now well established.

Moreover, it has been recognized (Sweeney, 1985) that one of the ways of assisting the growth and development of small industrial enterprises, stimulating the innovative and entrepreneurial vitality in an economy, involves two activities:

- Wide diffusion of the best technical practice.
- Creation of a rich and diverse flow of information that leads to awareness of a broad spectrum of technology and its creative application in all sectors of industry.

Most information needed for economic development passes through the informal personal networks that exist in every community (Sweeney, 1985). They exist not only at professional and business levels but also at local levels.

In Portugal at the present time, the providing of information to industry is scattered throughout a number of entities, including several governmental bodies, industrial associations, and R & D institutions (Petrie, 1986).

EXAMPLES OF APPLICATION OF THE EATPUT MODEL

Example 1

A Portuguese firm X (See Table 1) that was exporting plastic toys to foreign countries was having problems concerning some type of additives that resulted in the deterioration of the product when it was exported to countries with a hot climate.

The type of information that was lacking in this example was technical. The company needed not only documentary information but also some kind of specialized information and advice suggesting methods for changing the production processes. This could be achieved only by a trained information specialist (the human interface) with a sound scientific/technological background, who could read and analyze all the available information but mainly the relevant patent information.

Table 1. EATPUT Model of Example 1.

EATPUT Components	$Time_1$	$Time_2$
Event World	Colored toy exported	Loss of money
Acquisition	Eyes	Looking for new information sources
Transmission	Central Nervous System	Visits to trade exhibits, visit information center
Processing	None	Technical staff using library and databases to solve problem
Utilization	Recognition of loss of characteristics of toy	Use information to find new ways to prepare toys (e.g., exploit patent information)
Transfer	Loss of market	Recover markets and

Table 2. EATPUT Model of Example 2.

EATPUT Components	$Time_1$	$Time_2$
Event World	Exploration process started	Realize that alloy should comply with ASTN specifcation
Acquisition	Staff involved in the process	Marketing Department Manager
Transmission	None	Telex Information Center
Processing	None	Information staff used library, and resident scientist to identify answer; responded in 35 minutes
Utilization	None	Decision was made
Transfer	None	Directive issued to export batch

Example 2

A Portuguese firm Y (See Table 2) was about to export some steel appliances to a foreign country. Suddenly, it was realized inside the firm that the alloy (steel) must be in accordance with a certain ASTM specification. A telex was received in CITI/LNETI explaining the reason for the urgent need for this information and requesting details on the specification. In half an hour, the necessary information was provided after the standards collection was consulted in one of the departmental libraries.

In this case, the type of information required was factual (technical/scientific). The use of the standards collection was done efficiently by close collaboration between the information specialist and the material sciences specialist.

CONCLUSIONS

The examples of firm X and firm Y underline the importance of personal contacts as well as the need for decentralization in the provision of information to industry, irrespective of the size of the enterprise. The failure of the information systems described existed because the human interfaces needed to provide information did not exist or were not used.

This human interface should be provided by information scientists or information managers. They should behave like gatekeepers. Ideally, they not only should be knowledgeable about sources of information and how to use them, but also should be good communicators in the sense that they can tailor the information and advice they provide precisely to the needs of their clients and have the ability to win the client's confidence (Gibb, 1984).

The information scientist is a highly specialized worker who has the skills needed to teach other workers in an organization the techniques for improving productivity. Furthermore, he or she is trained to help different levels of management to communicate more effectively and knows how to use information technologies to manage information resources in an organization (Carbo-Bearman et al., 1985).

REFERENCES

Allen, T. J. (1977). *Managing the Flow of Technology.* Cambridge, MA: MIT.

Captial Planning Information. (1982). *Information and the Small Manufacturing Firm.* Edinburgh.

Carbo-Bearman, T., P. Guynup, and S. N. Milevski. (1985). Information and productivity. *Journal of the American Society for Information Science,* 36(6), pp. 369-375.

Gibb, J. M. (1984). Will Europeans ever learn to communicate? *Information Come of Age: Proceeding of the Institute of Information Scientists,* Rossendale, University of Kent at Canterbury, July.

Haygarth-Jackson, A. R. (1980). Industry and the research associations: Information provison to the users. *Nationwide Provision and Use of Information: Joint Conference,* Sheffield, 15-19 September, Sheffield, Aslib, pp. 22-29.

Hemer, J. (1984). *Information Networks Designed to Support Innovation in the Less - Favoured Regions of the European Community : Case Study in Baden-Wuerttemberg.* Fraunhofer Institut fur Systemtechnik and Innovationsforschung, October.

Launo, R. (1984). New concepts in the training of industrial information specialists: information resources management. *Seminar in the Training of Industrial Information Officers.* Budapest, National Technical Information Center and Library, pp. 47-50.

Petrie, J. H. (1986). *Development of Information Services for Industry. Technical Report: Review of Information Services for Industry and Recommendations on their Development.* Vienna: UNIDO.

Pipin. (1983). *External Information Provision for Small and Medium Size Industry.* The Hague: Bobin.

Pratap-Lingham, P. (1983). The design engineer: his information requirements. *International Forum on Information Documentation,* 8 (2), pp. 29-35.

Rosenberg, V. (1967). Factors affecting the preferences of industrial personnel for information gathering methods. *Information Storage and Retrieval* 3(3), pp. 119-127.

Rzecski, G. and D. J. Farrar. (1984). Information in the manufacturing industry: an assessment and proposals for improvement. *Aslib Proceedings.* 36(3), pp. 113-117.

Sweeney, G. P. (1980). Technical change in industry through the marketing of information. *Organization and Economics of Information and Documentation: Proceedings of the 40th FID Congress,* National Technological Library of Denmark, Copenhagen, 18-21 August. The Hague: FID, p. 115.

Sweeney, G. P. (1982). *New Enterpreneurship and the Smaller Firm.* Dublin: Institute of Industrial Research and Standards.

Sweeney, G.P. (1985). *Study of Information Networks Designed to Support Technological Innnovation in Less - Favoured Regions of the Community.* Dublin: IIRS.

Vickers, P. H. (1983). Common problems of documentary information transfer, storage and retrieval in industrial organizations. *Journal of Documentation,* 39(4), pp. 217-219.

GROUP REPORTS

GROUP REPORT: SYSTEM ISSUES

SYSTEMS

The function of the system group is to understand how the variables that influence the intermeshing of people, technology, and procedures can contribute to system breakdown and failure. To understand information system failure, a categorization of the differences that can exist among information systems needs to be established. This categorization should acknowledge that such a system exists in an environment - which John Spiers refers to "the universe of discourse." There are boundary layers between systems, the actions that are taken as the result of the sensing mechanism and the universe of discourse.

Systems either exist as entities of natural states, like organisms, or are created by humans. When information systems are created as surrogates of human information systems, they are designed. The design process normally follows an analytical process that identifies the objectives to be achieved. Thus, faulty analysis - which refers to inadequate assessment of the environments, and the insufficient account of sensing, transmitting, and processing functions - can lead to faulty design. Faulty design is considered to be a contributing factor to information system failure.

The ultimate purpose of an information system is to enable humans to deal with the exigencies of the environment. Thus, decision-making and problem-solving are the focal point of information system effectiveness; if the decision-making function and the total information system structure are incompatible with this function, we can say there is a misfit - a misfit that is instrumental in information system failure.

In essence, establishing the differential structure and objective of information systems, accounting for the design characteristics, and realizing the compatibility or incompatibility of the decision-making requirement in this design are considered important aspects of failure of information systems.

It is desirable to objectify certain aspects of information systems, thus allowing the development of hypotheses and related research toward their substantiation.

Two basic dimensions can be considered in this attempt at quantification, namely, measures of system performance, and the quantification of information quality, expressed in Figure 1. It may be speculated that the completeness of the design, and thus its completeness with respect to accommodation to failure and to diagnosis of failure capacity, is a function of the ability to specify the horizontal axis parameters. Note that a continuum of failure is identified, from full operation through levels of degraded operations, to a threshold (limit) of system functional failure.

Some of the qualitative information parameters have been identified as follows:

Information (DATA)	Information System
Validity	Transformation
Relevancy	Integrity
Exhaustiveness	Cost Effectiveness
Timeliness	Compatibility
Formal presentation	

NATO ASI Series, Vol. F32
Information Systems: Failure Analysis
Edited by J. A. Wise and A. Debons

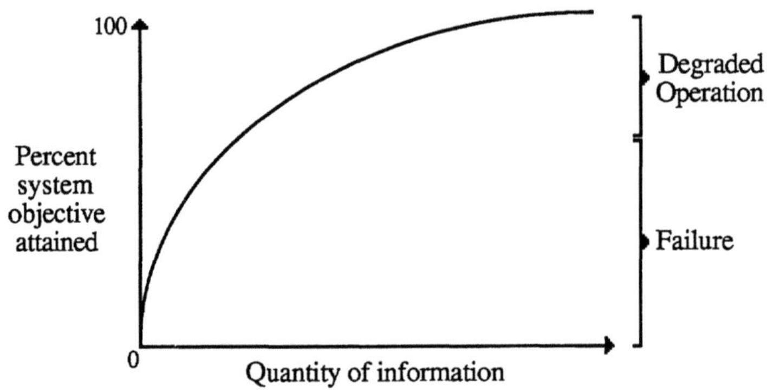

Figure 1. Hypothetical profile of the relationship between qualitative information parameters and information systems objective attainment.

PRINCIPLES

The following principles regarding information system operation are proposed as relevant to information system failure:

1. Information systems have specific objectives.

2. These objectives are contained within certain requirements (needs), i.e., decision-making, problem-solving, which are part of specific functions (management, services, control, etc.).

3. Information systems consist of subsystems, each with its own but related objectives.

4. Structurally, an information system is a flow system, and it inevitably degrades when the flow is disrupted. At the limit, the degradation process ends with failure.

5. Information needs also change. When these needs cannot be met by the information system, it can be said to fail.

6. A system can therefore be said to have failed when a) the system no longer meets its specific design objectives (internal failure), or b) the system objectives no longer meet the information needs (misfit failure).

7. Information systems should be designed in anticipation of failure and provide for graceful (i.e., homeostatic) degradation.

8. It should be assumed that any complex man-made system is supported by an information system and that failure of the system will be preceded by an information system failure.

9. The analysis of failure requires the reconstruction of events leading up to the failure. This requires a record of system activity. The record can come from the actual state of the failed system, or from a redundant record (audit trail), or from a combination of both. Not all design strategies for information

systems produce systems that inherently provide a complete record of activity. Failure of a system tends to destroy the system and hence the evidence required for failure analysis. The proper design of information systems should build a record of activity that can be used to analyze failure. The record must be protected against failure.

10. All real systems have a finite probability of failure. All systems are therefore vulnerable to failure, which is expressed as a probability. This probability of failure can be expressed as a monotonic function - dependent on increasing system complexity. The probability of failure of any system is directly related to the number of functions it performs.

CONCLUSIONS

1. It is possible to develop prevention and analysis methodologies. However, a singular methodology may not be applicable to all information systems.
2. There are certain basic principles regarding the nature of systems and their failure that should be understood by information system designers:

    - Identification of changing information needs and their relation to sub-system objectives
    - The development of indicators (barometers) that reflect mismatch between a desired information state and the current state.
    - The need for incorporating means for accounting of pre- and post-conditions leading to failure.

GROUP REPORT: INVESTIGATION TECHNIQUES

The purpose of this paper is to summarize the results of the Investigations Working Group and to introduce the papers prepared by the individual members of the group. Individuals of the group represented the countries of France, United Kingdom, and the United States, and originated from government, private industry and academia. The Investigations Working Group was chaired by Rudolph Kapustin, President of INTASCO Corporation.

THE BASIC QUESTION

The primary question of the Investigations Working Group concerned whether it was possible to establish a general framework and recommendation for the investigation of an information systems failure. Using a broad definition of an information system as composed of hardware, software, and human factors components, and with an equally broad based application to include space system failure, airplane accidents, information system failure, and manufacturing process failure, the following framework for the investigation was established.

ASSUMPTIONS

The group specifically addressed an investigation process that would be appropriate for a large scale, catastrophic information systems failure - such as the Space Shuttle Disaster, an airline crash, a chemical plant accident, a nuclear power plant failure, or a command and control early warning system failure. Additionally, the group focused on an approach that would be viable after the fact, i.e., after the accident that had taken place. Further, the group was strongly convinced that it is not a question of IF an information system fails, but WHEN and to what CONSEQUENCE it will fail.

THE INVESTIGATION COMPONENTS

Subsequent discussion describes the investigation process related to a large scale catastrophic information systems failure, including the structure of the investigation organization, the study of the information system, integration methods of obtaining evidence, and the investigation process.

INVESTIGATING ORGANIZATION STRUCTURE

Fundamental to the success and effectiveness of the investigation of the information systems failure is the organizational structure of the team charged with the responsibility of providing an explanation as to why and how the disaster occurred. The investigation team should be separate and apart from the organization being investigated. In the case of the early days of the Space Shuttle disaster investigation, it became apparent that a complete, impartial analysis could not be performed with NASA investigating itself. Similarly, the National Transportation Safety Board independently investigates airline, railroad, marine, pipeline, and surface disasters.

The organizational structure should be delineated with clear lines of authority and reporting responsibility. The people assigned to this organization should be completely relieved of other duties and responsibilities for the duration of the investigation. Additionally, except for certain staff functions, it is recommended that a matrix management approach be avoided in order to reinforce the lines of authority and responsibility of the investigation team.

The individual members of the investigating organization should have skills and backgrounds compatible with their areas of responsibility. In a disaster related to the failure of

NATO ASI Series, Vol. F32
Information Systems: Failure Analysis
Edited by J. A. Wise and A. Debons

a large information system, it would be extremely beneficial to ensure that the aggregate skills of the team represent an interdisciplinary team. In this manner, the analysis of the failure would cross all technical and human factors boundaries.

The need for regular meetings at various levels of the investigation organization provides the framework to share information, verify facts, and provide a method by which "mid course corrections" can be made in a controlled environment.

Finally, it is essential that adequate resources be provided to perform the investigation. In addition to an adequate level of dedicated staff, items such as travel expenses, computer and wordprocessing assets, and special technical expertise shouldbe provided.

STUDY OF THE INFORMATION SYSTEM

An effective, comprehensive study of the information system is paramount to the successful investigation of the cause of the failure. This study phase should not be confused with the logical deduction phase that concludes WHY the failure occurred. The purpose of the study is to focus the interdisciplinary team members on a detailed analysis of the components of the technical facts of the system.

An effective starting point for this study is to use the fault tree approach and/or the organization approach. For extremely large complex systems, pursuing both approaches simultaneously and later integrating the findings is necessary. During the study phase of the information system, the need to control the evidence of the failure is essential. This phase usually results in obtaining evidence from three primary sources: physical evidence, such as photographs, wreckage and debris; written documentation; and testimony from witnesses and other pertinent individuals. The Space Shuttle investigation resulted in over 100,000 pages of documents being submitted as evidence, hundreds of interviews, and hundreds of photos, film footage, radio transmission recordings, and tons of debris. In order to control and monitor the evidence, sophisticated computer systems were employed to retrieve, cross check, and inventory this material.

Specifically, control of witness interviews is extremely important. It is necessary not only to summarize the results of the interview (when it took place and who participated) but also to produce full text manuscripts or tape recordings of the interview. This level of detail is necessary in order to later integrate and cross check pertinent interview facts with written documentation. Similarly, the use of advanced automation facilitates control over the witness interview process. During the study phase of the information systems failure, it is essential that emerging, obvious conclusions do not serve to prematurely terminate the study in other technical areas. Full investigation and fact gathering should be done in all areas, and should later be discarded during the integration phase. Additional valuable evidence is also often obtained by employing the concept of a "skunk works" or a confidential evidence input means. Finally, the study phase of the failure investigation should conclude with a thorough cross checking of testimony evidence against written documentation of pertinent facts. Cross checking of physical evidence and written documentation is essential. An illustration might be a conclusion that a jet engine turbine blade was not manufactured according to the proper procedures for heat treatment.

INTEGRATION ANALYSIS

After the completion of the study phase, the gathering of facts and evidence, and cross checking facts across all the evidence, the integration analysis then serves to allow for the formulation of the reasons for the system failure. The incorporation of the fault tree analysis and associated evidence with the human performance characteristics derived from the organization approach begins the process of formulation of the reasons for the failure. Additionally, the results of the fault tree analysis, with evidence collected from design, manufacturing, and component reliability, lead to conclusions for the failure.

REPORTING

The recommendation for producing a factual and conclusive report of the failure analysis investigation hardly needs mentioning. In summary, it is essential to avoid the use of complex technical jargon unless it is absolutely necessary. The investigation report should stand by itself and not depend upon subsequent readings and references to clarify either facts or conclusions presented. It is extremely important to present the facts concerning the failure of the information system, separate and apart from the conclusions of the investigation. Since the facts and evidence prsented are objective and will stand upon their own merits, they must be separated from the conclusions and reasons for the failure, which are extremely subjective in many cases. The conclusions and reasons for the system failure should be separately presented, and the relevant facts used for substantiating the conclusion must be cited.

Finally, related to the specific preparation and publication of the failure analysis investigation report, it is recommended that the investigation team actually write the report rather than assigning the responsibility to technical writers. The reader is directed to the papers in this volume by the members of the investigation working group: R. Kapustin, R. Kehrli, A. Bilinski, M. T. Ho, H. Petrie, and M. Barnett.

GROUP REPORT: ORGANIZATIONAL ISSUES

Organizations depend on the information systems embedded within them. Malfunction of the information system often leads to organizational failure. The major concern of this group was the way in which the effectiveness and reliability of an information system could be improved. The group felt it would be helpful to develop a model of a well-functioning information system, and we therefore developed a list of characteristics for such systems. Although these characteristics must be treated as hypotheses, they are nonetheless valuable as a guide to future investigations.

ORGANIZATION STRUCTURE

The structural characteristics of the organization greatly influence the rate and nature of information systems failure. The organization should have clear and widely accepted goals. If there are divisive conflicts, certain members may even deliberately block or subvert the information system, as happened in the case of the W. T. Grant department stores. An organization which is unjust or which inequitably distributes its rewards may cause such subversion.

SYSTEM PROCEDURES

There are also a number of what might be considered sound information operating system procedures. Among these are the following:

A) Absence of primary barriers to communication, such as the rigid adherence to status or departmental function.

B) The presence of multiple independent channels for monitoring key environmental variables for avoiding simultaneous failure of monitoring channels.

C) The facility for bypassing standard channels if these appear blocked. The organization should encourage inquiry, critique and advocacy by creating an atmosphere for free flow of information. The organization should legitimatize reasoned and constructive dissent.

D) An information system designed to avoid catastrophic failure modes. If failure occurs, the system's performance should degrade gracefully. The organization should be able to continue even if the information system fails.

E) Information collection in such a way that the data collected can be valuable to use in future unanticipated ways.

F) A decision-making cycle rapid enough to cope with the rate of environmental change.

AN INFORMATION SYSTEM SHOULD BE ABLE TO IMPROVE ITSELF

A) An organization should draw upon its own cognitive resources and encourage external cognitive resources relevant to it. Internally, it should encourage contributions through reward systems, which may be financial,

NATO ASI Series, Vol. F32
Information Systems: Failure Analysis
Edited by J. A. Wise and A. Debons

positional, or symbolic.

B) Information flow should be monitored in terms of adequacy (lack of bias) and relevance (not overloading individual with unnecessary detail). A high-level, central individual should be given responsibility for monitoring overall flow and improving it.

C) Methods for anticipating failure are important. These include the following:

1) Fault tree analysis (see paper by Dhillon)
2) Gaming and simulation
3) "Incident" (near-miss) reporting systems and hotlines.
4) Mechanisms for learning of and from failures of other organizations
5) Recording system for data on past failures

D) Every failure can be an important occasion for improvement, if properly processed. Such processes include hindsight review and evaluation, and written reports on failure, which are circulated through the whole organization.

GROUP REPORT:
HARDWARE AND SOFTWARE

INTRODUCTION

This paper was jointly written by the members of the Hardware/Software working group. The group had two aims:

1. To identify the key issues relating to hardware and software in failure of information systems.
2. To make recommendations relating to the key issues.

Before considering these points further, it is appropriate to review the benefits and problems of using computer-based information systems.

Benefits of Using Computers in Information Systems

1. Computers provide quicker reaction to questions or demands than other systems do.
2. It is possible to process a large amount of data by fairly complex procedures within a short time.
3. The cost for the individual information demand is low if the system is used frequently.
4. The probability is very high that tests that have been carried out correctly in the past will be carried out correctly in the future.
5. Computer performance is not biased by personal interests, such as dedication to particular persons or fear of losing a job.
6. Computers don't get time off, and their results are reproducible.

Problems with Many Computers

1. The cost for initial analysis and system implementation is high.
2. System results may be misleading if the information system is not used properly or if it is poorly understood.
3. The system may provide wrong results because of internal failures.

Scope of Report

In the very limited time available, it was decided to review alternative approaches for classifying the key issues and then to pursue selected methods of classification in more detail.

It was also agreed to base the recommendations on some example questions so that the user would be using an inquisitive approach rather than a prescriptive one. It was recognized that the detailed questions would be incomplete but would give an indication of the problems involved.

Possible Classifications

The bases for classification considered were the following:

- The EATPUT model (a system function classification).
- The hardware/software system life cycle (a design and operations classification).
- The failure management procedure (a diagnostic and error recovery

NATO ASI Series, Vol. F32
Information Systems: Failure Analysis
Edited by J. A. Wise and A. Debons

classification).
- The information life cycle (an information generation and use classification).

LIFE CYCLE CLASSIFICATION FOR HARDWARE/SOFTWARE FAILURE ANALYSIS

The group recognized 11 stages in the life cycle of a system and identified some questions relevant to each. These stages are essential elements of software engineering methodology.

1. System Objectives

Are the objectives well-defined? Is evolution possible? Don't be too ambitious!

2. User Requirements

Are the different users identified? Is the information being displayed in a proper form? The form of dislay and the system operation consider the particular shapes and signs, etc., the user is acquainted with? Is it easy to operate the system? Does the system react in time? Are time constraints specified? Is the system robust against operator error? To what degree towards which errors?

Does the system have timely access to the relevant sources of information? Does the system inform those persons who are entitled to information and responsible for getting it? Does it perform a plausibility check on the correctness of its inputs? Is the probability of treating faulty input information specified? Have legal requirements been recognized and met?

The recommendations of the Organizational Group with respect to computer systems should also be considered here.

3. Specification

Are the specifications unambiguous and complete? Can the individual requirements be checked? Can the individual requirements be implemented? Is there any procedure that takes care of changes of the requirements? Is clearly stated what is outside the scope of the system? Are all requirements unambiguous? Is it clearly stated which requirements are of higher importance and which of lower importance? Are any reliability requirements specified? Are any recovery procedures, rollback procedures, graceful degradation, and the like specified?

4. Design

Is the design feasible within reasonable limits? Are the mathematical models used adequately? Is the design properly documented? Is appropriate care taken for redundancy and diversity? Are there enough reserves for future modifications? Are there appropriate failure self-diagnosing provisions?

5. System Construction

Is the system constructed according to the design? Are proper techniques built into the manufacturing process? Is proper quality control and audit enforced?

6. Validation, Verification and Testing (VV&T)

Is the VV&T complete and reproducible? Are test procedures updated synchronously with design modifications?

During the information system life cycle, the question of whether or not the specified functions have really been implemented is very important. Because of the size and complexity of modern information systems, simple testing through the developer will normally not be sufficient. This applies in view of the possibly high "responsibility" of such systems thatmakes their failure have serious consequences. In order to highlight the most important aspects of system VV&T, the group called attention to the paper by Ehrenberger concerning the role of software reliability tools in this area.

Attention is drawn to the correlation of "testing" and "fault tracking" (failure localization), which can conceptually be considered as an extended version of testing. Hence, starting with test, a three-level approach is supported. Hierarchically, at the top is a structuring process that assumes that a universal strategy can be adopted in generating test programs. At the second level is the generation of the test program, which adopts the structure imposed from the above (i.e., test program generator) and provides room for modifications from the level below. Finally, at the bottom is the execution of the test program, which also provides a mechanism for learning in all three levels above so that the overall system is set on a path where continuous improvement will be possible.

7. Monitoring

Is there an adequate monitoring tool? Is the monitoring function reliable enough? Is the monitoring data accessible?

8. Applications, Operational Use

Are the user needs satisfied? Are on-line help facilities provided? Are appropriate measures taken to prevent system destruction or system impairment through its users? Such measures are to be considered against voluntary and involuntary actions. Effects on frequently used system parts or data have to be considered and effects on seldom used parts or data.

The importance of user requirements and the related user interface in an information system was highlighted by the failing attempt to get relevant references with respect to "failure analysis in information systems" out of the databases made accessible by various Information Retrieval Services. The basic problem seemed to be the fact that the mental mode of the user and the one of the service supplier did not match. In the interface, the user also didn't find any help to guide him towards a better match between what he was expecting from the combination of certain keywords and what was meant by the same keywords at the input end.

Obviously, the system worked perfectly as a "data processing" system, but intelligence, or mimicked intelligence, is required to transform (combinations of) data (numeric or logical) into information. If the user had been able to access the reasoning pattern and underlying expertise of the person keywording the reference documents, he could have tried to rephrase his search in a more machine-friendly way.

Expert systems, or intelligent knowledge-based systems, to help the user in finding the most appropriate combination of keywords and databases, as a function of his specific but not always deterministically describable needs, would be a major help in solving some of his problems. One has nevertheless to be aware of the fact that actual state-of-the-art "expert systems" are domain-specific, while the possible users of an information retrieval service cover a wide spectrum of domains and disciplines. Although over the past years some relevant work has been going on with input to expert systems for information retrieval services (formulating search requests, cataloguing, library referencing work), it is still in its embryonic phase.

9. Modifications

Is there a check to ensure that modifications do not degrade performance? Is there a change control procedure? Is it possible to carry out minor hardware and software modifications during system operation? Is there an appropriate procedure agreed upon for carrying out modifications? Does it give due regard to possible side effects on the existing parts and their performance? Does it consider the reverification aspects?

10. Maintenance

Is the system maintainable? Many of the factors noted in Section 9 above are also clearly relevant here.

11. Replacement

Have the implications of a possible replacement of the system been considered?

It was recognized that the 11 stages above compose a conceptually serial model but that in practice there is a continuous need for iteration (e.g., in the specification, design and construction stages).

THE FAILURE MANAGEMENT PROCEDURE CLASSIFICATION ANALYSIS

Is in-built test and failure detection provided on the major hardware subsystems? Are software errors detected by an execution monitor? Is routing deadlock detected? Are the transactions verified? Can the hardware/software be reconfigurated automatically? Is formated structured contents checked for correctness (error detection/correction)? Is need to use/need to know checked? Is the use mode of the information system by the user checked for conformity with system goals? Are default values significant in case of unsuccessful query or search?

Information Life Cycle

Is there a check on format and content of information fed in ("garbage-in" evidence)? Is the information flow traceable among physical and virtual locations? When disassembled/reassembled, are the information fields checked w.r.t. to common origin and destination? Are the information feature extraction procedures fully documented with default values? Are acknowledgments implemented for all transactions?

For each stage there is clearly a need for tools to generate comparisons and for metrics to evaluate the results. The software engineering literature contains many references to such tools and metrics. (See Figure 1.)

Possible Contributions of Artificial Intelligence

Artificial intelligence engineering will allow for the incorporation into the information system of procedural rules and checks that were previously performed by humans or non-programmable firmware. Thus, they apply mostly to the failure management process and to the user/goal directed analysis of the information contents (e.g., in routing, distribution, verification or exploitation).

CONCLUDING REMARKS

It was observed that within the Advanced Research Workshop the predominant orientation was towards post-mortem failure analysis. This group, however, decided that for hardware and software failures it was more useful to consider an *in vivo* approach since it is clearly better to avoid a failure than to have a means of analyzing one. It is, however, practical to use the classification for post-mortem failure analysis.

In the real world, it is important to realize that even though systems are properly specified, there will be times when the information is found to be inadequate for the purpose intended.

For example, a recent search of a major specialized information database for papers on "Failure Analysis of Information systems" yielded 12 papers, of which only 3 were relevant. This highlights the need for improved search mechanism, perhaps using Artificial Intelligence. The system specification would then need to be modified to incorporate this major technological development or other unforseen advances. This is an indication of the need for including future possibilities in system objections.

It should be stressed that many important factors of hardware and software systems have

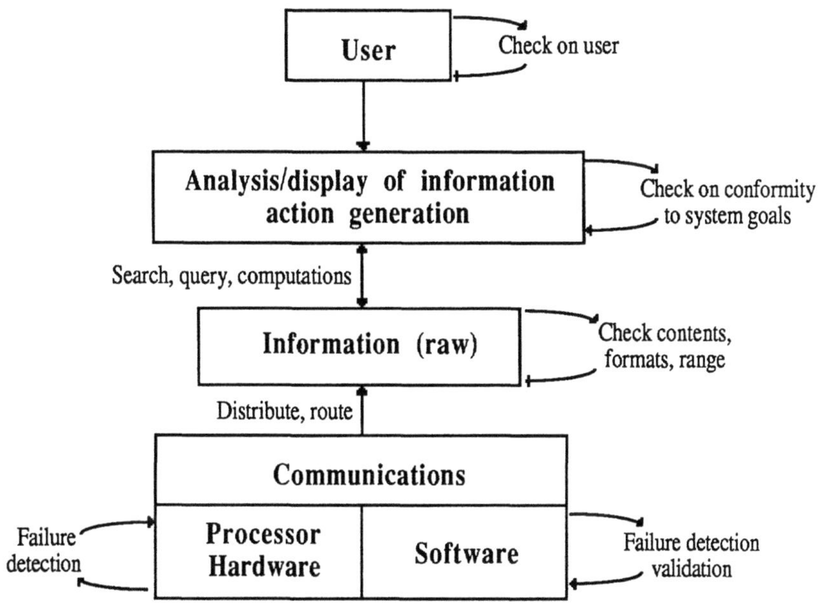

Figure 1. Levels and functions where hardware/software is implemented and may cause failures in information systems.

not been covered here (e.g., documentation, spares holding, training, etc.). These are omitted because of time constraints, and not because they are considered to be unimportant. Also, the information systems failures that have been happening in industrial organizations in most cases follow from the fact that the needed information was not transferred and/or disseminated to users where and when required in an adequate format.

The missing information might be available either as "formal information systems" built into computers or "informal information systems" available either internally or externally to the company. In order to avoid failures in information systems, it is necessary to monitor informal systems wherever feasible to ensure that they are in harmony with formal systems.

This task, at least for the time being (and unless some new innovations, e.g., in the field of artificial intelligence, appear in the very near future) might best be satisfied by information resources management specialists. The task of these specialists is to integrated the information resources available internally and externally (that is, from the continuously changing environment where thc company operates).

As a result of an efficient and effective coordination of the information available, one should expect not only an increase in the overall productivity of organization but also less likelihood of information systems failures.

This contribution is a brief and incomplete review of the opinions of the hardware/software failure analysis group at the ARW. It reflects the personal view of the members and not necessarily those of their organizations.

PARTICIPANTS

PARTICIPANTS

Prof. Peter Andow
Loughborough UniversityTechnology
Chemical Engineering Department
Loughborough LE11 3TU
U. K.

Dr. Mike Barnett
College of Maritime Studies
Newtown Rd.
Warsash, Hampshire
U. K.

Mr. Andrew E. Bilinski
The Pentagon
SAF/FMD Room 4E 975
Washington D.C. 20330-1000
USA

Prof. Gianni Casale
Institute Don Bosco
6 rue de la Sucrerie
B-7750 Warcoing
BELGIUM

Mr. Dereck Conway
c/o Dr. Gianni Casale
Institute Don Bosco
6 rue de la Sucrerie
B-7750 Warcoing
BELGIUM

Dr. Ana Maria Ramalho Correia
LNETI/CITI
Azinhaga dos Lameros Estrada
do Paco do Lomiar
1699 Lisboa Codex
PORTUGAL

Prof. Anthony Debons
University of Pittsburgh
749 LIS Building
Pittsburgh PA 15260
USA

Prof. Balbir S. Dhillon
University of Ottawa
Mechanical Engineering Dept
Ottawa, Ontario K1N 6E5
CANADA

Prof. Wolfgang Ehrenberger
Fachereich Informatik
Jungwirthstraße 12
D-8000 München 40
F. R. GERMANY

Prof. J. A. C. Empson
University of Hull
Department of Psychology
Hull HU5 2DW
U. K.

Mr. Halim Ergunalp
Telecam A. S. P. K. 28
Lüleburgaz
TÜRKIYE

Prof. Harold D. Foster
University of Victoria
P. O. Box 1700
Victoria, B. C. V8W 2Y2
CANADA

Dr Stein Gaarder
A/S VERITAS Research
Box 300
1322 Hovik
NORWAY

M. M. Trung Ho
Institute National de Recherche et
Sécurité
30, rue Olivier Noyer
75680 Paris Cedex 14
FRANCE

Dr. Erik Hollnagel
Computer Resources International
Vesterbrogade 1 A
DK-1620 Copenhagen V
DENMARK

Prof. Irving Janis
Univ. of California - Berkeley
627 Scotland Dr
Santa Rosa CA 94505
USA

Prof. Gunnar Johannsen
University of Kassel
Mönchebergstraße 7
D-3500 Kassel
F. R. GERMANY

Mr. Rudolph Kapustin
President, INTASCO
9926 Dellwood Avenue
Columbia, MD 21046-1024
USA

Ms Müjgan Kerman
Visnelik Mah. Kayihan Cad.
Erk Apt. No. 11/2
Eskisehir
TÜRKIYE

Mr. Randy R. Kehrli
U. S. Department of Justuce
P. O. Box 627
Washington, D. C. 20044
USA

Dr. J. G. Kretzschmar
S. C. K.
Boeretang 200
B-2400 Mol
BELGIUM

Prof. Donald Mason
58, Maeshendre
Waunfawr, Aberystwyh
Dyfed SY23 3PS
U. K.

Prof. L. F. Pau
Technical Univ. of Denmark
Electromagnetic Institut
Bygning 348 - 2800 Lyngby
DENMARK

Mr. J. Howard Petrie
Consultant
82 Green Lane North
Liverpool L16 8NL
U. K.

Prof. James Reason
University of Manchester
Dept. of Psychology
Manchester M13 9PL
U. K.

Mr. John D. Spiers
DCE Computer Consultancy Group
Chester House, Chertsey Rd.
Woking, Surry GU21 5BJ
U. K.

Prof. Gerrit Van der Veer
Free University
Postbox 7161
1007 MC Amsterdam
NETHERLANDS

Prof. Earl L. Wiener
University of Miami
Box 248237
Coral Gables, FL 33124
USA

Prof. Ron Westrum
Eastern Michigan University
929 5th St.
Ann Arbor, MI 48103
USA

Dr. John A. Wise
Westinghouse R&D Center
1310 Beulah Rd.
Pittsburgh PA 15235
USA

Prof. Faruk Yarman
Anatolia University
Bademlik Kampüsü
Eskisehir
TÜRKIYE

Dean Marshall C. Yovits
Purdue School of Science
P. O. Box 647
Indianapolis IN 46223
USA

OBSERVERS

Dr. Arv Larson
Booz, Allen & Hamilton, Inc.
430 East West Highway
Bethesda, MD 20814-4455
USA

Dr. Mchael Kaplan
ARI European Sci. Coord. Off.
223 Old Marylebone Road
London NW1 5TH
U.K.

INDEX

NATO ASI Series F

Vol. 1: Issues in Acoustic Signal – Image Processing and Recognition. Edited by C. H. Chen. VIII, 333 pages. 1983.

Vol. 2: Image Sequence Processing and Dynamic Scene Analysis. Edited by T. S. Huang. IX, 749 pages. 1983.

Vol. 3: Electronic Systems Effectiveness and Life Cycle Costing. Edited by J. K. Skwirzynski. XVII, 732 pages. 1983.

Vol. 4: Pictorial Data Analysis. Edited by R. M. Haralick. VIII, 468 pages. 1983.

Vol. 5: International Calibration Study of Traffic Conflict Techniques. Edited by E. Asmussen. VII, 229 pages. 1984.

Vol. 6: Information Technology and the Computer Network. Edited by K. G. Beauchamp. VIII, 271 pages. 1984.

Vol. 7: High-Speed Computation. Edited by J. S. Kowalik. IX, 441 pages. 1984.

Vol. 8: Program Transformation and Programming Environments. Report on an Workshop directed by F. L. Bauer and H. Remus. Edited by P. Pepper. XIV, 378 pages. 1984.

Vol. 9: Computer Aided Analysis and Optimization of Mechanical System Dynamics. Edited by E. J. Haug. XXII, 700 pages. 1984.

Vol. 10: Simulation and Model-Based Methodologies: An Integrative View. Edited by T. I. Ören, B. P. Zeigler, M. S. Elzas. XIII, 651 pages. 1984.

Vol. 11: Robotics and Artificial Intelligence. Edited by M. Brady, L. A. Gerhardt, H. F. Davidson. XVII, 693 pages. 1984.

Vol. 12: Combinatorial Algorithms on Words. Edited by A. Apostolico, Z. Galil. VIII, 361 pages. 1985.

Vol. 13: Logics and Models of Concurrent Systems. Edited by K. R. Apt. VIII, 498 pages. 1985.

Vol. 14: Control Flow and Data Flow: Concepts of Distributed Programming. Edited by M. Broy. VIII, 525 pages. 1985.

Vol. 15: Computational Mathematical Programming. Edited by K. Schittkowski. VIII, 451 pages. 1985.

Vol. 16: New Systems and Architectures for Automatic Speech Recognition and Synthesis. Edited by R. De Mori, C.Y. Suen. XIII, 630 pages. 1985.

Vol. 17: Fundamental Algorithms for Computer Graphics. Edited by R.A. Earnshaw. XVI, 1042 pages. 1985.

Vol. 18: Computer Architectures for Spatially Distributed Data. Edited by H. Freeman and G.G. Pieroni. VIII, 391 pages. 1985.

Vol. 19: Pictorial Information Systems in Medicine. Edited by K. H. Höhne. XII, 525 pages. 1986.

Vol. 20: Disordered Systems and Biological Organization. Edited by E. Bienenstock, F. Fogelman Soulié, G. Weisbuch. XXI, 405 pages.1986.

Vol. 21: Intelligent Decision Support in Process Environments. Edited by E. Hollnagel, G. Mancini, D.D. Woods. XV, 524 pages. 1986.

Vol. 22: Software System Design Methods. The Challenge of Advanced Computing Technology. Edited by J.K. Skwirzynski. XIII, 747 pages. 1986.

NATO ASI Series F

Vol. 23: Designing Computer-Based Learning Materials. Edited by H. Weinstock and A. Bork. IX, 285 pages. 1986.

Vol. 24: Database Machines. Modern Trends and Applications. Edited by A.K. Sood and A.H. Qureshi. VIII, 570 pages. 1986.

Vol. 25: Pyramidal Systems for Computer Vision. Edited by V. Cantoni and S. Levialdi. VIII, 392 pages. 1986.

Vol. 26: Modelling and Analysis in Arms Control. Edited by R. Avenhaus, R.K. Huber and J.D. Kettelle. VIII, 488 pages. 1986.

Vol. 27: Computer Aided Optimal Design: Structural and Mechanical Systems. Edited by C.A. Mota Soares. XIII, 1029 pages. 1987.

Vol. 28: Distributed Operating Systems. Theory und Practice. Edited by Y. Paker, J.-P. Banatre and M. Bozyiğit. X, 379 pages. 1987.

Vol. 29: Languages for Sensor-Based Control in Robotics. Edited by U. Rembold and K. Hörmann. IX, 625 pages. 1987.

Vol. 30: Pattern Recognition Theory and Applications. Edited by P.A. Devijver and J. Kittler. XI, 543 pages. 1987.

Vol. 31: Decision Support Systems: Theory and Application. Edited by C.W. Holsapple and A.B. Whinston. X, 500 pages. 1987.

Vol. 32: Information Systems: Failure Analysis. Edited by J.A. Wise and A. Debons. XV, 338 pages. 1987.

MIX
Papier aus verantwortungsvollen Quellen
Paper from responsible sources
FSC® C105338

If you have any concerns about our products,
you can contact us on
ProductSafety@springernature.com

In case Publisher is established outside the EU,
the EU authorized representative is:
Springer Nature Customer Service Center GmbH
Europaplatz 3, 69115 Heidelberg, Germany

Printed by Libri Plureos GmbH
in Hamburg, Germany